ALLE ZEIT WACH
1842

Brian Wynne

Risk Management and Hazardous Waste

Implementation and the Dialectics of Credibility

Springer-Verlag Berlin Heidelberg NewYork
London Paris Tokyo

Dr. Brian WYNNE
Director of Centre for Science Studies
and Science Policy
University of Lancaster
Lancaster, LAI 4YN
United Kingdom

IISA
International Institute for Applied
Systems Analysis
A-2361 Laxenburg
Austria

ISBN 3-540-18243-8 Springer-Verlag Berlin Heidelberg NewYork
ISBN 0-387-18243-8 Springer-Verlag NewYork Heidelberg Berlin

Library of Congress Cataloging-in-Publication Data
Risk management and hazardous waste.
Includes index.
1. Hazardous wastes--Management.
2. Environmental policy.
3. Risk assessment.
I. Wynne, Brian
TD811.5.R58 1987 363.7'28 87-26396
ISBN 0-387-18243-8 (U.S.)

Printed in Germany

Offsetprinting: Color-Druck, G. Baucke, Berlin; Bookbinding: Lüderitz & Bauer, Berlin

2161/3020 543210

Dedication

It was suggested that I should ask a well known person to write a foreword. However I could find few better words to express the spirit of what I am ultimately trying to say than those of Soren Kierkegaard, the nineteenth-century Danish writer and theologian:

> What is a poet? A poet is an unhappy being whose heart is so torn by secret sufferings, but whose lips are so strangely formed that when the sighs and cries escape them, they sound like beautiful music ... and men crowd around the poet and say to him: "Sing for us soon again" that is as much as to say: "May new sufferings torment your soul, but may your lips be formed as before; for the cries would only frighten us, but the music is delicious". And the critics come too, and say: "Quite correct, and so it ought to be according to the rules of aesthetics". Now it is understood that a critic resembles a poet to a hair; he only lacks the suffering in his heart and the music upon his lips. So, therefore, I would rather be a swineherd, and be understood by the swine, than be a poet and be misunderstood by men.

I therefore dedicate this work to analysts of public policy, that we might transcend the fate of the critics and begin to *understand.*

Preface

This book is one product of the project on Institutional Settings and Environmental Policies (INS) undertaken at the International Institute for Applied Systems Analysis (IIASA), Laxenburg, Vienna. INS began under my leadership in January 1983, and lasted until my return to Lancaster in May 1984. It had a brief but interesting organizational career which reflected the intense uncertainties, excitements, and frustrations of that turbulent period in IIASA's history. INS merged the previous Risk project with broader interests in technology, environmental policy implementation, uncertainty, and surprise.

As a scientist turned to sociology, my leadership naturally led us to try to analyze the *interactions* of technical knowledge and institutional realities in regulation. This was consistent with a general current – not then felt much at IIASA – that technical decision methods focusing on optimizing or "satisficing" public decisions as discrete events had become too abstractly concerned with decisions as *products*, and had lost sight of the importance of *processes*. The previous IIASA Risk project on facility siting had rightly emphasized the importance of decision processes. However decisions on siting – even controversial ones – are relatively well defined, and encourage the analytic assumption that all parties have the same problem – siting, and how to achieve it, stop it, or shift it. While it may be a reasonable starting point for well structured, specific decisions, this is not a generally realistic assumption. Indeed, work that does recognize the importance of processes often regresses into the abstract language of procedural rationality, as if universal procedural prescriptions for rational decisions can be made, regardless of the technological and institutional setting or *context*, and as if procedure handles a problem definition that is common to all parties.

Having been largely sheltered from this operations research based tradition, I had started with a different orientation, from the social analysis of scientific knowledge. From this angle *credibility* is always a problem, and experts are always tacitly negotiating different problem definitions, even whilst engaged in rational discourse about evidence, models, probabilities, etc. In practical "external" contexts, such as regulation and risk management, the analytical framework from the sociology of science that links knowledge *construction* with the processes of credibility negotiation or *persuasion*, becomes an important resource. When one considers more diffuse risk-regulation systems, where implementation of decisions is often distant from their birthplace, the related problems of credibility and competing problem definitions are even more significant – analytically and practically. Regulation and risk management use scientific knowledge as a central currency, but they are encountering increasing difficulties in achieving credibility – and hence effectiveness – for their standards and policies. The ensuing costs in economic, health, and general political terms are beyond reckoning. As is widely apparent, the *legitimation* of public policies, including regulatory methods and standards, is now a major preoccupation of their makers. Thus, how public knowledge is *constructed* to achieve persuasion; how this affects its other roles; and how technical knowledge and other institutional arrangements are related to achieve most effect are the basic concerns of this book. In taking an *institutional* focus, we also decided that we had to understand the underlying processes of institutional *credibility*.

It is at the level of public credibility of regulations, and of expertise generally, that this book aligns itself with the concerns about brittleness and resilience toward surprise and *viability*, which have been the basis of the policy research programs introduced by Buzz Holling, Bill Clark, and others. It is one of my regrets that we did not have the opportunity to explore properly the connections between the approach developed in the INS group, as reflected in this book and our other products, and that broader research program from systems ecology.

Being a comparative book, the relative effectiveness of different regulatory practices is an implicit concern throughout. However, although we do make evaluations and suggestions, we make no simple comparative judgments. The effectiveness of regulation is a complicated notion conceptually, and an elusive one empirically. To illustrate this book's evaluative orientation, we can return to the question of processes. Regulation is clearly more than a succession of separate "optimized" decisions; it *involves continuing relationships*; moreover these are developed in multiple arenas. Finding appropriate decision processes is far more complex than constructing rational procedural norms, because, as we see in real issues, different images of rationality and science are used as legitimating resources – as competing models of authority – by different parties. These imply different procedural arrangements. No procedure is neutral, nor is it ever completely issue-specific. A logical consequence is that we should be concerned to evaluate institutional processes and relationships of regulation. In the parlance of systems, we should regard decisions as open systems, not merely as a temporary device to find a more expansive reclosure, but because social systems are *inherently* open, and nondeterministic. In this necessarily more dialectical mode, attention shifts from sole preoccupation with "correct" decisions toward a more open-ended concern for socially feasible and robust decisions, and thus for the quality and integrity of policy relationships and interactions.

The argument in this book challenges convention by suggesting that the underlying approach of regulation in the developing climate of public justification reflects a misleading image of rationality and science. Whatever value this image may have had, it is now outliving its worth in a new context of public decision-making. Indeed, its use may be inadvertently corroding public credibility (credibility as far as the public is concerned) rather than nurturing it.

The central factor in this failure is the misconception that science is inherently and boundlessly uncertainty-seeking: scientific approaches will therefore ensure a comprehensive appreciation of uncertainties. However, science advances by the systematic limitation of its attention to *known* uncertainties within single frameworks stripped of their context. In public issues like risk management, the unrecognized context includes multiple social perspectives; these influence even technical terms like "risk" and "hazard" which are not merely imprecise, but subject to *inherent* ambiguity. Thus, the conventional regulatory language of rationality (derived from science) artificially reduces *structural* uncertainty and latent conflict to technical imprecision. Lack of control, deriving from ignorance and latent conflict, is concealed by the scientific language of manageable uncertainties. There is a curious paradox here, in that individual regulators, policymakers, and other experts privately recognize this syndrome, that the credibility of scientific approaches is almost inversely proportional to their degree of elaboration. The responsibility for *believing* in the myth (and thereby perpetuating it) is passed onto someone else - the public, the media, sociologists, naive scientists, or whoever. Yet none of these groups appear to believe it, either! They sometimes behave as if they do, but when pressed to act on it (e.g., by accepting a decision based on it, affecting them) they also renounce their faith.

The eerie conclusion is that we are all busy enacting a public framework of regulation driven by certain fundamental tenets of rational knowledge, that *no one* believes in. Now, there is always a certain tension between norms and actions. However the size and nature of the gap matters, and it now seems so large as to approach a collective hypocrisy – even institutionalized schizophrenia – on a grand scale. The public norms of rational control and decision in regulation seem to be little more than degenerate caricatures of reasoning. "Being pragmatic" as a (legitimate) norm has thereby been practically converted into a cynical lack of any greater purpose or hope than immediate survival, with the public language of rationality being left adrift to be elaborated as a Baroque facade of mounting structural dishonesty. One is reminded of the baroque fantasies played out in public life during the final years of the Habsburg dynasty.

This brief outline may put the situation too starkly. But I suggest the essentials are authentic enough. The myths of scientific rationality are no longer adequate to comprehend our experience of modern technology and the searching questions of its adequate control. We should not seek to throw away myths, but to develop more capacious and more human ones. These would be consistent with a critical social learning of the limitations of analytical control of technology via back-end regulation alone, and with a broader social interest in the purposes and *control* of (front-end) technological development.

After this confession of my ultimate concerns, what about the more practical dimensions of this book? What has all the foregoing moralizing to do with hazardous waste? Interdisciplinary works are notoriously difficult to orient to definable populations of readers. Systems analysis was supposed to transcend the reductionisms of single disciplines, but it has developed a recognizable reductionism of its own, toward formal quantitative models. Faced with this sterile scientism, sociologists often fall to the temptation of advancing their own, alternative reductionism. But this may do no more justice to reality than its opposite. This book offers a sociological perspective; but it is not a smoothly reductionist one. It does not claim to integrate the sociological and technical dimensions, or to reduce one to the other. All it claims is to hold them in something like proper tension. In so far as the reader's difficulty reflects that tension, I do not apologize for it.

The analytical framework threading the whole book is unashamedly interpretive – it tries to hold the descriptive and the normative in constructive balance. It goes beyond "the evidence", but does not wantonly neglect it. It is not a direct evaluation of policy options in the usual sense, but attempts to outline a new basic orientation toward the issues – to highlight some less obvious features of the present situation which may alter our taken for granted anchor-points for evaluation. This is potentially an important *practical* contribution.

I learnt at IIASA how pervasive and crippling is the lack of an institutional perspective in policy analysis. This is not a merely analytical matter, but more of an underlying moral philosophy. Along with its benefits, the scientific world view has brought to the understanding of public affairs a thoroughly corrosive duality, whereby meanings only come from either a monolithic objective truth of nature or a personal subjectivity. If we cast loose from the moral sheet-anchor of single facticity, it is therefore assumed that there is no stopping short of rampant subjectivity and anarchy. The dominant intellectual mood at IIASA then was of this kind, with an underlying belief that institutional analysis was really an encouragement of "antiscientific" subjectivism.

Though small, the INS project at IIASA actually incorporated three different strands. These were the analysis of methodological issues in policy modeling with Bill Keepin; hazardous waste, with Joanne Linnerooth and Mike Dowling; and Himalayan deforestation, with Mike Warburton and Mike Thompson. All of these (and more) diverse interests were united in our group's common concern to understand ways of defining and handling uncertainties and conflicting rationalities. We included as part of our domain how analysts, like ourselves, regularly become part of the policy *problem* by misconceiving and inadvertently misrepresenting these dimensions.

INS's place within IIASA is an instructive lessons about the above bifurcation. As an ex-"hard" scientist I assumed that a scientific approach to regulation meant actually looking – with a theoretically informed perspective – at what is happening. This meant looking at the back end, where regulatory approaches are actually implemented (or not), rather than assuming that the world behaves as in the abstract scholastic exercises of formal models of policy systems. This general approach, innocently advanced, was received with studied neglect and incomprehension, punctuated by sporadic eruptions of hostility. Ironically it seemed, "soft" social science was being rejected by so-called "hard" science as threatening, because it proposed too close a look at reality. Here was cause for reflection! The kind of answer we moved to in all three areas we examined was that abstract mathematical policy modeling – while fine and valuable in the proper situation – represented the artificial bifurcation outlined above. It tended to freeze a given institutional structure as natural and inevitable, and enshrine a given problem-definition corresponding with those (now invisible) institutional structures. In the act of commitment to a model structure, it then implied that any alternative approach was the onset of subjectivism. (It was not surprising that the throwaway, "quick and dirty" interactive modeling of Carl Walters, Mike Staley, and colleagues was our nearest intellectual companion at IIASA, and also suffered rejection symptoms). Indeed the tension of intellectual styles was not strictly between the technical and the social – but between a basic orientation that treated institutional questions as naturalistic (thus did not really treat them at all) and one that did not. Beneath this distinction lay the following questions:

Are institutional dimensions to be handled analytically in a quasi-technical manner which assumes that they obey naturalistic laws – even if these are within the expanded frameworks of "liberal" systems approaches, which may include evolutionary perspectives and discontinuities? Or are they to be recognized as *inherently* and forever open-ended, and therefore incomplete? Does the very lack of analytic tidiness and completion in the latter represent the arena of legitimate human freedom and responsibility?

The liberal wing of systems analysis appears to be ambivalent on this issue, and may confuse (its focus) "*adaptive management*" with (the emphasis of sociological critical theory) "*political learning*".

Whether conventional or liberal, the dominant IIASA intellectual style was arguably a reflection of the Institute's own institutional setting, and its relationship to typical clientele, "the policymakers of the upper reaches". Complex computer models of the applied systems vintage have often appeared to shelter such policy elites from the wider and baser realities that impinge on ordinary folk, and to exclude such questions from conceptualization. This is the very opposite of what is now needed.

Largely undebated methodological differences of this sort sat uncomfortably on top of some fundamental political and institutional issues. A thorough institutional analytical approach appears to encourage a fearful response because it does cast off from the comforting anchorage of universal, single rationality, including the more expansive liberal version. However, in the prehistoric age before "policy analysis" and "applied systems analysis" there did use to be, recognized as normal, an *institutional* dimension underlying the framing of issues, negotiations, and "decisions". Research alive to this social fabric recognizes the plurality available in factual accounts of the world (within the restriction that all accounts "work"); yet equally it reminds us that the *social institutions* of which we are part reduce to nonarbitrary patterns what would otherwise be a mess of countless subjectively chosen views of the world. These institutional frameworks of cognition are negotiable and malleable at the margins, and over the long term; but they are not immediately reconstructible, nor freely chooseable, except to the extent that migration is possible into alternative existing institutionalized frameworks. Solidly objective and elaborate institutional networks of norms, expectations, identities, traditions and forms of explanation severely restrict (but do not totally determine) the "caprice" of supposedly free individual will. Human beings are *social* animals. In neglecting or distorting the institutional dimension therefore, the rationalist approach to "policy" would effectively replace the responsibility and freedom of ordinary institutions and social actors by abstract and authoritarian "expertise". In the end, practical policy is likely to be best served by research that does not maintain the delusion that such institutional dimensions either do not exist, or – if they do exist – that they are only subjective distortions of rational commitments. Freed from these shackles, it might then foster analysis that does not preemptively shatter institutional processes into nothing but a kaleidoscope of fragmented, arbitrary, and competing individual values or wills. Policy analysis will first have to recognize and systematically address these institutional dimensions, before it can help in making such institutions and their policies resilient, and might we also dare to hope – just.

Acknowledgments

I feel a deep and genuine sense of privilege to have been able to work in an institution of unique importance and one whose role in the wider world underlines a point made in this book: in addition to whatever is *produced,* the quality of the constitutive human processes and relationships matter in themselves. My first debt of gratitude therefore goes to those people who first invited me, especially Buzz Holling and Howard Kunreuther, and those who made me welcome. I also owe a special debt to my friends and colleagues at Lancaster, who simply worked harder to cover my absence, without any extra help.

Many IIASA friends and colleagues were puzzled (to put it politely) that I actively encouraged the INS group to address three apparently disparate topics, on slim resources, during my short time in the IIASA sun. However, not only did the main work on hazardous wastes benefit greatly from the intellectual interaction with these sister topics on energy policy modeling and Himalayan deforestation, but I happened to have three extremely able and trustworthy friends working on them. Any "distraction" which I "suffered" was more than compensated by corresponding rewards to the work on hazardous waste. Both these sister topics have also produced work that has received international acclaim [see the special Issue of *Policy Sciences,* **17** (Nov and Dec 1984), *Journal of Applied Systems Analysis* Vol. 12 (April 1985), and Thompson *et. al.,* (1986) *Uncertainty on a Himalayan Scale*, Ethnographica, London]. It is a personal honor to recognize here the dedication to IIASA and all that it is supposed to stand for, shown by so-called "junior" colleagues Mike Warburton and Bill Keepin.

Short-term visitors like myself come and go at IIASA, but our existence is dependent upon long-term staff. I would like to pay tribute to those support staff who, through an especially difficult and uncertain period in 1983–1984, worked to try to maintain IIASA as a symbol of mutual understanding and trust between different peoples and political systems.

These challenges being long term, they do not yield to short-term secondments with the sense of impatience engendered by a time-scale of 15 months to start up and bring a complicated international project to a meaningful conclusion. Nevertheless, we were privileged to be able to develop collaboration with Hungarian colleagues whose personal hospitality, friendship, and professional cooperation I warmly acknowledge. Contacts were made with other colleagues in Eastern Europe but regretfully there was insufficient time to develop them. I hope this potential will be able to be realized in future in the risk and environmental policy fields.

In addition to Hungarian collaboration, we also benefited from collaboration and interviews, all of which took place in 1983 and early 1984, in The Netherlands, UK, USA, Austria, France, and the FRG. Space does not allow a full listing of all collaborators, but I would like especially to mention José van Eindhoven and colleagues at the University of Utrecht, Jan Dirven, then of the Ministry of Economic Affairs, and Eric Ferguson, the Secretary of the Netherlands National Member Organization of IIASA. In the FRG we benefited especially from the regular cooperation of Werner Schenkel and colleagues at the Umweltbundesamt, Berlin, and from Jobst Conrad's work, which acted as input to Chapter 6. Gary Davis, a former designer of the Californian hazardous waste legislation, was a very welcome visitor and collaborator, of greater significance than is indicated in joint authorship of one chapter here. I also want to thank Harry Otway of the Commission of the European Communities Joint Research Centre at Ispra, Italy, for personal and intellectual support, which included not only discussions on psychology and institutions, but also Ispra financial support for six weeks at the end of my IIASA stay. Collaboration with Harry extended to an IIASA–Ispra Workshop and subsequent book on risk and regulation [Otway, H. and Peltu, M. (Eds) (1985), *Regulating Industrial Risks,* Butterworths, London)]. From the UK, Ted Finnecy of the Harwell Environmental Safety Group was an especially helpful and rewarding colleague, who amongst other things read and commented in detail on all the IIASA Working Papers produced in the form of a first draft of this book in June 1984. I also warmly appreciate the thoughtful and detailed suggestions for revision of these early draft chapters made by Paul Kleindorfer of IIASA and The Wharton School of Decision Sciences, University of Pennsylvania. Val Jones also gave the book the benefit of her critical reading, and made several valuable suggestions. Various others, too numerous to mention, commented valuably on parts of the draft material. Needless to say, the use or misuse of their help by the named chapter authors leaves the authors, and overall myself, responsible for the errors of fact or judgment that remain.

Two other events at IIASA particularly helped to develop my thinking on this project. These were an INS informal summer study on regulation and expertise attended by Harry Otway (Italy), Jerry Ravetz (UK), Arie

Rip (Netherlands), Mike Pollak (France), Brian Campbell (Canada), Gary Davis (USA), Ted Radford (USA), Pal Tamas (Hungary), and Lech Zacher (Poland); and an IIASA International Forum on Science and Public Policy organized by Chet Cooper with help from Mike Dowling, also a member of my project.

The summer study and the Forum, especially discussions during the latter with Harvey Brooks, and informal discussions on both these and other occasions with Mike Thompson and Jerry Ravetz have been especially helpful in developing the approach to uncertainties and credibility expressed in this book. Jerry and Mike were also a source of regular encouragement, stimulation, and conviviality throughout my transfer to IIASA and back. With Mike Thompson I also shared – occasionally despairing, but always cheerful - ruminations on the ups and downs of a sociologist's life among the applied systems tribes. I cannot let the opportunity pass for expressing my thanks also to Chet Cooper, whose arrival at IIASA halfway through my term brought sanity and steadiness to mounting all-round disorientation. To Chet I also warmly record personal thanks for sending Mike Dowling, and telling him to batter my office door down if necessary in search of a job. In the event, seeing that my office was pretty busy Mike spared my door and, more subtly, buttonholed me in a local wine tavern! The end result was that Mike became the mainstay of our group, and a valued personal friend into the bargain. It is also a personal pleasure to acknowledge the invaluable role of Eryl Mädel, who soaked up more than she should have been exposed to, always with infinite patience, and who did everything from research to the most menial jobs with an air of calm common sense – whatever she may have been feeling! To Vivien Landauer I also owe personal thanks for teaching me about Austrian German, and office organization.

Finally, I want to express the deep debt I owe to my wife Cathy and our children, Matty, Tom, and Grum, who have not just tolerated on my behalf the unsettlements and uncertainty of domestic upheavals, but who turned it to positive effect together. We also as a family warmly appreciate the friendship and hospitality of the many Austrian people, especially the community of Biedermannsdorf, who made the risks that *we* took pay off so handsomely.

Contents

1. Introduction: A Conceptual Overview 1
Brian Wynne

2. Hazardous Wastes Risk Management and Environmental Regulation 23
Brian Wynne

3. Hazardous Waste – What Kind of Issue? 45
Brian Wynne

4. The Rationalities of Problem Definition: The Netherlands and Hazardous Waste Management 84
Brian Wynne and Dick Hortensius

5. The Listing and Classifying of Hazardous Wastes 114
Michael Dowling and Joanne Linnerooth

6. Government Responsibility for Risk: The Bavarian and Hessian Hazardous Waste Disposal Systems 150
Joanne Linnerooth and Gary Davis

7. Decentralized Regulation and Technical Discretion: The UK 195
Eryl Mädel and Brian Wynne

8. Hazardous Waste Management in Hungary 245
Ernö Kiss

9. Risk Assessment of Technological Systems – Dimensions of Uncertainty 269
Brian Wynne

10. Risk Assessment and Regulation for Hazardous Wastes 311
Brian Wynne

11. Risk Perception, Decision Analysis, and the Public Acceptance Problem 356
Brian Wynne

12. Summary and Conclusions 397
Brian Wynne

Index 445

CHAPTER 1

Introduction: A Conceptual Overview

Brian Wynne

1.1. The IIASA Hazardous Waste Study in Perspective

The production, conversion, or disposal of wastes pervades almost every human activity. Many of these wastes are harmful, or potentially so. They can also be valuable – and hence "goods" – depending upon different perceptions. Despite their ubiquity, wastes exist in a twilight zone where no clear, "natural" definition of them can be given, within wide margins of uncertainty and variation. They provide no service, and no one has an inherent self-interest in their safekeeping. Yet, overall, their damaging effects upon the environment and human health can be colossal. Major economic and social perturbations have been created by uncontrolled past toxic waste dumping, and adequate long-term management of hazardous wastes remains one of the most difficult challenges ever faced by regulatory authorities.

Current difficulties over radioactive waste disposal are serious enough, but they look simple by comparison with the problems of controlling thousands of different hazardous-waste-generating activities, waste types, handlers and transporters, diverse economic relationships, and disposal sites. On top of all this are still the environmental uncertainties of possible routes back to human or environmental harm. In hazardous waste life-cycles (unlike say, discharges from a plant), natural processes and human interactions are jumbled together in complex and widely variable ways, making a badly structured and, indeed, indeterminate behavioral-technical risk-generating system. It seems to be a testing ground for the institutional viability of regulation, where uncertainties and implementation difficulties feed on one another to threaten its public credibility.

This book draws the technical field of risk analysis toward the more messy organizational realities which make up technologies, regulatory institutions, and their practices. As risk analysis has perforce been dragged away from well defined engineering systems toward such decision-making arenas, approaches to risk seem to have divided into two sorts, both of them incomplete. Either they have remained within the same traditions of engineering reliability but have been extended to try to incorporate new relationships and uncertainties, without fundamental reshaping; or they have been sociologically defined – although often insightful, they have virtually neglected the physical and technical dimensions of the issues [1]. Our approach offers a novel treatment of this bifurcation of the technical and institutional dimensions of risk assessment and regulation. In tackling one of the more complicated environmental regulation problems, the study elaborates a fresh perspective on the general role and nature of scientific knowledge in public decision making and its practical enactment.

Risk analysis involves the scientific elucidation of damage mechanisms from different natural or technical processes, and the quantification of probabilities and consequences. As an aid to social management of risks, the analysis has to be integrated with evaluations and choices, usually under uncertainty and conflict. It is normally assumed that risk analysis takes place as a scientific process, free from the context of institutional interests and constraints, which enter at a later evaluative stage. Whilst this may have been a reasonable first-order approximation for a discrete engineering plant (which is where risk analysis developed), it is an inadequate foundation for the kinds of risk management problem, such as hazardous wastes and toxic chemicals generally, that have risen to prominence in the last decade or so.

Risk analysis aims to achieve replicability and control of a scientific sort. However, scientific progress and precision requires two simplifying principles:

(i) That the environment or context of a given relationship can be held in suspension.

(ii) That ideal conditions and entities (such as perfectly elastic solids or friction-free surfaces) really exist.

The "background" is frozen and stripped away, and the entities idealized, so as to "purify" the phenomenon of interest and render it manipulable and analyzable. This scientific orientation in risk assessment of complex technological networks has yielded benefits, but at the expense of other limitations whose extent and significance is only now becoming apparent. In particular, the focus on "risk" stripped of context and in terms defined by the analyst is increasingly being seen to lack foundation in the interpenetrating problem definitions, constraints, and concerns of real decision makers. What the analyst defines as external context may be part of the primary analytical problem. Our *institutional* approach is an attempt to take these contextual realities into account, without reducing the analytical framework to exclusively technical or sociological factors.

One can understand how the "context-free" fallacy has taken root. In addition to the general mythology that science is context free, there are reasons specific to the origins of risk analysis. Until the recent emergence of environmental risk analysis, the field of risk management was dominated by calculation of risks from well defined and monolithic problems, such as those of nuclear reactors and chemical plants. As risk systems, these were indeed relatively free of their context. A central problem for such "zero-infinity" type risks is how to model their extremely low probability processes and events in the absence of data based on experience [2]. Expert disagreement was attributed to the lack of data (reflecting an empiricist model of science). Even in such risk analyses, however, it was noticed in the previous IIASA study on the siting of liquid energy gas (LEG) facilities that uncertainty and expert disagreement over risks are not only due to lack of empirical data resulting in subjective disagreement on probabilities, but also to subtly divergent expert *definitions* of the precise risk-generating *system*. This was a clue to how to approach the more typical risk problems, which are not so well structured. Since expert disagreement and uncertainty had become (and remains) of great concern in every domain of technology and environmental policymaking, it seemed that a potentially useful way of developing the theory and practice of risk assessment was to examine risk-management issues where the risk-generating "technology" was manifestly less well structured than, say, LEG terminals or nuclear reactors, and more open to ambiguity in its precise definition.

We therefore decided to examine hazardous waste management, an important and growing risk management problem, which is ill structured and heterogeneous. Our aim was to clarify the forms of interaction between institutional settings, behavioral factors, and technical knowledge. The analytical framework of this book is made up of three linked fronts. The first is an integration of *implementation* into the framework for analyzing decision processes in risk management or regulation. The second front involves a basic revision of the conventional model of *scientific knowledge and uncertainty* embedded in dominant approaches to policy, regulation, and policy analysis. The third front involves a radically different interpretation of the origins and nature of *public perceptions* of "risk", and an attempt is made to follow through the supplementary argument that public reactions are reactions not to perceived risks as such, but to the institutional relationships that are part and parcel of technologies. These institutional relationships are normally excluded from attention because the analytical frameworks are structured by conventional notions of scientific rationality. An overall problem deriving from the way these three dimensions interact is the *credibility* (and hence the viability) of regulation.

1.2. Implementation

Our interest in implementation derived from a general curiosity about reality, as contrasted with the often rarefied and unreal perspectives of policy modeling and design. The latter seem to adopt the definitions of decisions that policymakers themselves naturally employ. However, there is an important distinction between "decisions" (e.g., regulatory standards or aims) and their practical enactment. Clearly, what substantive decision is made has an effect upon practices and outcomes, but it does not automatically *determine* them, especially where – as is normal for regulation – decisions are part of a continuing multiorganizational *process* of interaction, negotiation, commitment, and adaptation. Much environmental policy analysis – perhaps influenced by siting controversies – has tended to focus on decisions as definitive events, as outcomes in themselves, rather than to try to examine their relationship to practical outcomes. It is a moot point whether "decision" ought to mean what "decision makers" "decide", or what actually happens. Frequently, "decisions" are defined analytically as what "decision makers" do; but the definition of "decision makers" here has been influenced by the analyst's relationship only with central policymakers. This has consolidated a quasi-scientific "purification" of the behavioral arena to exclude the complications involved in "implementation". Downstream actors or "implementers", and associated organizational complexities, have thus been relegated analytically to the role of the merely mechanical (decision-less) enactment (or obstruction) of policymakers' "decisions" or rules.

The general view of policymaking as largely incremental, and decisions as embedded in *continuing* social relationships, is more congenial to us. Our concern is to bring implementation into more even analytical balance with more synoptic phases of regulatory policy-making, to treat it as an interacting decision-making forum of comparable importance and complexity. Implementation is more than mere external or secondary context to the "real" analytical arena. Our orientation was encouraged by the growing body of empirical work demonstrating that implementation is often different from that imagined in the policymaking phase. The widespread "enforcement deficits" or "implementation gaps" [3] identified by environmental policy analysts imply not only wasted policy resources, but worse, withdrawal of public acceptance from policy bodies and processes which manifestly fail to fulfil their formal goals and promises. Unrealistic assumptions that implementation is a supposedly trivial after-phase have often been built into policy decisions themselves, which subsequently fail to meet policy targets and promises. This undermines the credibility of the policy system, potentially entering a vicious circle of decreasing implementability. As a leading policymaker in the hazardous waste field has warned:

> ... the discrepancy between the daily practice of disposal and official political objectives and pretensions is obvious ... In many cases there is an obvious discrepancy between regulatory requirements and the actual means of meeting them. The enforcement gap is not only a permanent threat to the environment, it also affects the credibility of legal provisions in place [4].

This issue of regulatory authority and credibility was the ultimate interest underlying our work. It derived significant benefit from the comparative institutional analysis of national "regulatory styles" at the level of implementation. There is an international corollary of the realization that many "policy decisions" simply flutter in mid-air, as symbolic gestures [5]. In the growing number of issues with international implications, this encourages a cynicism and protective self-interest that threatens already fragile international policy negotiation and collaborative enactment.

The mounting problems of policy implementation are not due to *inadequate* technical knowledge, but to the institutional mechanisms for putting it to effective use, in contexts of conflicting organizational constraints, interests, and rationalities. Indeed, what counts as technically adequate is a variable determined by institutional factors. There are different *kinds* of technical knowledge and regulatory instruments whose effectiveness can only be evaluated in relationship to their institutional settings. Even within the same regulatory organization, the appropriate form of technical knowledge (or language) may be different at different levels, as we show later. Technical regulatory instruments (and the knowledge they imply) are *codes* for behavioral relationships, and can only have meaning and constructive effect in proper relationships with their surrounding networks of social relationships, perceptions, and interactions.

In its very framing risk and policy analysis ignores – or radically oversimplifies – implementation complexities, and treats failings as due to technical uncertainty or incompetence; this is hardly likely to help alleviate implementation problems. Indeed, it may inadvertently undermine public acceptance of technologies and their regulatory institutions. This idea is developed in the following sections.

In summary, our argument on *implementation* is that it is more than merely downstream decision enactment, but involves new organizational realities and rationalities which are artificially excluded from policymaking (and usually from policy modeling and analysis). These are more than merely "deviations" or "counter-implementations" of the optimal, but reflect objective local institutional realities whose pragmatic accommodation is just as important for regulatory stability. This is part of the "context" that scientific approaches unconsciously amputate.

1.3. Scientific Knowledge and Uncertainty

Risk assessment has developed as a more sophisticated, probabilistic scientific treatment of essentially the same regulatory problems as its predecessor, technology assessment, but under explicit conditions of uncertainty. However, the basic conception of uncertainty that structures risk analysis is fundamentally misleading, and is encouraged by a false underlying view of scientific rationality. As the difficulties, especially failures to achieve credibility, have multiplied, "scientific uncertainty" has been made the culprit, leading to intensification of scientific effort. In 1983 the then chief of the US Environmental Protection Agency (EPA) observed that "a climate of fear now dominates the discussion of environmental issues" [6]. Typically, he saw only one way out:

> Somehow our democratic technological society must resolve the dissonance between science and the creation of public policy. Nowhere is this more troublesome than in the formal assessment of risk – the estimation of the association between the exposure to a substance and the incidence of some disease, based on scientific data.
>
> I believe that part of the solution to our distress lies with the idea that disciplined minds can grapple with ignorance, and sometimes win – the idea of science. We will not recover our equilibrium without a concerted effort to more effectively engage the scientific community.

Likewise, when NATO's Committee on Challenges to Modern Society (CCMS) completed a major comparative study of national hazardous waste management policies in 1981, it noted the same public anxiety, linking it with implementation failures. But it expressed the same faith in science to overcome this and wider regulatory uncertainties:

> [There exist] obvious differences in political attitudes and philosophies about how hazardous waste disposal should be carried out. A priority task ... should be to develop a more common understanding of technical disposal requirements and to achieve some sort of harmonization of disposal standards and quality ... With regard to this objective, more efforts have to be made to put hazardous waste management on a sound scientific footing [7].

Implementation gaps and regulatory failings generally are thus attributed to technical inadequacy, an incomplete "sound scientific footing". On this reasoning, lack of public credibility is caused by uncertainty, which is caused by incomplete science. The self-evident solution is therefore to intensify and standardize the application of scientific knowledge in risk analysis, to provide more clear and uniform technical standards, more scientifically precise and universal legal norms, etc. All this assumes that the technical world is free of institutional context.

This underlying approach is very nearly universal in hazardous waste regulation (and elsewhere), and certainly dominates its public language and processes. It incorporates questionable assumptions about the nature of the problems and of prevailing uncertainties, the nature of scientific knowledge, the basis of public perceptions of the risks, and indeed of the appropriate nature of social authority. Our alternative orientation emphasizes that issues are made up of multiple interacting social groups, which have different cognitions of "the issue" at stake. These may incorporate different interests, so that divergent values or social assumptions may be built subtly into factual arguments and uncertainties about risks. Expert disagreement in risk analysis is also frequently due not to imprecision or ignorance, i.e., technical uncertainty alone, but to different *expert* definitions (not only lay perceptions) of what the risk problem is. These may partly reflect different institutional or disciplinary positions, and other assumptions which may imply behavioral judgments or prescriptions, i.e., indirect value commitments.

Whether at expert or public level, we call these divergent basic problem structures deriving from institutional realities, *structural* uncertainties. "Uncertainty" generated by such *structural* conflict of expert perception is fundamentally different from, but normally mistaken as, purely technical uncertainty or imprecision. Implementation is analytically important in this distinction of uncertainty types because it is usually most distant organizationally from the level of "scientific" input, i.e., where regulations are decided. Even at centralized levels different *cognitive* structures and not merely interest structures exist (e.g., expert disagreement), but implementation shows most sharply the operation of different substantive rationalities even within the same regulatory organization. What the conventional approach assumes to be lack of technical precision (scientific uncertainty) is often *structural* uncertainty or latent conflict between divergent perceptions and social rationalities. In its unawareness of these social dimensions *framing* rationality, we argue that "rational" methods in regulation and decision analysis effectively *reject* a necessary dimension of democratic negotiation between expertise and lay experience. By inadvertently dismissing a potential development of the relations of expertise with lay experience, they ultimately undermine the credibility and authority of expertise altogether.

To give a brief example, central regulatory managers normally have to relate to composite bodies (industrial lobbies, parliamentary scrutineers, the media, environmentalist groups, and international agencies) who want to know the rationale and see the overall "system" consistency of regulation. They may therefore define regulations as if they are clear and uniform, and as if key control terms such as "hazard" or "waste" are unambiguous and admit no significant loopholes. Local implementing actors, on the other hand, relate to local industrial plants and operators, local residents, and other particularistic conditions. They have to shape the practical interpretations of those very same regulations in ways that are viable in *their* setting, in negotiation with a different set of actors, interests, and constraints. This may frequently be the source of perceived "implementation gaps", as for example when a local inspector judges that he cannot enforce legal sanctions on a firm violating the regulations, because he knows his own legal department does not have the resources for a successful legal prosecution, or because he knows that the company is basically honest and diligent, but in temporary difficulties.

Whereas central regulators may defend the credibility of the regulatory process amongst *their* reference groups by projecting a universalistic, rigorous scientific image of regulation, local regulators may defend the viability of regulation by *ad hoc* accommodation to contingent local realities, thus potentially contradicting the centrally projected image of technical consistency, precision, and rigor. The most significant factor affecting implementation is not scientific uncertainty or incompleteness, but *structural* conflict of rationalities among the diverse legitimate parties in the regulatory arena. Under pressure for public justification and credibility, however, the language of science artificially reduces these into apparently one-dimensional technical uncertainties, which seem to be manageable by scientific methods – probabilistic risk analysis, etc. We call this the *bureaucratic processing of uncertainty.* It is supported by the myth that science is uncertainty seeking, whereas at a deeper level it is uncertainty rejecting.

The most central example of the practical and conceptual effects of this uncertainty fallacy is the dominant policy assumption that the key terms "hazard" and "waste" can *even in principle* be precisely defined. This is a major point developed throughout the book, that the main cause of inadequate hazardous waste management is believed to be the inconsistency and imprecision of these "technical" terms. But this assumes that they can be precisely defined (by more scientific effort), free of context. Our research shows that this is not so – they are rooted in their institutional context, and reflect their context. They are more than *transcientific* questions, because an infinity of extra data will not achieve their clear definition. "Hazard" and "waste" are therefore not just imprecise or statistically fuzzy – they are *fundamentally* ambiguous. Their intrinsic physical meaning is not given and objectively predetermined in nature; it is *always* incomplete, and has to be completed by *social* construction. This social construction has to be repeatedly achieved and repaired by partly tacit processes of negotiation in specific regulatory settings and waste life-cycle situations.

On a large scale, we find that the different national institutional settings resolve regulatory definitions of hazard and waste in different ways. And even within a single regulatory framework the precise practical definition of "fixed" technical criteria is an institutional process. The construction involves actors and groups with divergent interests, perceptions, constraints, and rationalities, and thus with competing favored definitions.

The viability of the regulatory process may therefore actually depend upon the very opposite of intensification of science; it may require that some imprecision and ambiguity of formal regulatory standards and definitions be maintained, as an adaptive arena in which the contending parties can interact, negotiate, and settle and renegotiate the practical meanings as they go along.

Indeed, the *lack* of technical precision may allow the intra-organizational "inconsistency" at different levels of regulation (as outlined before). This accommodates irredeemably conflicting social forces and demands without their ever being brought into immediate, destructive confrontation. This ability of the institutional domain to absorb intrinsic social contradictions, aided by appropriate formulation of scientific norms, may be an important buttress against regulatory paralysis, especially since regulation has changed from an administrative function to the more political role, traditionally performed by legislatures, of compromising between conflicting social values. The general insight to be drawn is that the technical language of regulation is a code for its social context. Contrary to the conventional view, it does not derive from a context-free scientific framework. Comprehensive, elaborate, and inflexible technical frameworks are likely to be established where *structural* (institutional) uncertainty is greatest. Conversely, when structural uncertainty is low (e.g., a relatively stable, high-trust social environment) precise, comprehensive, and inflexible technical language is unnecessary.

Here we can clarify the analytical distinction between prescription and description. For a badly structured issue like hazardous wastes, it may make no sense *descriptively* to employ standard, apparently precise models in risk analysis. However, if there is high structural uncertainty in the institutional setting, regulation will employ such precise and standardized technical language as a *prescriptive* code for controlling highly uncertain *behavior*, or for reassuring mistrustful and truculent social groups. In practice the analytical distinction between description and prescription is confused, and the institutional needs for prescription drive the criteria of technical description.

One could say that uncertainty does not exist in objective amounts; the threshold of perceived uncertainty is inversely proportional to the degree to which it is defined as a problem. This depends upon the amount of cherished commitment that it appears to threaten, which is a reflection of social factors.

These correspondences between technical and institutional uncertainties, and the roles of scientific knowledge in processes of regulatory credibility and *viability*, are a central concern of this book. As we show later, the specific mode of scientific knowledge in regulation is influenced by several factors, including its simultaneous use in more than one social role. One of these is its interaction with processes of public perception.

1.4. Public Perceptions – Credibility and the Politics of Anxiety

Policymakers, regulators, or risk managers face an increasingly difficult problem achieving credibility and authority, or *legitimation* of their policies, regulations, etc. As Peter Crawford, then head of OECD's Chemicals Management project, remarked in 1982:

> The era in which government decision makers can accept expert guidance without being able to point to, explain and validate the procedures followed is receding rapidly [8].

It has continued to recede, and the public *justification* of decisions has become a dominating dimension. The precise *decision rules* underlying risk assessments and regulations have been required to be more transparent. This has centrally involved science, and the authority of descriptions of scientific practice. As with implementation difficulties generally, the reaction has been molded by misconceptions about the basic nature of public skepticism about science [9]. The lack of expert consensus is blamed for the public's skepticism, the remedy being more and better technical knowledge, including more rigorous procedures for defining risks, and for converting analysis into practical regulation [10].

All this is founded on the premise that "exaggerated" public perceptions of risk originate in expert disagreement and in public ignorance and lack of intellectual rigor. However, we reject the idea embedded in conventional approaches, that "risk perceptions" are (subjectively distorted) perceptions of (objective) *risk*, which is supposed to be a single context-free dimension. We argue that it is *artificially abstracted* by analysts from people's multidimensional social experience of issues, technologies, and decision-making institutions. We propose instead that people rationally react to their (past and current) social experience of decision-making relationships, in which physical risk defined by analysts is inextricably embedded as only one element. For example, public reactions to "hazardous waste risks" and associated regulations or proposals are reactions to accumulated experience of "regulation" as a historical relationship. This includes control of public health and environmental damage, but also the intelligibility, competence, trustworthiness, and social identifiability of institutions, which themselves have several public interfaces. These social relationships are the grounding of "public perceptions of risk". In some areas of regulation they are already mediated by impenetrable scientific languages – they are *already* socially alienated.

As we argue in Chapters 11 and 12, past public "acceptance" should be regarded more realistically as passive quiescence, no more. Apparently declining public willingness to *trust* regulatory institutions may thus be a reflection, not of an inherently less-trusting public, but of escalation in what technological and regulatory elites *ask* of public credulity.

Therefore, the conventional response to regulatory problems, of intensifying the scientific content of regulation, may well *undermine* the authority of regulatory bodies because it may extend the existing social alienation on which, we argue, public risk perception is founded. We could say that the "risks" people objectively perceive and respond to are the *social risks* embodied in the social relationships of unintelligible control and apparently arbitrary power between risk receivers and risk managers. *By pretending that "risk perception" is perception of (physical) risks, and by obliterating the institutional dimension, conventional policy analysis may only add more fuel to the already amply fueled fire of public opposition.* The regulatory importance of social credibility and the assumptions outlined above mean that science has been drawn into a dual role in regulation. On the one hand is the ordinary practice of analyzing risks, their variances, determinants, and potential controls (we might call this the "descriptive" or "empirical" dimension), while on the other is the role of negotiating public credibility, i.e., of public justification. This involves a symbolic role, of projecting reassurance and persuasion that public expectations of trustworthiness and credibility have been fulfilled.

The first role only involves substantive statements, e.g., "the risk of activity X under conditions Y is Z: it can be reduced to Z′ under conditions Y′ ". However, the second role involves public accounts of the internal *process* of science, to claim credibility for the first kind of statement (and related regulations). This exposure of the internal practices of science has immense practical political implications for the credibility of science in public, and thus for the decisions and prescriptions that lean upon science for their authority.

To policymakers the "hazardous" aspect of hazardous waste is a relatively minor issue, and their main concern is to convince an uneasy public:

> Extensive efforts are needed to inform the public better on the technical criteria involved in hazardous waste disposal and on how far precautionary measures are taken to exclude or limit short term and long term risks. It has to be demonstrated to the public that disposal facilities are designed and managed properly so that people feel confident that things are done in the right way and in the interest of citizens themselves ... There are experiences well proved and persuasive disposal concepts and technologies existing. What we have to do is to sell them better to the public [11].

Largely as a result of this pressure to justify policy commitments to third parties, environmental risk analysis has implied the existence of developed scientific underpinnings that are often not there. To some extent this has been beneficial in forcing such scientific attention and development in areas such as pollution control technology. But frequently the elaboration of statutory regulations (and the political pressure behind them) has proceeded *as if the science were already developed,* and has created enormous pressure to produce the necessary "applied science" very hastily, and thus within strong constraints on the scope and depth of such research. The inevitable result has been the understatement of uncertainties and methodological limitations, the over-elaboration of scientific models, inadequate peer review, and even downright falsification [12]. Whether for "third-party" reassurance or to reduce institutional unpredictability, scientific knowledge has inevitably been shaped by its social role.

Both internal bureaucratic dynamics and external reassurance may require a belief in objective, rule-bound knowledge. But the full extent of this policy "forcing" of science, and of the ignorance, uncertainty, variability, and lack of normal scientific standards that underly it, is becoming increasingly, if incoherently, recognized by the public.

Unrealistic public expectations of science have been cultivated in the past by regulatory institutions, which have (not necessarily illegitimately) kept uncertainties and expert conflicts from public view. This may have been feasible when the scope of technologies and the speed of change were relatively limited. But such social filtering of experience is made less feasible by new conditions, under which uncertainties and limitations of method cannot be privatized by experts and other elites. As these institutional barriers between the public and uncertainties are broken down, daunting problems of credibility and communication are emerging, aggravated by the expectations engendered by past practices and social relationships in regulation.

Part of the problem is that even scientists themselves tend to underestimate the extent of informal judgments and unspecifiable decision rules within their own science; these cannot be formally justified, especially not in skeptical settings such as policy or legal arenas. When so tested, even maturely developed scientific knowledge frequently fails the test of credibility [13]. This not only damages specific policies or claims, but also the institutions' *general* credibility.

Our suggestion is therefore that prevailing approaches to regulation and their founding assumptions about science are in a cycle of self-destruction of institutional credibility. This is self-fueling as erosion of authority engenders more symbolic projection of images of formal scientific procedures and evidence, and ensuing (frequent) inspection leads to further public contradiction of this mode of authority.

In the hazardous wastes case there is an extra twist to this cycle. The symbolic reassurance discourse of science emphasizes standardized, universal risk and regulation frameworks. But the dispersed and heterogeneous nature of hazardous wastes life-cycles requires the particularistic, situation-specific, analysis of risks, in the descriptive or empirical mode. The pressure upon regulatory bodies to appear credible and to reassure may more sharply than for other issues conflict with accurate situational analysis.

Concrete examples of this conflict are given in later chapers. To illustrate the problem here we can quote the concern of a UK regulator over the publication of nuclear risk analysis criteria:

> Before any decision can be taken as to the wisdom of expressing policy in such terms, due regard must be given to the possible consequences of enshrining figures of this nature. One point is that if a certain probability figure is approved, any improvement required will be seen as illogical [14].

In other words, the symbolic credibility dimension forbids acknowledgment of the contingent property of current scientific beliefs, because credibility currently attaches to a universalistic image of scientific knowledge. Social analysis of science suggests that knowledge is more context-specific than is usually recognized, because there are many assumptions about contextual relationships that are assumed to stay constant or hold true in order for a given relationship to be valid [15]. These many "network" relationships are conditions for the validity of a scientific statement and limit its strict scope, but they are so taken for granted by scientists that they do not recognize their existence. These qualifications are therefore rarely fully stated, especially in the context of public debate, thus giving an exaggerated impression of universal truth to scientific knowledge. Of course, many such conditions, e.g., that Newtonian mechanics is only accurate for systems of velocity less than the speed of light, or for levels of energy transfer greater than subatomic energy quanta, are fulfilled over huge ranges of "natural" experience, so the qualifications become largely irrelevant and go "invisible". However, in systems of greater variability, complexity, and ignorance such as those involving behavioral relationships and unknown environmental processes, as with hazardous wastes, the universalist faith about science may be a bad model, and the tension between universalistic ("intrinsic risk") considerations and the actuality of situation-specific risk may be thereby exacerbated.

Because the foregoing orientation is complex, and unfamiliar, I will attempt to summarize, before introducing the practical issue of hazardous wastes and the structure of the rest of the book. The conceptual overview is central to the contribution we attempt to make, which is to analyze the significance for regulation of the new context of *legitimation* in which the scientific language and practice of risk analysis is increasingly embroiled.

(1) Policy and risk analysis has neglected the significance of *implementation*, which involves much more than the mechanical local enactment (or lack of it) of central rules. "Deviation" or "deficits" often entail correspondence with local factors; these are just as important as central considerations in making policies practically viable. Such local, situational factors are akin to the context of a relationship or problem usually assumed stable by science. Like science, risk analysis and other

"rational" methods in policymaking all unconsciously strip away this context and assume that it is, at most, marginal.

(2) Particularly where behavioral uncertainties interact with physical parameters, and ignorance is significant – as is true of most risk analysis problems (perhaps especially hazardous wastes) – this basic neglect of context is drastically misleading. From the "context" arise different definitions of the decision problems, different perceptions of what the primary risk-generating system is, and different kinds of relevant experience and expertise. These multiple rationalities represent latent *structural* conflict over the key technical terms, "risk", "hazard", and "waste". This *institutional* uncertainty in the heart of the issue is misrepresented as technical imprecision. Implementation difficulties, technical imprecision, expert disagreements, and public opposition are all taken to be ultimately remediable by more precise and rigorous scientific standardization.

(3) Because of the intensifying need for *justification* of regulations to achieve public credibility, universal and precise scientific models are disseminated whilst the underlying scientific ignorance and uncertainties escalate. The tension is therefore sharpening between the scientific roles of symbolic reassurance and of empirical analysis of actual, situated risks. As expertise is increasingly used to justify regulatory decisions, it is increasingly challenged, leading to public examination of the internal scientific *process*. This is also found to fall short of (unrealistic) formal, rule-bound images of scientific practice, leading to further deterioration of regulatory credibility.

(4) Public reactions are falsely defined as "exaggerated" psychological perceptions of risks defined by the established experts. These artificially abstracted physical concepts of risk bear only distant relationship to the concrete institutional relationships of technology – social risks – which are arguably the authentic and legitimate sources of public perceptions and reactions. These divergent public definitions of the relevant problems are distantly indicated, but not at all adequately represented, by the elaboration of risk attributes, which people are thought to perceive and weight differentially in "subjective" reactions to risk. Current conceptions and uses of science as the basis of risk management artificially, if inadvertently, conceal the full uncertainties and ignorance. This precludes mature social learning through regulation and instead encourges inconsistent and often extreme public reactions as the intelligibility gap between the public and policy elites increases.

In these circumstances it seems there are only two overall directions left. We can try business as usual, with more desperate patching up as we go along, with rising polarization, instability, the devaluation of public discourse into "public relations", and social embrittlement of public decision processes generally [16]. The sense that in this direction we have to run faster and faster even to stand still on the sand is aptly summed up by Greenberger's observation that we are using scientific expertise more and more, for it to be believed less and less [17]. Whatever abstract expertise may be intensified in this process, overall there is no mutual learning.

An alternative is to seek out the elements of current practices and approaches where critical dialogue of frameworks and mutual learning does seem possible, and to build upon these. The particularly critical need seems to be to explore new forms of dialogue and learning, and new institutional relationships between "experts" and "nonexperts". This should not be taken as anti-expertise. To the contrary, if expertise means legitimate authority, it is to restore a context in which expertise can exist. I would echo the observation of Harvey Brooks, that:

> ... perhaps the principal lesson from our experience with the interactions of experts and laymen in public policy decisions with high technical or scientific content is the need for greater introspection into the non-technical values and preferences that affect both the selection of evidence and its interpretation by all the participants, both laypersons and experts [18].

Brooks' thoughts were directed to "big" decisions, as one-off *events*, but when reoriented toward the kind of *process* decisions or relationships involved in typical regulatory settings, they provide the idea of a gradual elicitation of fundamental frameworks held by different parties, at the deeper level where values, basic problem definitions and clusters, and descriptions of the world, all meet. Some thoughts on the kind of institutional relationships and moral-cognitive heuristic (the equivalent of the role that "rationality" now plays, but something less authoritarian and narrow) needed to give effect to such principles are given in Chapter 12. A step in this general direction has been the interactive modeling workshops initiated by Holling and colleagues [19]. A common experience of these has been that during five intensive days well away from the shop, policymakers have been fired by fundamentally new insights into the problems they confront, only to find the inspiration relentlessly melt away as they are re-enclosed in their usual organizational world. Some kind of analogy to the interactive workshop process of joint exploration and mutual appreciation, but on a more public and more mature and regular basis, will eventually have to be found.

The analysis as outlined above indicates why I believe the developments in the practical experience of regulation in the last decade require a new look at scientific rationality in public arenas, because of the rapid development of the legitimatory or justification dimension. It has always been there; but I believe it is now of a different scale, quality, and significance where, starting from conventional approaches to risk assessment, we can see the need for a new orientation.

It should by now be clear to the reader why this book is about Hazardous Management of Wastes. In Chapter 2, I explain why hazardous wastes is such an apt analytical focus as well as an important empirical one.

1.5. The Chapters in Outline

Having given an overview of the issues that this book addresses, we now outline how the book is structured. First, however, a word on strategy.

The cross-national comparative work on which the book is founded was not a large-scale effort with several fieldworkers operating to the same methods and question schedules in every country. Nor did we see any value in a rigid comparative framework; we were using comparative opportunities to help structure a more analytical problematique about the interactions between technical knowledge and uncertainties, institutional structures, and regulatory credibility. The research was based on our own interviews, some commissioned interviews, IIASA meetings, documentation, and data collection, supplemented by collaboration, especially in the Netherlands and Hungary. This collaboration was limited by the existing commitments of colleagues working for us on their own funds.

The chapters are thus organized as much around analytical subthemes as descriptive case studies of specific countries. However, the central issue of each chapter is illustrated by empirical materials from a leading country, with supplementary comparative materials from other countries. Comparative analytical discussion is woven into the chapters where it naturally arises. To define a problem more deeply may ultimately be the most important practical contribution.

In Chapter 3 we analyze some key features of the hazardous waste issue that seem to exist in whatever cultural–institutional system it finds itself. Most of these properties add up to an issue that is more poorly defined than most – not merely technically uncertain, but institutionally uncertain and indeterminate, on top of a great deal of technical ignorance. How different societies define the "technical" boundaries and structure of "the" hazardous waste risk analysis and regulation problem reflects institutional factors in that society. Although they are influenced by different settings, as we see in subsequent chapters the heterogeneity and extensive behavioral–technical life-cycle of hazardous wastes create important *general* conditions for the proper role of risk analysis in relation to local institutional relationships. Other properties of hazardous wastes are also identified that have a marked effect upon the interaction of these dimensions with public credibility of regulation.

Chapter 4 describes the processes of issue definition into formal regulatory frameworks in the Netherlands, and the practical relationship between such formal prescriptive systems and the institutional processes of enacting them. The relatively elaborate formal precision of the Dutch regulations, and the contrast with the vagaries of their implementation, is interpreted in terms of issue-specific factors worked out in a particular and (for Europe) unusual political–administrative culture.

Chapter 5 analyzes the factors that underly what is regarded as *the* central problem for rigorous national regulation of hazardous wastes, as well as for international control – making the definition and classification of hazardous wastes more precise and standardized. After reviewing the different technical approaches that have been taken toward this problem, the chapter analyzes the way in which such hazardous waste classification lists were constructed, especially in the USA, the FRG, and Austria. A point of prime importance throughout the book emerges, namely the inevitably *social* constitution of technical frameworks. This is because the detailed choices of method, criteria, discriminations, etc., were found to depend upon local administrative purposes and *institutional* needs (including legitimation or public justification), as well as upon more universal scientific–technical factors. This has important practical implications for attempts to stem international loopholes by intensifying the scientific effort to find precise, universally meaningful hazard definitions and classifications.

One of the most important issue properties described in Chapter 3 is the diffuseness and dispersion of hazardous waste life-cycles. Chapter 6 examines one concrete approach to the unusually severe risk management problems that this presents; this approach is reflected in the regulatory systems of Hesse and Bavaria in the FRG. These states (like Denmark and Sweden) effectively reduce the indeterminacies of multicentric, multiactor definition, movement, and "control" of hazardous wastes by "condensing" the hazardous waste life-cycle from its "cradle" at production to its "grave" at final disposal or destruction, into one, comprehensive organization for regulation and management. We call this the government ownership or "absorption" of risks, which would otherwise be at the mercy of autonomous behavioral, perceptual, economic, and technical interactions. Important institutional initiatives include combinations of public and private investment (though increasing public input to capital programs), and obligations on all waste generators to register and transfer their waste to the ownership of the public management authority. Public investment has been followed elsewhere, more recently in the Netherlands, but the FRG's export restriction has occasioned a great deal of controversy. Its merits and drawbacks are discussed in the light of recent developments.

A sharp contrast in the institutional approach is then examined in Chapter 7. The UK has devolved regulation to local county council or district council authorities who have ultimate responsibility for the provision of treatment and disposal (T&D) capacity, and responsibility for interpretation of central technical principles. This has led to inconsistencies in standards, disposal costs, and enforcement, and has obstructed capital investment in new and better T&D facilities. Controlled codisposal of some kinds of hazardous waste in municipal landfills is a central crutch of UK policy, to the chagrin of a substantial body of international opinion. The institutional decentralization, technical flexibility, and virtually total private enterprise T&D of the UK system are discussed in the context of policy development in the UK and Europe.

The Dutch and UK cases are the most clear illustrations of an important general point that technical knowledge is a code for *behavioral* relationships and expectations. Where there is informal trust among the relevant *institutions*, (e.g., the UK in the past, but this is changing) elaborated technical codes are unnecessary. Conversely, where there is institutional fragmentation (here Dutch political culture bears some resemblance to that of the USA), elaborated, inflexible, and precise technical norms effectively stand surrogate for *institutional* uncertainty.

Chapter 8 is an account of the Hungarian system of managing hazardous wastes. As well as providing information not hitherto available, the chapter also underlines the point that political–economic factors of industrial structure and financing, and government–industry relations, are more fundamentally important in risk regulation than technical risk analysis *per se*. Disconnected from organizational realities the latter is meaningless. It is interesting to note too that the wider institutional relations of government and industry in Hungary probably have more potential for fundamental, upstream regulatory solutions than do most market economies.

Chapters 9 and 10 turn from case studies of national processes of regulation toward risk analysis and management, the requirements of it, and the constraints upon it in the light of earlier chapters. Chapter 9 elaborates the distinction between two fundamentally different kinds of uncertainty, whose confusion has important and damaging effects. This chapter's underlying framework allows an examination of the tension between standardization of scientific risk criteria and more variable, situation-specific optimization. This is a special case of the more general point that expert risk analyses often differ not due to *imprecision*, but to the different underlying risk problems they have framed. These may well embody institutional assumptions. A key observation of Chapter 10 is that hazardous waste belongs to a class of policy issues whose uncertainties embody multiple, contradictory social definitions of the problem, and even widely divergent constructions of central technical terms like "hazardous", and of apparently straightforward data such as hazardous waste arisings and movements (cf. other fields such as oil reserves, or deforestation rates and fuel wood consumption).

Chapter 9 argues that the dominant approaches continually misperceive what is in reality behavioral–institutional uncertainty, as if it were technical uncertainty. Thus, at a strategic level the need to work on diminishing or constructively living with the former is obliterated in the automatic intensification of scientific work on the latter. More specifically, the greater basic importance of *institutional* uncertainties than is recognized *generally* in risk and policy analysis (not only for hazardous wastes) implies that institutional strategies such as those analyzed in Chapter 6, may be more valuable than usually recognized. We argue that social–institutional uncertainty generates and maintains technical uncertainty. The dominant rational decision-making approach in policy assumes the opposite, that institutional uncertainty or conflict is generated, or anyway, *allowed*, because of unresolved technical uncertainties.

In general, our inversion of the conventional framework of rational public decision making restores a rightful emphasis on the quality of social relationships in decision making – to negotiation and mediation as well as discovery and declaration; to *process* as well as products. However, this involves more than embracing the frameworks of "procedural rationality" which have evolved from the conventional framework in response to this point [20]. This argument is elaborated in Chapter 11. Notions of procedural rationality are built upon the belief that rational procedures can be developed, but for *given* and discrete problems that are assumed to be shared by all parties. Our institutional relations framework is also about *process*, but it includes the essential point that the process must recognize the legitimacy of, and allow for negotiation between, diverse definitions of the fundamental issue(s) at stake. Defining a problem is also an expression of values, and to assume a given definition is already to impose authoritarian restrictions, however "liberal" or "rational" subsequent procedures may be. In risk issues at least, events and decisions are part of a stream of relationships and experiences. It is reasonable for people to judge these institutional relationships when they are asked to accept risks under conditions of uncertainty. No issue is an island.

In Chapter 11 we show how public "perceptions of risks" are being falsely interpreted by essentially the same dominant approach, which takes a single scientific framework for granted without appreciating its own context-free artificiality. This has led to the elaboration of technical decision-analytic management methods instead of more negotiative, open-ended, and mutually appreciative *social relationships* of decision making.

Finally, in Chapter 12 we attempt to integrate the insights from the more empirical chapters with the more theoretical discussion of Chapters 9–11. This involves an extensive analysis of the *legitimation* dimension of regulatory decision making, as problems of institutional credibility and public reassurance have escalated markedly in the 1980s. Science as risk analysis is caught up in both these intertwined dimensions – descriptive and symbolic action. The tension between them, analyzed in Chapter 10, has been amplified by the need for public justification. The symbolic projection, however, is not a disconnected "secondary" shell of false consciousness; it shapes and constrains the empirical discourse, limiting what it can and cannot analytically recognize, and what meanings can be invested in empirical analysis. The justificatory dimension tends inevitably to condense and obliterate real situational variances in a universalistic language of risk assessment and regulatory implementation. Especially for hazardous wastes, this does violence to the realities of variable risks, and of local regulatory implementation, and thus paradoxically it undermines its own attempts to reassure the public. Attempts to climb out by intensifying and elaborating conventional "rational" scientific methods only dig the pit deeper. We conclude the book by outlining a more constructive general alternative, based upon notions of adaptiveness of institutional relations, broader social learning, and more open-ended, critical, and pragmatic notions of rationality.

Chapters 9–11 therefore work through the complicated and interacting themes: of *implementation*, which accentuates the importance of cross-cutting rationalities and *institutional* uncertainties; of the tension between these unrecognized uncertainties, and the assumption that implementation gaps are due to *scientific* or *technical* imprecision alone; the contradiction between the dual roles of science, in description and justification; and the *dialectics of credibility* problems set up by the unceasing attempt to use a misleading model of science in regulation – even to respond to public concern – in ways that are fundamentally insensitive to the institutional determination of practical rationality.

Hazardous waste has been an important practical issue in which to work out these ideas, but they would be useless if they could not be shown to have some corresponding practical implications. Some of these are described in Chapter 12. The general analysis of the book tilts the balance: in favor of greater "condensation" of waste life-cycles by some form of institutional "ownership" of risks; and in favor of greater intervention to regulate "upstream" waste production, rather than back-end disposal which treats waste production as an impenetrable "black-box".

Notes

[1] Most readers will be familiar with these different approaches. The first is typified by D. Okrent *et al.*, the second by Douglas, M. and Wildavsky, A. (1982) *Risk and Culture* (University of California Press, Berkeley, CA); a third strand, the extension of cost–benefit analysis and other market economic solutions, is represented by Lave, L.B. (Ed) (1982), *Quantitative Risk Assessment in Regulation* (The Brookings Institution, Washington, DC). As is evident in Chapter 11, I regard most economic approaches as fundamentally similar to engineering approaches in their characterization of social relationships and institutional dimensions of risk management.

[2] Weinberg's original (1972) concept of "transcience", to describe those many risk-analysis problems that can be formulated as scientific questions, but are never definitively answered as such, was a valuable characterization of a key property of such problems, namely the inherent impossibility of gathering definitive, controlled data. However, an unfortunate consequence is the implication that scientific consensus is determined by empirical data alone, or to put it the other way, that lack of consensus is only due to lack of empirical data. See Weinberg, A.M. (1985), Science and its limits: The regulator's dilemma, *Issues in Science and Technology*, **II**(1), 59–72. For one of many discussions of the theme, see Majone, G. (1984), Science and transcience in standards setting, *Science, Technology and Human Values*, **9**(1), 15-22. Since modern understanding is that, as Majone expresses it (p. 16), "empirical evidence only reflects the theoretical and metatheoretical assumptions that guide the experimenter's work," it is simply tautologous to say that scientific consensus is determined by empirical data.

[3] US Congress, Report of Hearings on the Hazardous Waste Control and Enforcement Act, 1983, before the Transportation and Tourism Subcommittee of the Committee on Commerce and Energy, March 1983, Serial 98-32, Washington DC. Commission of the European Communities (1982), *Present Situation Concerning the Disposal of Toxic and Dangerous Wastes in the Community, Implementation of Directive 78/319/EEC, and Future Actions*, Directorate-General for the Environment Consumer Protection and Nuclear Safety, B (DG.XI) XI/338/82-EN, Brussels. See also Risch, B. (1983) The activities of the European Community on hazardous waste, in J.P. Lehman (Ed), *Hazardous Waste Disposal*, pp. 123–136 (Plenum, New York, NY). Waddel, T. (1981) *Integrated Waste Management in Europe: A Tentative Assessment,* Report to the German Marshall Fund and Institute for European Environmental Policy, Bonn. Downing, P.B. and Hanf, K. (Eds), (1984) *International Comparisons in Implementing Pollution Laws* (Kluwer-Nijhoff, The Hague).

[4] Wolbeck, B. (1983) Political dimensions and implications of hazardous waste disposal, in Lehmann, *op. cit.* [3].

[5] See, e.g., Majone, G. (1985) International regulation of risks, in H. Otway and M. Peltu (Eds), *Regulating Industrial Risks* (Butterworth, London).

[6] Ruckelshaus, W.D. (1983) Risk, science and democracy, *Issues in Science and Technology*, **1**, 19-38. An earlier version appeared as Ruckelshaus, W.D. (1983), Science, risk, and public policy, *Science*, **221** (9), 1026–1028.

[7] NATO Committee on Challenges to Modern Society (CCMS), Pilot Study, Final Report, Brussels, 1979. See also, Lehman, *op. cit.* [3], which is a product of the same NATO program.

[8] P.J. Crawford, paper delivered to the International Conference on Toxicology and Regulation of Chemicals, Bellagio, Italy, August 1983. See also OECD (1982), *Managing Chemicals in the 1980's* (OECD, Chemicals Division, Paris).

[9] Wynne, B. (1982), Institutional mythologies and dual societies in the management of risk, in H. Kunreuther and E. Ley (Eds), *The Risk Analysis Controversy: An Institutional Perspective* (Springer, Berlin); Rip, A. (1985) Experts in public arenas, in Otway and Peltu *op. cit.* [5].

[10] For example, US National Research Council Committee (1983), *Risk Assessment in the Federal Government: Managing the Process* (National Academy Press, Washington, DC).

[11] Wolbeck, *op. cit.* [4].

[12] The US legal case of Industrial BioTest in 1983 involved the exposure of fraud, falsification of data, and other malpractices on a large scale. Much of the data were embodied in international regulations before these became apparent.

[13] Bailey, G. and Hawkins, R. (1983), *The Future of Rubbish: Waste Management Options for the Environment Reviewed* (CPC, London).

[14] Charlesworth, F. and Gronow, W. (1967), A summary of the experience in the practical application of siting policy in the UK, *Proceedings of a Symposium on Nuclear Reactor Siting*, pp. 7–24 (IAEA, Vienna).

[15] Hesse, M.B. (1980), *Revolutions and Reconstructions in the Philosophy of Science* (Harvester Press, Hassocks, UK). Lodge, P. and Law, J. (1984), *Science for Social Scientists* (Macmillan, London).

[16] For a somewhat similar case, see Daneke, G.A. (1983), An adaptive-learning approch to environmental regulation, *Policy Studies Review,* **3** (11), 7–12.

[17] Greenberger, M. (1981), Humanizing policy modeling: Confronting the paradox in energy policy modeling, in J. Gass (Ed), *Validation and Assessment of Energy Models* (US Dept. of Commerce, National Bureau of Standards, Washington, DC).

[18] Brooks, H. (1984), The resolution of technically intensive public policy disputes, *Science, Technology and Human Values*, **9**(1), 39–50.

[19] Holling, C.S. (1978), *Adaptive Environmental Assessment and Management* (Wiley, Chichester, UK).

[20] For some examples of these stretches of the conventional framework, see Grauer, M., Thompson, M., and Wierzbicki, A. (Eds) (1985), *Plural Rationalities and Interactive Decision Analysis*, Proceedings of an IIASA International Conference in Sopron, Hungary, August 1984 (Springer, Berlin). The first part of that book, on cultural bases of rationality, contains a solid collection of articles consistent with the orientations of this book. (Parts of Chapter 11 of this book were presented at the Sopron conference, but failed a production deadline for the proceedings).

CHAPTER 2

Hazardous Wastes Risk Management and Environmental Regulation

Brian Wynne

2.1. Introduction

In 1979, the Dutch waste treatment company, UNISER, was found to have illegally disposed of large quantities of hazardous wastes by defrauding its customers, which included most of the Dutch chemicals industry [1]. A member of UNISER's board who was jailed, was also a member of a government expert advisory committee. The only law under which the company could be prosecuted dated from the 1870s, and was originally designed to prevent the sale of contaminated meat. The credit for eventual prosecution went to local authorities who had persisted doggedly through years of evasion, lack of support and even informal obstruction from central government. Despite the existence of formal regulations from 1981, the lack of domestic treatment facilities meant that until now the main method of Dutch hazardous waste "control" has been to export it to other countries. Some of these countries allegedly have unacceptable disposal practices, such as uncontrolled landfilling next to national frontiers. Others (e.g., Belgium) have objected to certain Dutch export practices and have demonstrated the inherent fragility of a system dependent upon large-volume exports by imposing bans on their wastes.

In the USA, the original 1976 legislation to control hazardous wastes, under the Resources Conservation and Recovery Act (RCRA), was widely recognized to be the most elaborate and comprehensive environmental protection framework ever. Yet the timetables for implementing regulations were so badly breached, and the regulations (even when they came) so full of loopholes, that some analysts judged that there would have been less of an environmental risk problem had there been no legislation at all [2]. The US Congressional Office of Technology Assessment's 1983 review of the "Superfund" program to clean up past uncontrolled toxic waste dumps actually concluded that despite this program, more superfund toxic sites needing remedial cleanup would be created under RCRA's "control regime" than Superfund would clean up [3]! The same report conservatively estimated that more than half of the toxic waste generated by US industry – over 40 million tonnes (mt) – was escaping control. Large sectors of US toxic waste disposal have been found to be in the hands of organized crime. Superfund's allocation of $1.6 billion was estimated to be anywhere between 10 and 100 times too small to deal with discovered past toxic waste dumps. In another case, a waste disposal company operating in Denver, Colorado, was officially found to have been deceiving the EPA and the Colorado State authority between at least 1980 and 1982, yet the courts judged that no legal penalty could be applied under RCRA or anything else [4]. These and many other flagrant loopholes in RCRA led the US Congress to reauthorize and extensively strengthen hazardous waste control legislation in 1983 [5].

Two mammoth legislative programs, supported by endless policy analysis, have therefore seen the exacerbation of the problems. This offers an invitation, not to say an obligation, to examine the issues afresh.

In the UK, which often boasts of the longest established framework for environmental protection, a 1985 review of hazardous waste regulation by the newly established Hazardous Wastes Inspectorate (HWI) produced one of the most outspokenly critical official reports ever seen from a UK government agency [6]. It noted that

> ...a considerable number of undesirable disposal practices have been witnessed. It may be inferred that malpractice at hazardous waste sites is not uncommon. ... standards are anything but consistent or satisfactory, ... all too many major hazardous waste landfill sites have been seen which exude an atmosphere of total dereliction and decay.

The regulatory system appeared to encourage the "cheapest tolerable option", and could not be expected to command public confidence. About 150 incidents of *discovered* illegal disposal are recorded every year, and some places were recognized officially as "cowboy country", where enforcement officers had to proceed in groups for protection against physical assault. The HWI was especially severe on what had been officially claimed as the long-standing linchpin of hazardous waste control, namely site licensing to restrict uncontrolled dumping. So appalling was this system found to be that in many cases the HWI could not even understand from the license conditions whether a site was meant to be a hazardous (in the UK, "special") waste site or not! Even in the UK's traditionally quiescent climate of environmental protest, public antagonism has grown toward hazardous waste practices – ironically, this has so far been most intensely focused on the high-technology, supposedly cleaner, sector of waste treatment and disposal. However, cracks are beginning to appear, almost literally, in the waste management practice of landfill codisposal of hazardous wastes with municipal wastes, on which the UK is more dependent than any other industrial country.

Despite a vigorous defence of the extensive dependence on this practice, justified by a widely referenced scientific research program in the 1970s [7], criticism of unacknowledged uncertainties and biases in this program have begun to emerge, and groundwater contamination from landfilling has been found. Some UK water authorities have expressed concern at this state of affairs, and have had to close some aquifers as supplies because of contamination [8].

In the EC, which issued a toxic waste Directive in 1978 requiring all member states to implement certain controls, and to submit reports of progress, about half of the member states were being threatened with legal action by the Commission of the European Communities (CEC) because they had not fulfilled their responsibilities, even many years after the deadline. In addition, the inherent difficulty of even establishing a coherent overall policy perspective has been underlined by a major conflict within the EC as to whether wastes should be treated as a fully commercial commodity with free trade movement across frontiers, or whether they should be seen as part of environmental protection and subject to locally financed and managed public–industrial treatment and disposal (T&D) services. This issue is developed later.

In most countries, hazardous waste management is in a state of internal flux and public strife. International controls are crippled by problems and divergence in national approaches. The "disorderly state of affairs" reported by one European regulator in 1982, with "discrepancies between the reality and spirit of regulations" [9] could be an appropriate description for the general situation, after about ten years or so of legislation in most countries.

There is also a growing concern that the problems in developing countries may become far worse due to the rapid growth of chemical industries in some regions (e.g., in India, at nearly 20% per year, between 1975 and 1985), the lack of environmental control expertise, infrastructure and resources, and the potential interest of firms in industrialized countries in exporting their most expensive (and thus, most toxic) wastes to such cheap options.

Figures compiled by Harvey Yakowitz at the OECD suggest that approaching 400 mt of hazardous wastes are produced worldwide, though the definitions employed are extremely variable and uncertain [10]. Roughly 5 mt arise in less-industrialized countries, 20 mt in Western Europe, and 15 mt in Eastern Europe. These figures conceal rapidly developing volumes of hazardous waste arising in particular geographical *and* technical areas, and new patterns of waste trading, e.g., between less-industrialized countries, both of which put special strains upon proper controls. Indeed, international trade in wastes seems to be increasing in all dimensions, between industrial countries, between less-developed countries, and between industrial and developing countries. In addition to the fact that local waste management directly affects international movements and problems, industrial–world regulatory frameworks are being used as models for the less-industrial countries, so there is an extra impetus to examine carefully their technical and (especially) *institutional* characteristics, interactions, and implications.

A central and unique dimension of the hazardous waste issue is highlighted by the belief of experienced environmental policy staff in international organizations that much toxic waste *legally* allowed to be exported to developing countries never even arrives, because cheaper T&D prices to the customer and a larger profit margin to the handler are achieved if the waste is quietly discharged into the ocean from tankers en route. By definition, no one knows the full extent of this. But such "active" diversion only exemplifies the *general* point, that whereas "dispersive" pollution is distributed by *natural* processes, hazardous waste is distributed by human and natural processes together. Organized crime cannot intervene – nor indeed can *legitimate* actors – to extract profits from the distribution of sulfur particles in aerial discharges. The scientific uncertainties about natural processes in environmental risk analysis are daunting enough. But to these are added the even greater uncertainties created by the intervention of strategizing human agents, with diverse interactions, motives, perceptions, and values. Many of the challenges in even adequately *conceptualizing* the hazardous wastes issue turn on this point, whose full implications have not hitherto been properly recognized.

Conventionally, environmental regulation is seen as a process of back-end *cleaning up* around industrial production. It has been taken to be a marginal adjustment problem; technically, of processes, and economically, of resultant prices. With many hazardous wastes, like some other major pollution issues, such as acid deposition, this is no longer true. Adequate regulation of these issues may involve major shifts of production, or at least, major increases in production costs. A reason why this is particularly true of hazardous wastes is that it is a "last frontier" environmental problem. That is the earlier moves to control dispersive (air and water) pollution, instead of internalizing the full costs upstream in production, have "unloaded" into the less costly, largely unregulated, adjacent back-end option of indefinitely storing or dumping the *concentrated* toxic wastes that accumulate from reducing dispersive emissions. An important implication is that there is less room for compromise and fudging the conflict between minimizing production (including waste) costs and charges, and maximizing environmental and public health protection. To put this another way, there is even more pressure upon regulatory bureaucracies to interpret the ignorance and the many large uncertainties in the estimations of risks into lesser uncertainties and, thus, marginal costs.

Yet because the waste is not merely dispersed and diluted, but "packaged" and moved around as such by human agents there is a greater range of more complex behavior to be controlled, and less chance of concealing the lack of control from public experience (as witnessed in many episodes, such as Times Beach and Love Canal in the USA, the Seveso dioxin-contaminated waste barrels in Europe, etc.). Hence the public trust of regulation is more fragile and brittle for hazardous wastes than for other issues.

Even as a clean up or containment exercise, let alone anything more *positively* interventionist, hazardous waste regulation has been swamped by the rapid growth of the chemicals industry during the 1950s and 1960s. Because of the roundabout historical route to hazardous waste control via earlier dispersive emission control and then the *ad hoc* storage or dumping of concentrated hazardous waste, the full implications of this *past* growth and lack of anticipatory control in the industrialized countries have only recently become apparent. Although many countries legislated in the mid- or late-1970s, actual regulations did not appear until the early 1980s, and have had to be continually revamped to try to keep up. Indeed, the target is not merely moving, but is rapidly changing shape. A key factor now is not only growth in production and its environmental risk after-effects, but structural change and diversification, from bulk chemicals toward greater numbers of more diverse, esoteric, and more active (and hence, usually, more toxic) "fine chemicals" for more refined uses.

Set against this rapid growth, diversification, and "toxic intensification" of waste production, the development of corresponding effective regulatory institutions and methods has been lethargic. Despite extensive activity and some progress, *effective* regulation, especially at the international level, lags well behind the growth in scale and sophistication of waste types and life-cycles. Formal legislation is often undermined in practice by discretionary decisions in favor of deregulation, gross ignorance about actual waste arisings and life-cycles, or resource starvation for enforcement. An overall result is a developing tension between the values and expectations created by legislation for risk control, and the evident lack of substance of these formal policy promises. This only encourages public cynicism, as it is made to look like symbolic hand-waving only.

We do not suggest that hazardous waste regulation is unique among environmental risk management issues in suffering failures. But regulation is qualitatively more complicated for hazardous wastes: they are "managed" by autonomous human beings, not nature alone. Thus, although "implementation gaps" are neither unique to hazardous wastes nor to the last few years, the momentum of expectations generated by modern environmental initiatives, combined with specific structural features of the hazardous wastes issue make it more practically acute and analytically challenging than the conventional "dispersive" pollution risk problems whose control, in many respects, sired it.

One major factor in the complex *behavioral* life-cycles of hazardous wastes is that they are traded, and this can take place between different economic and regulatory systems. This affects the significance and the strategy of comparative analysis.

2.2. International Dimensions and Local Interactions

International movements of hazardous wastes have been increasing dramatically – more rapidly than have the volumes of wastes generated. Indeed, in Europe between 1982 and 1983 wastes shipped across national frontiers virtually doubled. Although some of this could be an artefact of reporting, a significant real increase has undoubtedly occurred. Somewhere between 2 and 3 mt of hazardous waste is thought to cross European borders every year (about 10% of the total industrial waste produced) in consignments of about 20 t each. This means about 100 000 frontier crossings per year of declared waste to be managed, documented, and checked in the EC alone. These occur between national, even local, state systems with different waste definitions, legal requirements, and technical standards. In addition there are an unknown number of transfrontier movements of "recycleable" toxic materials, a significant amount of which undergoes uncontrolled disposal. Some countries rely very heavily upon exports – estimates for the Netherlands range from 35% [11] to 60% (see Chapter 4). Some also act as large conduits for waste in transit – until recently the Netherlands took about 25% of the industrial waste from the FRG on its way to sea disposal. A major international movement of hazardous wastes has also developed from Western Europe to the GDR; for example trade from the FRG to the GDR doubled between 1982 (140 000 t) and 1983 (275 000 t) [12]. This is now said to be declining.

The uncertainties in even these rough figures are very large, however – a point whose implications are discussed in detail in later chapters.

Ironically, the lack of implementation and the inconsistency in *national* and local policies and costs, spill over into increased *international* transport, and make the lack of effective regulatory frameworks and stable agreements at this level all the more costly and vulnerable to major disturbances. There is therefore a very strong interaction between local dimensions (e.g., standards and T&D costs) and international ones.

A growing list of international bodies have become concerned with the problems caused by inadequate regulation of cross-border movements of hazardous wastes. These include the EEC, which enacted a 1984 directive on cross-border waste transport [13], OECD, UNEP, WHO, IRPTC,

UNECE, and the UN Special World Commission on Environment and Development. The intractability of risk management problems in the hazardous waste field is indicated by the fact that the international insurance business, which might in principle be able to use its financial influence to form a regulatory alliance with regulators to help strengthen national and international control, has shied away from the area (as is its right to do) because of the colossal uncertainties that it sees there. However, regulatory bodies do not have the freedom to walk away from the problems. Furthermore, it is not only environmental sectors of governments that have to try to coordinate, but industrial, agricultural, construction, and service sectors. In the bodies dealing with international pollution issues, the usual national conflicts of interest groups and institutional uncertainties are complicated by basic differences between national regulatory styles.

Regulatory bodies, local and international, have to find a balance on an increasingly sharp knife-edge between international concern about hazardous waste trade and local concern about standards and facilities, which nevertheless have international ramifications. In operations research language, the feasibility space for effective regulatory solutions appears to be relentlessly diminishing. The apparent intractability of many local waste disposal conflicts, especially siting difficulties for new T&D facilities, creates pressure to export. Pragmatism thus requires the belief that international regulatory regimes can adequately control increasing volumes of hazardous waste. Evidence to the contrary then undermines *local* confidence and practices, which in turn exacerbates the international problems.

Hazardous waste management has joined a group of environmental issues – such as acid rain, carbon dioxide, and other RATS (radiatively active trace substances) emissions, and marine pollution – which are major transnational pollution issues. Yet for all its complex international dimensions the *local* complexities of hazardous waste are *also* daunting – arguably more so than its partners in the "transnational" pollution risk category.

Because of certain unique properties, hazardous waste may be described as the black sheep issue even in this problem family. At least with acid rain there are some uniformities – there is pretty good agreement, for example, at least as to what sulfur is (even if there is disagreement about its causal role in creating damage). There is no such agreement even as to what "hazardous waste" is (and *also* disagreement about its role in creating harm). Furthermore, although on the face of it an essentially valueless material, it is packaged, concentrated, and traded – it is passed between human agents and institutions which define it in different ways. Some may define "waste" as "goods". Recall the estimated 100 000 border crossings of hazardous wastes per year in Europe. In OECD countries, on average, it is estimated that a hazardous waste cargo crosses a national border once every five minutes, 24 hours a day, 365 days a year. These are heterogeneous cargoes, wastes and risks, involving varied materials, relationships, and rationales. *Social mechanisms* transform and channel the environmental disposition and risks of hazardous wastes more significantly than, say, for acid rain. As described in Chapter 3, these and other factors introduce qualitatively new challenges to regulation, and to policy analysis.

The growth in scale and the complexity of the international aspect of hazardous waste management naturally raises the idea of a supra-national control regime. As already indicated in the Introduction, the predominant response – almost a reflex reaction – to failings is to intensify the scientific and technical basis of regulation. The assumption is that this is tantamount to clarity, unification, and coordination. Research to help handle the international problems tends to fall within this assumption. Whilst not wishing to deny the importance of clarity and realistic international coordination, we have adopted a fundamentally different approach in our research. Although this is international, we perceived that much of the existing work at this level – either policy analysis or policy development – already begged important questions, especially about the relationship between different local technical regulatory frameworks and their institutional settings, and thus what could be expected of a supranational framework. Our international research was therefore *comparative* national research rather than supranational in premise. The predominant assumption is that free international trade in all wastes is legitimate, therefore, regulation simply has to cope, whatever shape or volume trade reaches through market forces. We did not take this for granted. Given the pressure on underdeveloped international frameworks caused by *local* inadequacies, we decided that it was necessary first to see whether comprehensive local approaches could not adequately handle the problems, bearing in mind the public acceptance and implementation dimensions. We therefore had to compare different local approaches to implementation and *legitimation*, as well as technical practices, as a necessary prior input to deliberation about supranational initiatives.

Large-scale potential perturbations of policy systems, international trade, and other arrangements are created by some of the key structural properties of the hazardous waste issue. Yet these structural properties are analytically unclear even within single regulatory systems; their interactions with conventional uses of science in risk analysis, implementation processes, and public reactions need to be clarified at this level first.

In this work these interactions are systematically analyzed for their national, technical, and institutional variations. Since international regulation of hazardous waste is widely regarded as urgent, yet in practice is still embryonic, we have attempted to return to first principles. In particular, we believe that moves to establish effective *international* regulatory regimes should be based upon a better understanding of how characteristic *national* or *regional* institutional factors influence the definition, meanings, and diverse uses of what is in theory the *universal* resource of scientific knowledge and technical methods for regulation. It is therefore appropriate now to review existing work at the international level, and then discuss what comparative institutional analysis of regulation might offer.

2.3. Previous Work and the Present Study

Several international organizations have become involved in policy research and development for hazardous waste management. This has involved combinations of research, expert meetings, policy negotiations and feedback from specific actions, such as the EEC's 1978 toxic and dangerous waste directive [14]. It is worth outlining this work to show how the IIASA study is complementary to it.

The largest study has been that of the North Atlantic Treaty Organization (NATO) Committee on Challenges to Modern Society (CCMS), which began in 1973. A first phase was completed in 1977, and a subsequent phase in 1981 [15]. Later work has focused on the problems in reclaiming contaminated land. However, the main effort was more or less completed before any of the countries then legislating or about to legislate policies for hazardous waste managements had been able to gain experience of the formidable *implementation* problems, which have since become so prominent. At that stage the general assumption was that if loopholes existed in regulation, these could be closed by making the legal definitions of key technical terms more scientifically accurate and precise, and by investing more effort in local enforcement.

The NATO study examined regulatory and/or technical options, but being an early effort was naturally not sensitive to the full impact of ignorance and uncertainties, and their interactions with different institutional arrangements, in risk-defining methods and processes. A further limitation of comparative insight was that the study was confined to liberal market economies, and although there are many interesting and important differences of approach amongst these, the opportunity for comparison with a centrally planned economy did not exist.

One prophetic observation stands out from the 1977 *Final Report* of the NATO study, but its prospective insight was not followed up in further research. It was noted that if implementation of hazardous waste legislation was not more effective, then public credibility of the legal regulatory frameworks would be threatened. We would go further and add that the credibility and viability of the governing institutions themselves – not only their regulations – is undermined by the ineffective and uneven implementation that has been widely experienced since the NATO CCMS report and the enactment of most hazardous waste legislation in the mid- to late-1970s.

OECD's Council adopted a formal recommendation on a comprehensive waste management policy in 1976 [16]. Its Waste Management Policy

Group began in 1974, and has been compiling data and acting as an information exchange for several years. More recently its level of activity has increased and diversified, gathering data worldwide, as far as possible, on waste arisings, movements, and costings. In addition important policy analytic work has been conducted, especially on several aspects of the transfrontier movement problem (resulting in a formal policy in 1984 [17]), but also on problems of past waste dumps, the economics of hazardous waste regulation compliance, and the liabilities of different parties in the hazardous waste life-cycle [18]. An important project has been the attempt to develop a single cross-referencing scheme, which would relate the different hazardous waste classification systems of OECD countries, of the EEC, and in international agreements, such as the London and Oslo sea-dumping conventions [19]. Some kind of workable cross-referencing is required in order to track hazardous waste movements across regulatory frontiers and to close the many loopholes for diversion into uncontrolled disposal. Having explored the harmonization of hazard classification systems to the full, however, attention is now returning to the intrinsic limits of harmonization, and the *origins* of the international transfrontier problem. These lie in the lack of consistency in institutional processes, practices, and costs between individual nations or subnational regulatory authorities. This is why our study was focused there.

Although its own in-house policy analytic resources are slight, the EEC's practical experiences as an international regulator of hazardous wastes are important as an input to policy research. A brief digression into the practical experience of the EEC is therefore in order here.

The EEC, with whom the OECD has coordinated its policy development, first took action on wastes in 1975 [20]. A framework Directive was filled out later by more specific measures. These enactments were on waste oils (1975), PCBs (1976), and a more general framework for "toxic and dangerous wastes" (1978), involving a list of 27 categories of substances to be included under special control. In 1984 a Directive was also issued consistent with that of the OECD on transfrontier shipments of hazardous wastes within the Community [21].

The main difficulty for the EEC has been that, to obtain agreement between member states on a regulatory framework, it has had to limit the specificity of requirements and to allow significant national discretion in implementation. Thus, the original Directive *proposal* from the Commission to the Council of Ministers contained provisions for a more closely defined list of hazardous wastes, to license transporters, and to specify legal liabilities, all of which were omitted from the *finally agreed* Directive [22]. This politically necessary, interpretive license for implementation, however, has undermined the aim of improving overall control. Tightening up in some countries has also led to increased transfrontier flows to (legally or illegally) escape local increases of costs. A recent report on the implementation of the 1978 Directive [23] shows a discouraging level of success, and criticizes as a main culprit the weakness and ambiguity of the original Directive. The same was admitted by a Commission regulator [24], but with the realistic extra comment that had it *not* been ambiguous there would never have been agreement on a Directive at all. In the absence of a strong and consistent political mandate, and without any significant scientific expertise and policy research of its own to call upon, the Commission's regulatory effect in this area is hampered by reliance for information and expertise upon delegates from member states, who naturally have a certain vested interest in defending their own national autonomy or approach.

Indeed, the 1984 "transfrontier" Directive occasioned controversy within the EEC because the principles of prenotification of intention and permission to ship waste, registration, and inspection at frontiers, etc., etched national customs boundaries even deeper, against the strong current of European opinion wishing to do away with internal economic boundaries altogether. The hurried finalization of the Directive in the very nationalistic political aftermath of the missing Seveso dioxin-contaminated waste barrels (1983) added to the difficulties, and the whole affair became tangled up with internal EEC conflict with the FRG, who (under pressure from the German Green Party) were introducing new federal amendments to the waste law (see Chapter 6), to include a ban on imports to and exports from each of the German *Länder* (the regulatory authorities), except under special exemptions. This approach is viewed as contravening the free trade principles of the Treaty of Rome and provoked the Commission into threatening to take the FRG authorities to the European Court in order to revoke the import–export ban policy [25]. Some countries would be left with virtually no "regulation" at all if they were not free to export; others depend on imports to maintain the load factor and economic viability of domestic waste T&D facilities.

The whole issue of transfrontier shipments and the rigorous methods of waste classification for registration and monitoring is therefore seen to be directly linked to the broader issue of coordinating *local* waste arisings with appropriate treatment facilities. This issue is dealt with in later chapters. The reason for describing the EEC position here is to indicate how its process of policy development has been overwhelmingly influenced by inter-member state trade and politics, and very little by considerations of overall optimality, or by expert analysis responsible to the Commission itself. The practical experience of extreme difficulties in controlling the transfrontier movement of wastes in the EEC also underlines the importance of better comparative understanding of the *local* variations in practice and approach, which are largely the cause of the international problems.

The UN Environment Program has taken considerable interest in hazardous waste management since 1980. UNEP has attempted to provide a framework for ensuring that less-developed countries do not repeat the mistakes of industrialized countries, which now find that the economic and political costs of old, uncontrolled waste dumping far exceed the costs of controlled treatment and disposal from the start [26]. Part of this concern has caused UNEP to attempt to gather reliable data on the production and movements of hazardous wastes, especially across international borders and to poor countries desperate for foreign currency and with no technical resources to deal properly with such wastes.

Recently, a special UN Commission, the World Commission on Environment and Development, has taken on the hazardous waste issue, like UNEP in a mainly North–South framework. WCED's work, coordinated with that of UNEP, OECD, and the UN Economic Commission for Europe, is attempting to take a fresh look at the role that different kinds of uncertainty play in undermining policy initiatives, as well as to actually generate new, practical initiatives [27].

A major concern of all international bodies has been the lack of an internationally agreed definition of a hazardous waste. This cripples even the compilation of basic data. Attempts at conventional forms of regulation, such as registering and controlling all movements and disposals of hazardous waste, fall at this first hurdle because no workable, universal definitions of the crucial terms exist. Even so, such bodies continue to seek a formula that would act as a single internationally meaningful framework into which all national and regional definitions and hazard classifications could be fitted. This objective has been the cornerstone of all attempts at regulation; where difficulties of harmonization and implementation of international agreements or directives have been encountered, the blame has been attached to imprecision in the key terms, especially the definition of a hazardous waste. The assumption has tended to be that this ambiguity could be eradicated by more careful analysis and *technical* precision. However more recently, some experts have begun to appreciate that these apparently purely technical terms always of necessity incorporate behavioral assumptions, and reflect local, particularistic administrative concerns and other institutional factors [28]. The fact that these are not random, but are systematic, and amenable to cultural and institutional analysis, has not been developed until the present work.

International policy initiatives, therefore, such as UNEP/WHO's 1983 collaborative document, *Hazardous Waste Management – Policy Guidelines and Code of Practice* [29], offer important policy principles, especially for less-industrialized countries; but these lack consideration of the further problems of implementation that may arise in varying local institutional circumstances. Many of the very severe problems of implementation that hazardous waste policy has encountered have to do with the widely different perceptions, purposes, and strategic definitions being used by diverse actors in the hazardous waste field. Even common technical definitions can take on widely different practical meanings when transformed from policy into implemented reality.

With the possible exception of the UNEP and WCED work, all of the previous international policy analyses of hazardous waste management were, naturally enough, conceived and/or conducted at a time of innocence about the implementation difficulties yet to come. That innocence has rapidly and, in some cases, painfully evaporated. We now know that making policy *reality* rather than myth, especially at the international level, involves far more than making the crucial terms and definitions more precise. But the complex interactions between technical and local institutional factors, which make (inconsistent) implementation such an important part of policy*making*, have yet to be understood. This is where the IIASA study is targeted, as a complementary effort to those already in existence.

Thus, in summary, two important points of difference between our research and other internationally oriented work should be stressed. First, it is conducted from the relative luxury of a less directly engaged, more analytical level. Second, it seeks to examine the origins of international problems and pressures in *local* practices. International policy bodies tend, instead, to focus on the international dimensions of the regulatory problem: they take commercially autonomous waste export trading for granted, then attempt to control it at the "back-end", or on its margins. The technical harmonization and consistency of local costs and implementation practices needed for regulation at the international level is intrinsically limited by national institutional factors. Therefore, comparative research on *domestic* policies and institutional processes *is* research on the *international* policy options too.

2.4. Cross-Cultural Analysis of Regulation

It is widely thought that the regulatory systems of the UK and the USA are poles apart, especially on the basic question of landfill. The relaxed UK acceptance of codisposal (of toxic wastes with domestic wastes) is contrasted with the supposedly zealous restriction of landfill, initially in California, but since the 1983 RCRA reauthorization, in the whole of the USA. Yet when we examine the details, this picture is shattered. The concentration limits for landfill codisposal on a list of waste constituents for the US system are found to be about the same as those in the UK. In some cases the UK limits are *more* restrictive (two examples are given in *Table 2.1*).

Yet the perception about the UK's laxity is widely held, even amongst some regulators and policy analysts. The point is, however, that these apparently precise technical norms play very different practical (and symbolic) roles in each system. The UK concentration limits are advisory – and they are scattered throughout the labyrinth of advisory documents and

Table 2.1. Examples of UK and USA concentration limits for landfill [30].

Constituent	*USA*	*UK*
Cyanides	1000 mg/l	1000 mg/l
PCBs	50 ppm	10 ppm

circulars, up-dating notes, etc., in that system. The US figures are mandatory, and they are now listed clearly for all to see in the legislation itself. Further, the US *projects* an anti-landfill image, whilst the UK's symbolic projection is forthrightly in favor of landfill.

Thus we begin with a widely held belief that the US and UK landfill policies are vastly different; we then see that technical norms are in fact about the same; but even closer inspection reveals that these norms mean very different things in each regulatory system – they support different practices and relationships, and embody different uncertainties. One cannot, however, conclude that because the UK norms are advisory they are more lax: they may allow more discretion – but discretion is also explicitly built into the US use of the limits. The UK may also be, in practice, more lax because of the symbolic projection. In the (large) range of uncertainty between clearly allowed and clearly forbidden waste compositions for landfill (the precision of the figures is misleading), pressure to *allow* landfill is more likely in a symbolic climate of presumption in favor of it.

This brief example is introduced to demonstrate the importance – and the complexity – of comparative analysis of *political cultures,* and to extend this into the processes of implementation, enforcement, and legitimation or justification, not regulatory "decisions" alone.

There are several ways of approaching cross-national studies of regulation and risk management. Cross-national research seems most obviously relevant for issues where there are pressures toward international harmonization, and important loopholes due to local inconsistencies [31].

Interest in cross-national comparison of risk management began in the mid-1970s from two starting points. The first arose in a combination of geography and political science, for example, in comparing coastal zone management in California and France [32]. This developed into environmental policy analysis using the conceptual equipment of political science. The second starting point was in science and policy studies, where the observation was made that, despite the assumed universality of the underlying scientific knowledge, different regulatory decisions were being made on the same problem in different countries. This was manifestly more complicated than the conventional wisdom that scientists provided the objective facts to society, which then exercised its (variable) values of *acceptable* risk upon the facts. The first studies focused upon USA–UK contrasts; for example, the USA banned aldrin/dieldrin and DDT for alleged carcinogenicity, whereas the UK reviews cleared them [33]. Analysis showed that although, in principle, the science base may have been the same, what scientists *perceived* as the *facts* in the science varied strongly between the two countries; scientific and policy "facts" could not be clearly distinguished, with scientists centrally involved in both domains.

These and other comparative studies found clusters of elements that tend to be found together, influencing the conversion of science into practical risk management or regulation, in consistently different ways. These patterned clusters were called characteristic styles or cultures. Thus, the USA was characterized as an "adversarial" political culture or regulatory style, the UK was called "consensus" or "consultative", and the FRG was called "corporatist", [34].

Comparative studies have varied widely in their empirical scope, conceptual depth, and policy ambitions. Many have been largely descriptive [35], whereas others have kept to abstract political theory, trying to generate theoretical models of policy implementation processes. Some have suggested policy prescriptions merely by giving accounts of national discrepancies, for example, in the implementation of pollution control regulations [36]. Others have argued that such "discrepancies" are the natural outcomes of autonomous cultural systems, which exist *sui generis* [37].

In interpreting comparative studies it is worth remembering that, even without *formal* regulation, regulation of a sort is already taking place via the ordinary marketplace, cultural norms, consumer attitudes, existing social, legal, and perhaps professional obligations, public and media attitudes, wider bargaining with related parties, etc. Formal regulation may enter into these preexisting dimensions, but it does not supersede them; it shapes, and is in return shaped by them. The point is important because it emphasizes that form is not necessarily identical to outcome or substance, and that apparent strictures or laxity in regulatory standards may conceal the opposite in effects. Equally, it cautions against simplistic plucking of standards or other indices out of context and comparing them superficially. This raises the question of what indices to use as a measure of regulatory "outcomes". The growth of comparative studies of regulatory *implementation* has sharpened these questions [38]. There are therefore tremendous problems involved in the real comparison of practical outcomes, and – if desired by policy – in standardization. One of the values of comparative research may, indeed, be to demonstrate just how inaccessible such ambitions are, and to ask whether there is a policy need to find means that rely less upon such convergence.

Another important objective of comparative research is to explore the reduction of transaction costs involved in elaborate superstructures of formal regulatory organizations, by reintroducing more direct relationships between actors partly regulating themselves. These may be market or quasi-market arrangements, or they may be informal cultural traditions and social networks, where regulators blend as advisors and colleagues with those they regulate. A large literature has developed on this general issue, some of it in open admiration of the British approach, which, in comparison with that of the USA, economizes enormously on processes of expertise, review, and complaint [39]. The role of natural knowledge (scientific expertise) in negotiating *social* credibility suggests an inverse relation between institutionalized authority and the volumes of (disputed) scientific expertise in circulation. The work of Douglas, Wildavsky, and Thompson is in this general vein [40].

It is not easy to draw simple policy conclusions from such work, however. Analysis at the political–cultural level may indicate that "wasteful" competition between all parties in an "adversarial" culture, like the USA agencies, is part and parcel of the *general* culture of competitive social relationships necessary for a dynamic economy. Thus, the counterpart of more economical and consensual regulation might be a noncompetitive economy, as indeed the UK is widely thought to be. Jasanoff's (noncomparative) analysis of innovation and regulation in biotechnology in the FRG is suggestive here [41]. Comparative analysis may show the limited viability of cultural transfer of specific regulatory mechanisms, taken out of their original context. Comparison can show where flexibilities and rigidities exist in a given system. This may show up needs for the negotiation of new institutional possibilities. For example, the fact that full sharing of production information between regulator and regulated can occur with benefit in Hungary may cause it to be more seriously considered in places where "upstream" production intervention is seen as an important need, but where it has hitherto been regarded as impossible. Like transfer of technical methods required however, care is required because such conditions may only prevail in one system by invisible support from other institutional arrangements. Thus, the Hungarian arrangement may only be possible because of the lack of competition at the level of large firms.

Good comparative research on regulation will always be seeking to identify such discriminations, the boundary conditions on the effectiveness of specific mechanisms or procedures, and the interconnectedness or otherwise of specific regulatory instruments. This applies whether they are technical approaches (e.g., how precise should a standard be?), institutional instruments (e.g., an expert advisory committee or a public hearing?), or broader institutional strategies (e.g., the balance between market forces and public intervention in regulation).

The complexity and degree of detailed discrimination necessary is shown by the case of concentration limits for hazardous wastes in the UK, Belgium, and the Netherlands. At one level, the systems are similar because of this property, and differ from the rest of Europe. However, Belgium has not implemented its regulations at all, and the UK and the Netherlands use the limits in very different ways. The regulations play completely different institutional roles, involving different organizational responsibilities and relationships, and have different effects on the material control of wastes.

Generalizations can be made, for example, about the familiar, litigious, adversarial nature of US regulation and the paternalistic, discretionary character of British ways of converting science into regulation. However, within any system there is always significant variation. The particular histories of given issues and regulatory bodies, their surrounding reference groups and formal responsibilities, the rates of change in the issue climate, etc., affect both the style and content of actual decisions within the same overall "culture" [42]. Mingled with these factors will also be the inherent properties of an issue (for example, is it extremely high technology or not, international or local, etc.).

Whereas the tendency has been to suggest that given regulatory styles or political cultures are all-embracing and monolithic in their imprint upon all issues within their range, a matrix framework would be more appropriate. This would have an axis embodying *issue characteristics* orthogonal to elements of overall national regulatory style. This might, at least, indicate where incompatible or compatible combinations of issue property and regulatory option are likely to arise. For example, it is difficult to envisage nuclear power being regulated by decentralized municipal authorities; in other cases, extreme scientific uncertainty in the issue has been incompatible with legal trial of the regulations, and moves toward negotiation have been encouraged by that issue property, even in an "adversarial" culture.

In the case of an issue like hazardous wastes, therefore, with strong but poorly defined interactions between local practices and international regulation problems, comparative analysis may help to distinguish which general issue parameters are matched by which optimal modes of regulation. For example, their boundary limits of effectiveness might be analyzed, or inconsistencies in technical instruments might be examined to see whether they could be compensated by other instruments at some other point in the system. The heterogeneity of the hazardous waste problem seems to match best with more *local* responsibility for risk management. But the extent of its life-cycle, stretching across several local areas, as in the rising international trade in wastes, might militate in favor of the opposite. One of these dimensions of the issue may be more curtailable than the other; for example, shifting the balance of decentralization or centralization, or curtailing the extensiveness and *behavioral* uncertainty of waste life-cycles. As another example, an imprecise hazard classification system may not be so bad if monitoring and inspection or transporter licensing is rigorous, or if self-policing is strong. Discretionary standards may be acceptable so long as strong mechanisms for informed open, local, case-by-case review are available. Whether such supporting conditions can be fulfilled will vary from one regulatory culture to another. Such comparative exercises will not discover regulatory answers, but it should help to clarify which trade-offs are the important and feasible ones in a highly imperfect regulatory world.

To return to the matrix approach, examination of the structure of an *issue* across cultures will clarify the extent to which international consistency is *necessary*. If it appears to be necessary, cross-cultural *comparison* of the regulation of specific issues can help to elucidate, practically, how far it is *realistic* to push international technical harmonization before one begins to imply cutting into the sovereignty and internal consistency of administrative styles, and thus risk purely symbolic, nonimplementable gestures rather than effective policies. For example, the attempt to establish precise concentration threshold values of hazardous chemicals and to define legally regulatable wastes uniformly across countries in a trade area such as the EEC would encounter opposition from countries that, *as a general decision-making style*, not only on the single issue in question, give established scientific advisors large areas of discretionary judgment in policy decision making. To change the regulatory style on one issue, so as to try to make it *internationally* consistent on that issue, might not only threaten established local institutional arrangements for that issue; it might also upset the more general consistency of *domestic* decision-making style, or power relationships that have been established within it.

This brings us to the question of how much elements of regulatory style, such as "adversarial" or "consensual" modes, are bound up in public legitimation processes for regulatory bodies. I examine this again in Chapter 12, but it is worth anticipating that discussion briefly by recalling the example with which we began this section, of the widespread perception that the USA regulates landfill far more severely than does the UK. Whatever the real case – and the evidence points to variability and unacceptable landfill practices in *each* system – these cultural perceptions are clearly of *practical* political importance. They must relate to the styles of *legitimation* – the discourse of credibility negotiation around regulations – in each system. The main point is that forms of legitimation are bound up in the *substance* of different risk analyses and decision and outcomes. They are not only *post hoc* persuasion exercises. This is true even when apparently similar technical norms exist in different regulatory regimes or political cultures. As Hawkins has put it, regulation is done in a moral, not a technical world [43]. I would only qualify this by emphasizing that it is done in both; but the negotiation of shared moral perceptions, which Hawkins rightly described as necessary to effect practical compliance, and the validity of technical instruments, are both constituted in the *cultural* and institutional settings of different national regulatory systems.

If the push of cross-national issues toward consistency were strong enough, inconsistency might have to be accommodated in the local political culture. For hazardous wastes, there is already a strong concern to achieve international consistency, so the value of comparative research is more direct than it might otherwise be. Finding different approaches, contexts, and technical standards in different countries, we still need to distinguish what is universal, and what is a function of local factors. In addition to its increasingly international dimensions, however, the hazardous waste issue has other distinct structural properties of importance. We turn to these in the next chapter.

Notes

[1] Hortensius, D. (1984), *Case Study of the UNISER and BOOY Clean Episodes in the Netherlands,* IIASA draft; see also Chapter 4.

[2] Getz, M. and Walter, B. (1982), Perilous waste, in D.C. Mann (Ed), *Environmental Policy Implementation*, Ch 3 (Heath, Lexington, MA).

[3] US Congress, Office of Technology Assessment (1983), *Technologies and Management Strategies for Hazardous Waste Control* (OTA, Washington, DC). Also, *OTA Review of Experience of the "Superfund" Program,* February 1984.

[4] *Hazardous Waste Control and Enforcement Act 1983*, Bill HR 2867, US Congress, Report 98–198, 17 May 1983 (Washington, DC).

[5] US Congress (1983), *Report of Hearings on the Hazardous Waste Control and Enforcement Act, 1983*, before the Transportation and Tourism Subcommittee of the Committee on Commerce and Energy, March 1983, Serial 98-32 (Washington, DC).

[6] UK Hazardous Waste Inspectorate (1985), *Hazardous Waste Management: An Overview,* First Report, Department of the Environment, (HMSO, London).

[7] Department of the Environment (1978), *Cooperative Program of Research on the Behavior of Hazardous Wastes in Landfill Sites*, known in the UK as "the Brown Book" (HMSO, London).

[8] Harris, R. (1985), UK Severn–Trent Water Authority, *Leachate Attenuation in Aquifers: The Story So Far*, paper presented to conference on Good Landfill Practice, Castle Ashby, UK. See also, London Environmental Data Services (ENDS) Reports: *Do Landfills Pose a Threat to Britain's Aquifers?* No. 129, October 1985; *Chlorinated Solvents in Groundwaters – A UK Survey*, No. 127, August 1985; *Chlorinated Solvent Pollution Closes Groundwater Supplies*, No. 118, November 1984; *Trace Organics in British Aquifers – A Baseline Survey*, No. 129, October 1985; report of a survey (of same title) by WRC Environment, Medavenham Laboratory, UK.

[9] Commission of the European Communities (1982), *Present Situation Concerning the Disposal of Toxic and Dangerous Wastes in the Community, Implementation of Directive 78/319/EEC, and Future Actions*, Directorate-General for the Environment Consumer Protection and Nuclear Safety, B (DG.XI) XI/338/82-EN, (CEC, Brussels). See also Risch, B. (1983), The activities of the European Community on hazardous waste, in J. Lehman (Ed) *Hazardous Waste Disposal*, pp. 123–136 (Plenum Press, New York, NY).

[10] Yakowitz, H. (1985), *Global Aspects of Hazardous Waste Management*, background paper for expert meeting at the World Commission on Environment and Development, Geneva. See also background papers for OECD conference (1985) on International Cooperation concerning Transfrontier Movements of Hazardous Wastes, Basel, Switzerland (OECD, Paris).

[11] *Ibid.*

[12] *Ibid.*, and personal communications with Ministry of Interior, FRG.

[13] EEC (1984), Council Directive on the Transfrontier Movement of Toxic and Dangerous Wastes, 84/631/EEC, *Official Journal*, L 326, December.

[14] EEC (1975), Council Directive on Waste, 75/442/EEC, *Official Journal*, L 194, July.

[15] NATO CCMS Pilot Study (1978), *Final Report* (NATO, Brussels). See also Lehman, *op. cit.* [9], which lists (p. 3) the 12 reports of the CCMS Study, and which is itself the proceedings of a 1981 symposium under CCMS auspices.

[16] OECD Council Act C (76) 155 (final), September 1976 (OECD, Paris).

[17] OECD Council Act C (83) 180 (final), February 1984 (OECD, Paris).

[18] OECD, *Identification of Responsibilities in Hazardous Waste Management* (OECD, Paris).

[19] This pertains to environmental (and other) policy generally. Jenkins, W. (1978), *Policy Analysis: An Organizational Approach* (Martin Robertson, London); Mann, D.C. (Ed), (1983) *Environmental Policy Implementation,* (Heath, Lexington, MA); Mazmanian, D.A. and Sabatier, P.A. (Eds), (1984) *Effective Policy Implementation,* (Heath, Lexington, MA); Hjern, B. and Hull, C. (Eds) (1982), Implementation beyond hierarchy, *European Journal of Political Research,* Special Issue; Hjern, B. and Porter, D.O. (1981), Implementation structures: A new unit of administrative analysis, *Organizational Studies,* **2**, 211–227; Mayntz, R. (1983), The conditions of effective public policy: A new challenge for policy analysis, *Policy and Politics* **11**(2), 123–143.

[20] *Op. cit.* [16].

[21] EEC (1975), Council Directive on the Disposal of Waste Oils, 75/439/EEC, *Official Journal,* L 194, July; EEC (1976), Council Directive on the Disposal of Polychlorinated Biphenyls and Polychlorinated Terphenyls, 76/403/EEC, *Official Journal*, L 108, April; EEC (1978), Council Directive on Toxic and Dangerous Waste, 78/319/EEC, *Official Journal*, L 84, March; EEC (1984), Council Directive on the Transfrontier Movement of Toxic and Dangerous Wastes, 84/631/EEC, *Official Journal*, L 326, December.

[22] The original draft proposal of the directive was provided by the Commission of the European Communities Directorate-General XI, Waste Division, dated 1974, Brussels.

[23] Haigh, N. (1984), *EEC Environmental Policy and Britain,* especially Ch. 8, Waste (Environmental Data Services, London), for the Institute for European Environmental Policy, London and Bonn.

[24] *Op. cit.* [9]

[25] I am grateful to members of the European Commission staff and the German authorities, for accounts of this conflict.

[26] See, e.g., UNEP (1983), *Industrial Hazardous Waste Management,* joint publication of the International Register of Potentially Toxic Chemicals, and the Industry and Environment Office) (UNEP, Paris).

[27] WCED, expert meeting, Geneva, May 1985.

[28] E.g., Wolbeck, B. (1983), Political dimensions and implications of hazardous waste disposal, in Lehman, *op. cit.* [9]. E. Finnecy, private communication, and IIASA seminar on "Hazardous waste management – definitions and implementability", IIASA, Laxenburg, Austria, March 1984.

[29] UNEP–WHO joint study (J.W. Huismans and M.J Suess, Eds), 1983.

[30] See, for example, Solesbury, W. (1976), Issues and innovations in environmental policy in Britain, West Germany, and California, *Policy Analysis*, Winter; Sabatier, P. and Mazmanian, D. (1973), Implementing coastal zone management laws in California and France, in P. Downing (Ed), *Implementing Pollution Laws: International Comparisons* (Kluwer, Boston, MA).

[31] It is useful to remember also that the hazardous waste issue is treated as important not just because there are loopholes, but also because they have *escalating* negative effects (e.g., via public reaction) on the regulatory system and on the regulated when they are breached. Regulatory inconsistencies between nations in other issues would not necessarily produce any significant reaction, except by parties wishing to use the example of stronger or laxer regulation elsewhere as a tactical weapon in arguing for what they wish in their own system.

[32] Sabatier and Mazmanian, *op. cit.* [30].

[33] Gillespie, B., Eva, D., and Johnston, R. (1979) "Carcinogenic risk assessment in the USA and UK" *Social Studies of Science*, **9**, 283–298.

[34] See, for example, Brickman, R., Jasanoff, S., and Ilgen, T. (1985), *Controlling Chemicals: A Cross-National Study of Policy and Politics*, (Cornell University Press, Ithaca, NY); Wall, G. (1976), National coping styles: Policies to combat environmental problems, *International Journal of Environmental Studies*, **9**, 239–245; Vogel, D. and Khun, A. (1983) The Comparative Study of Environmental Policy, paper presented at a Conference on Cross-National Policy Research, Berlin, December.

[35] See, for example, Enloe, C. (1975), *The Politics of Pollution in a Comparative Perspective: Ecology and Power in Four Nations* (David McKay, New York, NY); Knoepfel, P. and Weidner, H. (1983), Implementing air quality control programs in Europe: Some results of a comparative study, in P. Downing and K. Hanf (Eds), (1984) *International Comparisons in Implementing Pollution Laws* (Lumer-Nijhoff, Boston, MA).

[36] See, for example, Lundqvist, L. (1974), Do political structures matter in environmental politics? The case of air pollution control in Canada, Sweden, and the United States, *Canadian Public Administration*, Spring; Lundqvist, L. (1980), *The Hare and the Tortoise: Clean Air Policies in the United States and Sweden* (University of Michigan Press, Ann Arbor, MI).

[37] Thompson, M. (1983), Postscript: A cultural basis for comparison, in H. Kunreuther, J. Linnerooth *et al.*, *Risk Analysis and Decision Processes: Liquefied Energy Gas Facility Siting in Four Countries*, pp. 232–262 (Springer, Berlin).

[38] See, for example, Downing, P. and Hanf, K. *op. cit.* [35]; Hjern, B. and Porter, D.O. (1981), Implementation structures: A new unit of administrative analysis, *Organizational Studies*, **2**, 211–227; Mayntz, R. (1983), The conditions of effective public policy: A new challenge for policy analysis, *Policy and Politics*, **11**(2), 123–143.

[39] See, for example, Vogel, D. (1983), Cooperative regulation: Environmental protection in Great Britain, *The Public Interest*, **72**, 88–106; Vogel, D. (1980), *Coercion Versus Consultation: A Comparison of Environmental Protection Policy in the United States and Great Britain*, paper presented to the Annual Meeting of the American Political Science Association, Denver, Colorado.

[40] See, for example, Douglas, M. and Wildavsky, A. (1982), *Risk and Culture: An Essay on the Selection of Technical and Environmental Dangers* (University of California Press, Berkeley, CA); Wildavsky, A. (1982), *Doing More and Using Less: Utilization of Research as a Result of Regime,* unpublished mss.; Thompson, *op. cit.* [37].

[41] Jasanoff, S. (1985), Technological innovation in a corporatist state: The case of biotechnology in the Federal Republic of Germany, *Research Policy*, **14**, 23–28.

[42] Wilson, J.Q. (Ed) (1981), *The Politics of Regulation* (Basic Books, New York, NY).

[43] Hawkins, K. (1984), *Environment and Enforcement: Regulation and the Social Definition of Pollution* (Clarendon Press, Oxford, UK).

CHAPTER 3

Hazardous Waste – What Kind of Issue?

Brian Wynne

The structure of an issue – even a "physical" environmental issue – is not given and ordained by nature, but is organized by social processes. This seems obvious at the margins of an issue: for example, what size of producer is to be regulated as a hazardous waste generator. But the internal structure is also socially defined: for example, how "hazard" is classified, what counts as "waste", or whether the main problem is to reduce waste generation or keep existing treatment and disposal (T&D) facilities operating at full capacity.

However, this analytical insight should not seduce us into sociological reductionism. A given issue does have intrinsic physical and technical characteristics – and perhaps some organizational parameters – that are common to whatever social context in which it arises. A troubling point about hazardous waste is that one of its intrinsic features seems to be the great freedom of social actors *within the system* to define the problem and its key terms. In this chapter we explore the complex balance between the social bounding and construction of "the" hazardous waste issue, and the intrinsic properties with which the processes of social construction operate.

Regulators, of course, are key problem constructors, though as indicated in Chapter 1 there are systematic divergences even within regulatory authorities, especially between center and periphery. These are more than divergences of interest, but more of fundamental frameworks. The challenge for regulatory policymakers is to impose a problem framework, of risk analysis and management, which structures the behavior of the different regulated parties in acceptable ways, while not interfering with other freedoms that are regarded as legitimate.

For perspective we begin by noting the arbitrariness of the conventional boundaries of the hazardous waste issue. Then we note its relationships with adjacent issues, and different possible strategic orientations toward its regulation. We then examine some idiosyncratic properties of hazardous wastes, which make their management peculiarly difficult and analytically interesting. These structural properties interact strongly with our central analytical themes – implementation, uncertainties, and credibility – and their implications are developed through the book.

3.1. The Social Construction of Natural Problems

Although all policy issues are socially defined, amongst issues with prominent technical dimensions hazardous waste does seem to stand out as especially sensitive to institutional factors.

The largest, most elaborate regulatory framework for hazardous wastes ever devised, namely the US Resources Conservation and Recovery Act (RCRA) of 1976 [1], was, as its name suggests, originally intended as a response to the 1974 energy crisis, and hazardous waste management was a minor part of the bill. This was dramatically overturned in 1978 with the discovery of Love Canal [2]. In every country practical hazardous waste regulation is strongly shaped by its origins in ordinary garbage disposal, whatever the formal content of more recent legislation. It is always worth digging beneath the current structure into the historical emergence of such problems, to see how *unnatural* such problem boundaries may be, and to understand the social influences that shape and constrain them. Exploring in this way can help to understand practical options and constraints, and the time scales on which these operate. Defining hazardous waste is a hazardous task, and ultimately rests on institutional values and commitments as well as innate properties of materials.

Let us look briefly at just some of the problem boundaries:

(1) Like squeezing a balloon in one place, for it to inflate in another, hazardous waste volumes and types are directly affected by regulation of atmospheric or aqueous dispersal of pollutants.

(2) "Hazardous" waste should strictly include municipal, household waste, because the toxicity of many domestic wastes – batteries, cleaning and polishing fluids, cosmetics, medicines etc. – is higher than that of some industrial wastes. The toxicity of some leachates from municipal landfills is at least as high as that from controlled, "hazardous" waste landfills [3]. The greatest source of environmental cadmium is thought to be from batteries thrown out in domestic waste. Municipal waste incinerators are also known to emit dioxins with their aerial discharges, possibly to worse levels than toxic waste incinerators. Yet municipal waste is excluded, not for technical risk, but pragmatic

reasons. To control municipal wastes as hazardous would not only swamp resources, but would probably create enormous political reaction too [4].

(3) Radioactive waste disposal falls under entirely separate legislation, interest groups, and institutions. Although there are analogies, especially in terms of siting difficulties, hazardous waste management normally excludes radioactive wastes [5]. The contrast is discussed later.

(4) Hazardous waste should include all domestic sewage, which is usually itself toxic in organic matter and heavy metals. The US Environmental Protection Agency's (EPA) standard test criterion to define wastes as hazardous was relaxed by a factor of ten because the original contamination criterion would have included most domestic sewage, again swamping the system [6].

(5) Agricultural wastes are generally excluded from hazardous waste regulation, yet they can be highly toxic. In the EEC, roughly 40 times more agricultural waste is generated than "hazardous" industrial waste. Of course, such materials may be seen as goods instead. In Hungary the "hazardous waste" population to be regulated suddenly increased by a factor of two or more when animal carcasses from the meat industry were included in the regulations.

(6) Most systems exclude mill and mine tailings simply because they are so colossally voluminous. It also seems to offend common sense that something should be neutral when in the ground, but when *extracted*, even if a toxic component (e.g., a metal) has been removed, it is regarded as toxic waste. Often, of course, residual contaminants may be more environmentally mobile (e.g., radon and radon daughters from uranium mining and mill tailings) and thus more hazardous than when in geological strata. This kind of ambiguity caused a controversy in the USA, where oil drilling muds and brines were initially defined by EPA as hazardous wastes, only eventually to be excluded because the implications for the oil exploration industry were so horrendous. Dow Chemicals also (in the end successfully) argued against having brine from which they had extracted bromine defined as hazardous waste when they proposed to pump the less toxic brine back from where they had extracted it [7].

(7) Many common natural substances would be defined as hazardous waste if the definitions used in regulations were strictly enacted. "Phenolic substances" (e.g., wood) and "biocides" (e.g., salt) are categories of toxic and dangerous waste in the EEC's 1978 Directive. Some regulatory definitions of hazardous waste by the property of "reactivity" include the emission of methane and hydrogen sulfide. Without discretionary freedom to ignore this criterion when pragmatically necessary, all bulk domestic wastes would have to be regulated as hazardous on this count. To specify precisely and comprehensively the

appropriate conditions of inclusion or exclusion from control would be absurdly elaborate, if not downright impossible.

(8) Many hazardous wastes are exempted from national regulatory systems of registration because they are dealt with in-house by their producer. This excluded proportion of all hazardous waste is estimated to be as high as 96% in the USA, and over 70% in several other countries. It is thus a major factor in the social narrowing of the official problem definition.

(9) Nearly all hazardous waste regulations exclude small-volume waste generators, which markedly reduces the population of generators to be regulated; but in aggregate this excluded sector may produce large amounts of hazardous waste, moving around unregulated. The amendments to the US RCRA in 1983 reduced the threshold for small producers to be regulated from 1000 kg per month to 100 kg per month. The original threshold was claimed to have *excluded* 90% of the hazardous wastes generated in the USA from regulation [8], though this is contradicted by official views [9], which state that the 1000 kg threshold regulates 99% of the hazardous waste (which is not, of course, 99% of the generators). Others have pointed out that the 695 000 estimated generators who collectively produce 1% of all hazardous wastes in the USA may nevertheless be a disproportionately large contributor to improper disposal [10].

(10) Many substances defined as "goods" are wastes being stored indefinitely, for theoretically possible future recycling. This is a notoriously difficult loophole. It was partly in response to this ambiguity that the 1983 amendments to the US RCRA obliged the EPA to include underground storage of petroleum products and hazardous substances: this was also a major expansion of the population of regulated materials.

For a further perspective on the selectiveness of regulatory attention, it is useful to remind ourselves of some general figures: [11]:

(1) There are about 7 000 000 known chemicals.
(2) Approximately 80 000 are in commercial circulation.
(3) Approximately 1000 new chemicals enter commercial use each year.
(4) Using the *total* of world laboratory resources, about 500 chemicals per year could be testable for toxicity (at colossal expense).
(5) One test, for carcinogenicity alone, can involve 800 test animals and 40 different tissue specimens per animal for pathology examinations; that is, 32 000 specimens. This needs approximately $500 000 and 3.5 years to perform.

(6) There are approximately 14 000 food additives and contaminants. Many natural components are also thought to be toxic.
(7) The US EPA hazardous substances list contains approximately 500 substances. In practice, only 114 of these are regulated.
(8) The UK and EEC hazardous substances lists contain about 30 items.
(9) The US EPA's EP test for hazardous wastes covers only 14 chemicals.
(10) In the USA, the Occupational Safety and Health Agency (OSHA) had, in 1984, regulated no new chemical for six years.
(11) In the EEC in 1981, the total waste generated was estimated to be 1.8 billion tonnes, of which about 20 mt was legally defined as "toxic and dangerous" for formal regulation [12].

Even this rough indexing indicates just how severely any framework for hazardous waste reduces the full scope of the potential issue to manageable proportions. This reduction creates artificial boundaries and internal categories and structures, which have little to do with objective, natural dictates, and much to do with social factors, pragmatic necessities, and administrative purposes. An example is the way that agencies are created around a given social function – e.g., agriculture and soil – thus creating inconsistent but often overlapping jurisdictions. It is reported that during the development of the RCRA regulations at the EPA a pesticide waste standard was proposed that would have rendered a sprayed apple, that could be legally sold and eaten, prohibited from disposal on a municipal landfill [13]! In this case the anomaly was spotted and rectified. The process of reduction also leaves many central aspects of the problem – even apparently precise technical matters – open to contradictory definition. As is shown later, neither "hazard" nor "waste" have intrinsic natural meanings free of social and institutional determination. This is true between *different* national or state systems, but it is also the case between different groups and authorities within the *same* system.

In practice, different countries create *roughly* the same frameworks to attempt to regulate this ill-defined field of risks. At first sight the resultant differences may appear small when set against the enormity and vagueness of the target. But in reality they inevitably leave large loopholes and inconsistencies, which are rightly a major concern. It is not only that systems differ in their initial definition of precisely what counts as a hazardous waste for regulation, but they also differ in who they make responsible for such identifications; what they require, of different parties, once a waste is so defined; at what stage of the waste life-cycle and for what purposes, and with what degree of precision it needs to be so defined; and in various further respects. We will look later at these differences and the implications they have for risk analysis.

3.2. Strategic Orientations for Hazardous Waste Management

Historically, the developing framework of environmental regulation has followed relatively self-evident empirical categories – specific media, such as public hygiene, occupational groups, food, water, air, and soil, or, occasionally, specific agents, such as pesticides. Different reasons for regulation, different interest groups, different economic and other constraints, different legislative initiatives and implementing institutions, and different technical specialist inputs have tended to compartmentalize policies, criteria, regulations, and practices into these separate fields. Each has developed its own momentum, purposes, and traditions. Sometimes there is coordination between overlapping agencies, but often there is not. Yet actions and approaches in one field may have strong effects upon neighboring areas, and in their full life-cycle toxic chemicals may pass through several such uncoordinated regulatory fields, allowing loopholes, confusion, and the wastage of already scarce regulatory resources. Furthermore, each sectoral approach has been marginal and reactive to upstream production decisions and practices.

The hazardous wastes problem is conditioned by the fact that it eventually came to regulatory attention largely through ordinary domestic waste management. Since this was always of low priority compared to air, water, or food pollution, it was subjected to explicit attempts at control later than all these other environmental areas. Cleaning up *dispersive* emissions to air and water to meet tightening regulations for these media left increased volumes of more concentrated, hazardous materials, such as filter sludges, condensates, dry-precipitated dusts, slags, and so on. These could not be handled by domestic garbage systems. It was the dramatic impact of previous regulation in the conventional areas upon volumes of toxic sludges, concentrated liquids, etc., that eventually stimulated industrial and regulatory concern about what to do with hazardous (packaged or concentrated) wastes.

These intermedia switches are the product of unsystematic sectoral decisions and of arbitrary historical sequences of selective attention. A "chemical life-cycle" regulatory framework, for example, would be more proactive, and analyze projected chemicals through production, products, uses, waste cycles and dispositions across media. Waddell refers to the syndrome of "solving one environmental media problem at the expense of another medium," and in particular, "a tendency to generate solid/-hazardous waste problems [by] solving air and water discharge problems" [14]. This is one of the reasons why hazardous waste arisings have been thought to be growing faster than the general economic growth in the last decade or so. A sense of the potential scale of this perturbation is given by noting that a typical, medium-scale, coal-fired power station burning medium grade sulfur coal would yield hazardous solid waste of one square mile of elemental sulfur one foot deep, if all the sulfur were electrostatically precipitated out from the normal annual aerial emissions [15].

Risk processes in themselves are basically continuous across the socially constructed boundaries of legislative framing, administrative competence and attention, technical specialization, and political–economic interest. There is no guarantee that regulatory boundaries as established bear any relationship to some optimal, overall environmental or economic norm; indeed, there are many reasons to suppose the opposite. Yet, for better or for worse, these (often contradictory) socially institutionalized boundaries and divisions have developed and evolved as the concrete reality in which policymaking and implementation have to perform. These institutional processes and structures have substantial, objective [16], causal effects on events.

The arbitrariness of the institutional patterns that carve up natural systems is sharpened by conflicting pressures, e.g.:

(1) Intermedia continuity and movements of pollutants, which coincide with the political pressure toward integration and centralization of pollution management.
(2) A dynamic of fragmentation within regulatory bureaucracies as a rational response to their organizational environment. Internal bureaucratic groups specialize toward separate problems, expertises, and external interests, and thus define separate boundaries round their fields of attempted control [17].

There is no preordained formula that balances such conflicting institutional tendencies; compromises evolve *ad hoc* with shifting, and often contradictory definitions and boundaries. Formal legal definitions and regulations are therefore highly unlikely ever to end up totally consistent between different regulatory systems, nor are they ever likely to overcome interpretive variances and uncertainties that can undermine regulatory effectiveness.

We can now see that, even from a central policy standpoint, there are at least three fundamentally different, primary definitions of hazardous waste management as a policy problem:

(1) As an extension of conventional pollution abatement, risk management, and environmental protection.
(2) As a problem of industrial innovation and investment.
(3) As a problem in changing industrial processes.

3.2.1. An extension of conventional environmental protection

As historically the last area to be regulated, hazardous wastes are not only swollen in volume but also leave less room to maneuver between different possible options for proper treatment and disposal. Thus, physical prohibition (e.g., discharge limits) or ambient performance standards as regulatory instruments for dispersal into air, water, etc., left some room for the autonomous choice by an industry as to how to meet such standards. Furthermore, the economic implications of controlling and internalizing such externalities were artificially suppressed precisely by creating the hazardous waste problem. The *economic* costs of hazardous waste control therefore become less marginal and loom larger over mainstream production decisions than those of conventional pollution control. The logic of hazardous waste management is thus to become more positively interventionist, implying external *direction* of industrial practices and choices. As a leading figure in the field, Berndt Wolbeck, put it, "A sound policy should not confine itself to saying what should not be done, but should, above all, indicate what should be done" [18]. Indeed, in the 1984 amendments to the FRG's Federal Waste Law, there are included stipulations about the proper treatment technology for specific wastes. The UK is also considering more specific regulations on this point [19], and the US Congress RCRA reauthorization [20] expressly excluded landfill for a range of specified wastes.

This naturally implies more potential conflict between industry and regulatory bodies. It also raises the shadow of a greater political burden on science, as the technical justification for stronger (and more contested) interventionist directives. This is the context of risk analysis in regulation, where science is one of the few resources of authority (in addition to formal legal sanctions) available to regulatory bodies.

A further implication of the risk management perspective is that the problem area should be capable of definition in terms of risks to human health and valued environmental entities, and in comparative terms where possible. Yet, the hazardous waste issue is, in important respects, simply underripe for conventional risk management – it may be dominated by dimensions of sheer *ignorance* and indeterminacy, rather than by definable risks. This important difference is obscured by the social process wherein organizations artificially redefine such radical uncertainties as manageable probability functions [21]. It may be significant that whilst regulatory bodies responsible for the issue may portray it as manageable in terms of risk definition, others more free to choose, such as insurance companies, are voting with their feet and steering clear of this area of risk coverage, at least until it becomes better defined [22].

3.2.2. A problem of industrial innovation and investment

Although conventional dispersive pollution control needs technological development (e.g., filter systems), the management of hazardous wastes requires that a whole new industrial field be created and maintained. When most industrial countries enacted legislation on hazardous or "special" wastes in the 1970s, few of them had much existing industrial facilities or infrastructure for collection, transport, storage, treatment, and disposal of hazardous industrial wastes, beyond what was available for ordinary garbage. Therefore, an urgent problem was to develop such an industrial network. The US EPA's aim in implementing the 1976 Resources Conservation and Recovery Act (RCRA) was to stimulate the growth of such a T&D industry via private investment in what were assumed to be relatively well known technologies. The theory was that the regulations would create a large, attractive "market", that is, a defined population of hazardous wastes legally requiring licensed treatment; and free enterprise would do the rest. However, as Kragg describes [23], this failed to happen for a complex set of reasons. Foremost is undoubtedly the unusually direct connection between the details of regulations and the scale of the potential market of wastes thus defined. The uncertainties surrounding regulatory definitions created crippling uncertainties as to the scale of the market for possible investors. These were exacerbated by the variability of waste mixes, physical forms, and chemical compositions, making for unpredictable variations in the required treatment conditions. Not only that, but the degree of regulatory intervention necessary to fulfill regulation was too stifling to a fledgling industry in an already unstable technical and commercial state.

This paradox, that regulation requires industrial innovation, *which is itself jeopardized by requiring regulation*, is unusually sharp in the hazardous wastes case. It does not really exist for dispersive pollution; here, regulation often involves development of control technology, but not of a whole industrial network that itself also needs extensive regulation [24].

Other countries, such as the Netherlands and UK, adopted approaches similiar to those in the USA and encountered the same difficulties. The Netherlands, for example (see Chapter 4), enacted formal legislation to define and regulate chemical wastes (including prohibition of landfill) *before* establishing an industrial treatment and disposal infrastructure. It was persuaded by industry that such an infrastructure already existed, but of the three central elements, one (the AVR incinerator) developed serious technical and financial difficulties, one (UNISER) was closed down after prosecution for fraud, and the other option was export. The government presumption that, once controlled wastes were legally defined the market would ensure private investment in T&D, was undermined by a combination of uncertainties, and the freedom to use cheaper export options. Other regions or countries, such as Bavaria and Hessen in the FRG (see Chapter 6), and Denmark, adopted very different approaches.

From this industrial innovation perspective the problem of further analyzing and discriminating risks from different T&D sites and methods, etc., is less important than the organizational problems of evolving and protecting an effective industrial infrastructure for waste management. The degree of refinement of the risk-analytic framework that is considered to be necessary and feasible is, in practice, more crude than formal risk management approaches imply.

There is, therefore, a deep conflict in the issue: between necessary industrial innovation, which usually requires *protection*; and regulation, which involves interference and restriction. Different actors in the overall arena define this problem in different ways, as is made clear in Chapter 4. Indeed, in any T&D facility siting or safety conflict, the regulators' prime objective is usually to keep the T&D industry in business. A crucial factor in the management of this balance is the hazardous waste classification list (see Chapter 5), whose technical specifications may substantially affect the market of declared wastes for T&D. The key point, however, is a conundrum: Hazardous waste regulation needs industrial innovation of a fragile industry whose own maintenance requires that it not be regulated too closely.

3.2.3. A problem in changing industrial processes

In this perspective the critical focus is further upstream in the production–waste cycle. Despite the prominence given to resource recovery, recycling, and waste reduction in the public language of hazardous waste management policies, the reality is that rates of recycling have in some places languished, even fallen back in recent years [25]. Claims of success are made, but by and large progress, if it exists, has been patchy and unimpressive overall. Undoubtedly, a critical problem is the inevitable lack of stability of volume and quality of the input "raw materials" – which are someone else's wastes – to recovery plants, which may be highly sensitive to input composition. It is generally agreed that here, technical and commercial possibilities are severely undermined by institutional factors, including contradictory perceptions of the materials in question.

Recovery and recycling overlap with, but are different from, innovation in waste-reducing production technologies, since the latter are self-evidently in-house to the waste generator, whereas recovery and recycling may well be through transfer to some other operator(s) in the "waste" life-cycle. The production–innovation approach has found its sharpest conceptual expression in proposals to integrate notions of "true" waste costings in strategic planning and in the choice of new processes and products in industry. This is consistent with the integrated "chemical life-cycle", cross-media approach mentioned before.

While waste reduction and recycling is in theory convergent with proper waste T&D, in practice there are several intractable conflicts:

(1) Investment in adequate "downstream" T&D facilities creates a vested interest against successful "upstream" waste *reduction*, because of the need to run T&D facilities at full capacity.
(2) Waste reduction is encouraged by increasing the costs of proper T&D, but this is only effective if the waste life-cycle (see later) does not "leak". In reality, increased costs of a T&D option tend simply to push the wastes elsewhere, e.g., into illegal disposal or (legal) exports to cheaper alternatives.
(3) Encouragement of waste recycling requires loosening restrictions on waste definitions, movements, and exchanges, so as to allow entrepreneurial wisdom the scope to identify valuable resources in others' wastes. Again, a mechanism that encourages upstream innovation also facilitates downstream laxity.

Recovery and recycling imply that the materials are goods, whereas T&D implies that they are wastes. If they are defined as the former they are exempt from regulation, yet the discrimination is ambiguous, slippery, and dependent upon who is defining it. A galvanic sludge containing copper, for example, might be "stored" uncontrolled against speculation on the future price of copper or against further accumulation into an economically viable tonnage. Many solvents move around unregulated as "recycleable" materials, only to be eventually burned or dumped in improper ways. The implications of this are further explored in Chapter 10. The impossibility of tightening this boundary by precise definition of waste is illustrated by an expression of legitimate concern on behalf of the UK reclamation industry:

> The UK reclamation industry has a turnover of some 2.5 billion pounds sterling amongst 1000 firms involving 20000 employees; there is a 1.25 million pounds sterling credit as a result to the UK balance of trade. We have already seen how difficult it often is to distinguish between process materials and waste materials to be processed: excessive documentation based upon fictional descriptions must not impede the free movement of these materials [26].

Whereas one sector in the system sees "fictional description" of wastes as a main cause of lax practice and unacceptable *environmental or health* risk, another, equally active sector sees "fictional description" of wastes as an unacceptable commercial risk to a legitimate industry.

To summarize, there is a basic three-way conflict. Successful T&D creates a demand for wastes, which undermines recovery and waste-reducing production innovations. Successful external reclamation also undermines back-end T&D, and production–innovation. Encouraging upstream innovations by demanding tighter and more costly *legal* T&D risks pushing wastes into export or illegal dumping. Organizing a *regulatory* industry commercially creates wicked conundrums.

Ultimately, a risk assessment framework for the *waste-reducing* strategy can probably only be developed via a chemicals' life-cycle approach. The OECD's program on toxic chemicals regulation, for example, has explicitly incorporated the adequate disposal of wastes as part of an integrated, anticipatory control framework covering the entire life-cycle of a chemical, from inception (as a good or a waste, regardless) to grave [27]. This would be a systematic way of reflecting what is happening anyway – but as yet unpredictably and inconsistently – as sporadic public resistance and unevenly multiplied back-end costs force industry selectively to bring waste considerations from the margins into the center of upstream investment and technology development calculations.

The aim of this section has been to demonstrate some deep strategic conflicts and dilemmas within the structure of hazardous wastes management. These are prior to the concrete workings of conflicting frameworks brought by the diverse actors within hazardous wastes life-cycles. It does not seem possible to reconcile these different strategic definitions of the core problem, so our interest should be on the ways in which the different institutional structures in different national systems negotiate different practical compromises between them. We then need to ask how these strategic definitions proliferate or transform under the influence of diverse actors and interests in the field – its ill-structured, heterogeneous behavioral nature – even within a single regulatory setting.

In the background of these questions remain those about the interactions between scientific knowledge, technical uncertainties in risk analysis, and the deeper institutional uncertainties, which the technical language obscures. Analyzing the inherent structural properties of hazardous wastes performs some necessary groundwork for clarifying these interactions.

3.3. Structural Properties of Hazardous Wastes as a Regulatory Problem

We have identified six main issue characteristics that have crucial effects upon rational strategies for its management, and for its analysis. These interpenetrate, but it is useful to describe them separately – the extreme heterogeneity of the field; the extensive and indeterminate behavioral–technical life-cycles of wastes; the tension between intrinsic and situational risks of wastes; the multiple and often competing points of possible intervention, decision, and analysis for regulation; the levels of professionalism; and the uniquely inverted exchange relationship for materials and payments when wastes rather than goods are involved.

3.3.1. Heterogeneity

Potentially hazardous chemical wastes arise in a huge range of industrial activities. Indeed, chemicals are used in almost every human activity and, inevitably, eventually become regarded as waste. Most, if not all, hazardous waste legislation expressly excludes not only domestic and other nonindustrial wastes that may contain hazardous chemicals, but also specifically exempts small waste generators (typically $<$ 10 tonnes per year) and many other wastes, such as oil drilling sludges, various refining slags and mining tailings, etc. (see earlier). Even so, with the framework so whittled down that some estimate it to legally exclude more than 90% of all hazardous wastes, the US RCRA, for example, is still supposed to regulate the activities of over 60 000 separate waste generators, producing and transferring to a free market thousands of widely different chemicals in different consignments and unknown mixtures. Regulation has to cope with great uncertainties in these figures, despite repeated analysis. Early attempts to survey the US problem produced even greater estimates of generators, and reducing the generator threshold to 100 kg per month multiplies the regulated population to about 200 000 (and to zero, up to about 700 000) [28]. These thresholds are very flexible in practice, and single generators may also have multiple sites and operations. There are also approximately 50 000 waste transporters in the USA, performing an estimated 200 000 million tonne-miles per year of waste haulage in about 1.5 million vehicles [29]. The treatment and disposal industry in the USA, let alone waste generators and transporters, involves an estimated 75 000 facilities of many different types. In addition, there are over 4000 officially reported hazardous waste storage facilities of varying kinds, storing about 600 000 tonnes of diverse hazardous wastes. The movements, transfers, mixing, storage, and treatment of these wastes is all performed by countless, uncoordinated decisions and arrangements operating under normal decentralized market principles.

The regulated actors range from international giants, such as DuPont, to tiny family businesses with one operation. There are also many new industries, emerging rapidly in areas like microelectronics and biotechnology, that are small, but numerous, and use small quantities of highly toxic and little-known materials. The extreme differentiation of scale in the field is emphasized by the fact that 4% of hazardous wastes in the USA was generated by 84% of the generators. There are countless industries creating chemical wastes with less than 100 employees, yet 30% of all chemical sales (in all market economies) are achieved by only 25 giant companies. One giant hazardous waste T&D company, Safety Kleen, began as a solvent collector from small dry-cleaning firms with an average pick-up of 7 gallons [30]. This also indicates why a key expertise in hazardous waste T&D has been in transport and collection systems, and why middle stages, such as collection and transfer station, mixing, storage, etc., complicate the regulatory problem.

A similar portrait of extreme heterogeneity, albeit on a correspondingly smaller scale, can be painted for other countries. Japan is estimated to have about 9000 T&D plants. In the UK there are thought to be over 100000 different "special" (hazardous) waste producers; but these special wastes have to be segregated from the wastes of between 1.5 and 2 million producers of industrial wastes overall. The UK also has a register of over 5000 T&D facilities (mainly landfills), but it was unable to determine how many of these were genuine "special" waste facilities, because their legal, site license conditions were so vaguely worded (see Chapter 7). Transporters and intermediate waste handlers, or brokers, are unlicensed.

Thus, both hazardous wastes themselves, and their generating industries, transporters, brokers, and T&D operators, are extremely dispersed, heterogeneous and ill-defined. Under these circumstances it is easy to understand why, even after major efforts to obtain information in the last decade, ignorance, uncertainty, and indeterminacy predominate, especially in defining hazardous waste for registration of its types, producers, volumes, movements, handlers, and final disposal. Apparent details, such as quantity thresholds or concentration thresholds for producers of toxic constituents, can influence the regulated population enormously. So, too, can autonomous and variable interpretations of the regulatory definitions by waste producers and handlers themselves. Even in a supposedly coordinated system like the EC, for example, inconsistent standards and methods create an extra heterogeneity. For example, the FRG defines any waste containing hazard-listed chemicals as "special", whereas the Netherlands, UK, and Belgium use concentration thresholds. Even the latter are inconsistent, however, being based on very different approaches. Thus, cyanide wastes are legally hazardous above 50 mg/kg in the Netherlands, but above 250 mg/kg in Belgium. A dioxin content above only 1.2 ppm is regarded as hazardous waste in the UK, but above 50 ppm in the Netherlands [31]. Even within single national systems, marked variations exist between authorities on, e.g., site-licensing standards, transport controls, and even hazard classifications. In practice this means that even the *same* material can change character as it moves from one handler to another, or one area to another, without any physical or chemical transformation occurring.

In addition to this general uncertainty and irreducible indeterminacy, there may be genuinely different *optimal* regulatory strategies for different kinds of waste and different kinds (e.g., size) of generator. However, they are all combined under an undifferentiated policy umbrella of "hazardous waste management", in which the extreme problems of identifying and discriminating the *organizational* targets for regulation are directly exacerbated by the *conceptual* impossibility of precisely defining the relevant *materials*, and vice versa.

All this offers a striking contrast to the structure of the nuclear waste problem, with which chemical waste is often compared. Relatively speaking, nuclear wastes involve a very narrow range of well known radionuclides in relatively narrow bands of composition. Furthermore, they arise in a limited number of well known sites and are transported in a small number of consignments to very few treatment or storage facilities. It is easy to keep tabs on the arisings and movements, and those arisings are, in chemical and hazard terms, rather uniform and easily supervised. The largest waste producer, the nuclear industry itself, is relatively monolithic, with few centers of decision, control, and responsibility [32].

Thus, whereas regulators of nuclear waste or spent fuel at least know what is produced and moving where, chemical waste regulators are in a completely different position. In important respects the regulatory problem for chemical wastes is more akin to regulation of nuclear proliferation than of nuclear wastes; there is a need to define, detect, and control unknown, perhaps clandestine, activity.

Those problems of heterogeneity and regulatory ignorance of the basic elements of who and what is supposed to be regulated are highlighted by the contamination of groundwater from the rapid growth of microelectronics, led by the mushrooming of many small firms [33]. These discard small volumes of new, highly toxic chemicals, (e.g., as surface cleaning or etching agents), but they are often too small to carry in-house chemical expertise, and they produce batches of waste possibly too small and variable to be worth collection for recycling.

The question of industrial structure is important because it strongly affects the strength of self-regulatory and external regulatory supervision. Uncertainty about industrial structure also makes it difficult to evaluate the assertion that production threshold exemptions exclude a vast number of generators, but only a small proportion of wastes, from regulation. Claims on this vary. OECD and EPA data [34] contradict the view of others that the larger proportion of hazardous wastes falls outside of regulation by this loophole. The great difficulty is that "special" or hazardous wastes are a (variable) subset of a much larger population of wastes, which cannot all be regulated just in case they contain hazards. The lack of a clear-cut technical or organizational basis for distinction leaves a pervasive insecurity as to what proportion of what *should* be controlled is, in fact, left uncontrolled. The extreme heterogeneity amplifies this *institutional* uncertainty. Whatever the real scale of the problem, policymakers have indicated that the countless, unknown small generators are the greater source of regulatory anxiety. In this regard, it may also be relevant that some countries or states, such as Denmark and Bavaria, are making strenuous efforts to establish effective, regulated collection and treatment of even domestic hazardous wastes, which begin at each household as small volumes of highly toxic materials, but may end up on municipal sites in large volumes. In Britain it is now estimated that cases of soil pollution by mercury and cadmium (amongst other things) from domestic waste sites may be as bad as from industrial waste sites [35]. It is also necessary to remember that the reassurance of small volumes *per se* in aggregated figures may be misleading if legally uncontrolled small volumes of highly toxic waste are locally concentrated in human exposures.

A question raised earlier is whether there is, in fact, a unitary regulatory field. Can standardized regulatory approaches based on this assumption realistically tackle the issue? Is the same regulatory framework as appropriate for a transient "firm" of two or three individuals stripping and disposing of asbestos wastes, as it is for a major international chemicals company? Even between established and respectable companies the differences are very wide. Small firms, by definition, have less control over their economic and social environment, and are likely to be less concerned with the long term pay-off or side effects of present decisions. Large companies may be more secure, but with greater forward commitments at risk; they have more control over their immediate environment, and may have a national image and international trade to nurture and protect. They are likely to have a longer term planning and decision-making horizon, and will also want consistency and predictability across a wider horizontal area. These cultural aspects of decision making, and various organizational and economic factors, such as how internal accounting and responsibilities are allocated, affect hazardous waste regulatory relationships. They are highly variable.

Regulatory bureaucracies tend toward fragmentation into necessary specialist competences and approaches, dictated partly by the innate differentiations in a given issue [36]. The heterogeneity of hazardous wastes accentuates these tendencies as does the legacy of its origins in fragmented local municipal waste disposal. Another factor is the multipolar character of the hazardous waste "life-cycle", which contains different nodes of possible risk control.

3.3.2. Waste life-cycles

A defining characteristic of hazardous waste is that it is not dispersed and diluted in the environment. It is concentrated and needs handling and treatment before eventual destruction, containment, and/or dispersal. The immediate significance of this is given by *Figure 3.1*.

In order to emphasize the contrasts between hazardous waste and conventional environmental regulation, we can outline the possible stages in the life-cycle of any single "packaged" waste:

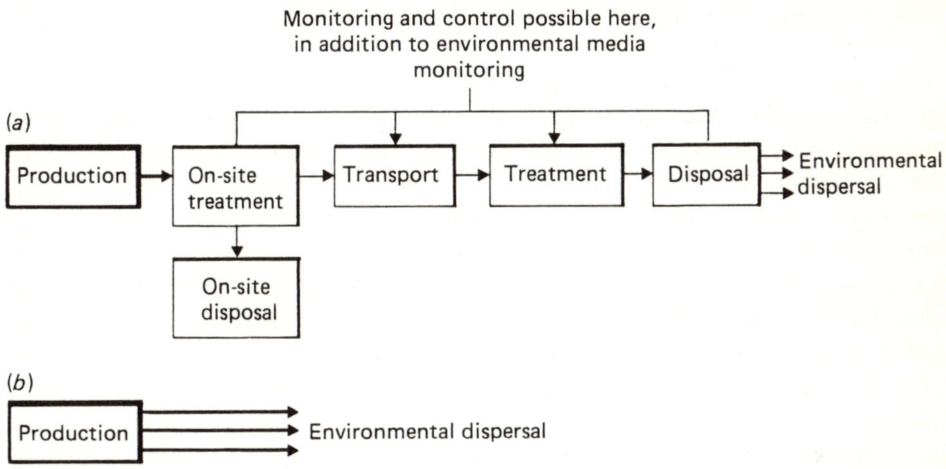

Figure 3.1. Differences in terms of potential emissions and risks between packaged hazardous wastes as in (*a*) and conventional dispersive pollution as in (*b*).

(1) In-plant process generation.
(2) Mixing with other in-plant process streams.
(3) Factory-gate arisings, "packaged" into transport consignments.
(4) Collection and transport.
(5) "Interim" storage (notorious as *de facto* deposition sites).
(6) Mixing, repackaging, and reloading at a transfer station.
(7) Treatment and transformation (deliberate or inadvertent) – new by-product wastes.
(8) Further collection and transport.
(9) "Final" treatment or preparation for disposal (e.g., for incineration or landfill).
(10) "Final" disposal – destruction; abandonment; monitored but irretrievable deposition; retrievable deposition (e.g., in an ex salt mine).

Through these stages a single waste may also change several times, not only in composition, form, and value, but also in ownership or control; and its defined nature and status may thus change, in risk, economic, legal, and other terms, as it moves from one agent to another, especially across different regulatory systems. Indeed, since wastes are often mixed or transformed at transfer stations, and treatment processes produce concentrated residues which begin life as a new waste, life-cycles are more accurately described as branching networks rather than discrete linear chains.

Consider, for example, the case where a producer transfers a waste to a broker, paying for T&D. The broker may see opportunities for recovery of constituents, say silver residues in photographic wastes. He therefore sells it to a reclamation company. The "waste" is transformed into a "good" whilst remaining chemically the same, and exits from regulation. If a handler dehydrates a sludge to reduce transport costs, he may unwittingly create a hazardous waste from an unregulated one, by reducing the flashpoint if it contains flammables. Another difficulty concerns wastes that have several toxic constituents, e.g., an acid containing heavy metals. The waste should rationally be defined as acid for the transport phase, but heavy metals for final disposal, reflecting the most important hazard at different stages of the life-cycle, even if the material stays chemically the same.

These properties of indeterminacy and chemical or institutional transformation create severe difficulties for "cradle-to-grave" documentary systems of notification, which are supposed to be a central means of waste control.

When combined with the aspects of heterogeneity and diffuseness already outlined, these complex *institutional* life-cycles of wastes have several important consequences, especially upon the relationship between regulation and implementation and the interaction of technical and institutional uncertainties.

Conventional pollution regulation involves the dispersion of potentially hazardous materials directly from the source plant into the relevant environmental medium. Assessment has thus been of the relevant *natural* transport and transformation mechanisms by which the material returns to damage people or significant media. Control has thus usually been exercised by specifying ambient environmental standards or discharge limits, which industry has to meet by its own means. But hazardous waste leaves the plant on a wagon, under the ownership of an economic agent, who may mix it with other wastes for convenient and economic transport, or storage, trade it as a good, pass it on to other transport, storage, or treatment, and so on, until final disposal by destruction, containment, and dispersal eventually occurs. These final dispositions, such as incineration or landfill, as well as intermediate storage, transport, etc., even if licensed and controlled, still *also* involve conventional environmental dispersal (e.g., flue gases from incineration, or groundwater from landfill leaching), and possible natural reconcentration. All the usual uncertainties of natural mechanisms are therefore still involved, but with a qualitatively new dimension between waste production and eventual *natural* dispersion.

The important *extra* dimension of uncertainty for hazardous waste is the complex and indeterminate *behavioral* arena between source and final disposition. It is, for example, because it is not dispersed, but packaged and handled, that the ambiguity exists as to whether hazardous waste is a waste or a (recyclable) resource. This distinction cannot possibly be defined by central authorities because it is a legitimate and necessary factor of enterprise, innovation, and thus trade between free industrial actors. On grounds of overall efficiency they have to be given latitude to define a "waste" as a "good" from within. Yet this is a major loophole in hazardous waste regulation. This intrinsic ambiguity is analyzed further in Chapter 9.

The more direct and extensive regulation of behavior inevitably implies more conflict between industry and regulator, because, being packaged, the waste being transformed and transferred is, in principle at least, more traceable to its source. If it is found to be in the wrong hands or wrong place at the wrong time, the original waste producer may be liable, even if acting in good faith he was let down by an incompetent or illegal operator further down the life-cycle. Unlike, say, 10^{10} anonymous molecules of sulfuric acid found to be acidifying a Scandinavian lake, a container of acid residues with Punters Chemicals Ltd. printed on it that is being carelessly stored when it should have been neutralized and carefully disposed of, will get Punters into trouble.. Specific cases like this (in addition to Hoffman–LaRoche's notorious Seveso wastes) have been experienced, and have also caused sharp reaction. It is because of the significance and complexity of these networks of actors in hazardous waste life-cycles that legal liability issues have been a focus of regulatory concern, so far without resolution.

The intractability of controlling specific behavior within waste life-cycles (as well as emissions at final disposal points, as in normal environmental legislation) is illustrated by looking at one part of the system. At the international level, for example, recall that in 1983 there were about 100 000 border crossings of hazardous wastes consignments within the EC, or one every five minutes. This is only at international transfer points. The transaction costs of rigorous supervision of all movements (within as well as between national systems) would be so enormous that it could not be done, except by adding extra restrictions that preempt some of the heterogeneity, such as limiting allowed border crossing points, or reducing the population of transport firms to a few large organizations. But such interventions more closely interfere with industry's autonomy and costs in ways that are more pervasive and less finely tuned to individual cases than conventional dispersive pollution controls. They also still have to be enforced.

There is also an important implication for public reactions created by the complex institutional dimension in the life-cycles of hazardous wastes, especially cross-border hazardous waste flows. The conventional cross-border pollution issues, such as acid deposition, involve diffuse, mainly atmospheric (or oceanic) natural processes. They cannot be switched on or off by political fiat. There is thus a relatively uniform, slowly changing deposition of pollutants over large areas, and this occurs via natural processes. The issue of acid rain has been gradually rising to policy status for some time, and whilst its public recognition may have been triggered by specific local observations, its general growth as an issue has been evident for several years.

Indeed the same point of difference applies *within* borders too. There has, for example, been no equivalent for conventional, dispersive pollution to the regulatory decision to buy up a whole town in Times Beach, USA, which had been totally contaminated by dioxin sprayed in waste oil onto dust roads all over the town. Acid deposition and other dispersive pollution occur more evenly, and (once dispersed) by natural processes not human action.

Hazardous waste is different for several reasons:

(1) Since it is concentrated, it can cause more intense local damage. Dispersive pollution can sometimes naturally reconcentrate, but hazardous waste is *already* usually concentrated, in human hands.

(2) Since it is moved directly by human agents it is:

 (a) More likely to be subject to unpredictable movement and disposition.

 (b) More likely to occasion sharp and dramatic negative perception.

 (c) More able to be rapidly stopped (e.g., if political will is exercised) with consequent sudden disruptions.

All of the factors in (2) played a role when the 41 barrels of waste containing dioxin from the Seveso plant were lost in early 1983. This incident and its effect upon public perceptions has led to restrictions on the cross-border movement of hazardous wastes. It has also led to changes of policy that posed a sudden and dramatic threat to systems dependent on export. Thus, the region of Wallonia in Belgium announced a decree in June 1983 banning imports of foreign waste. The EC's 1984 transfrontier directive, galvanized into being after the Seveso barrels episode, is regarded by some as a hastily conceived, backward step toward restriction of the free international movement of wastes within the EC, and the controversial FRG ban on exports and imports, which has had direct ramifications in other countries, was a result of the Green Party's growing influence upon decision making.

The perception of the *severity* of events, such as the Seveso barrels, may be multiplied by the fact that they are *directly* under human responsibility (and hence perhaps more threatening), whereas, sulfur deposition, for example, is indirect in that the operator at the coal plant source did not actually direct its emitted sulfur to a given deposition point. Thus, the political/regulatory and economic impact of an incident with hazardous waste, especially if it is a transborder incident, is likely to be more severe and unpredictable than the impact in conventional transborder pollution. Furthermore, sanctions are more concrete. Transborder traffic in air pollution is difficult to stop by the importing nation, but far less difficult with trucks or freight cars importing hazardous wastes. These factors are multiplied by the fact that wastes are inherently more likely to go astray than hazardous goods; and they are probably enhanced by previous experience of uncontrolled hazardous waste dumps. That is, past human practices in hazardous waste disposal are *concretely experienced*, unlike, say, past laxity in aerial emissions discharges. As a policy issue, therefore, the *amplitudes of unpredictability* and risk perception in hazardous waste management are intrinsically greater than those of otherwise parallel issues: this is also a function of its strong *behavioral* as well as natural life-cycles.

3.3.3. Intrinsic versus situational risks

The multistage life-cycle for hazardous wastes means that there are multiple points at which the life (and thus hazard) of the waste can be more or less independently affected. This suggests a need for coordinated, multiple, but distributed, regulatory interventions and for a framework of risk analysis to match this institutional reality. However, the most basic point of attention remains the first phase of the waste life-cycle, namely hazardous waste *arisings*. At this point data are needed on what is being produced, where, and by whom. Also needed is a universally applied precise *identification* mechanism for defining hazardous wastes, as a prerequisite for obtaining meaningful (i.e., centrally consistent) data and material control.

However, centrally collected knowledge of waste arisings may not be an inevitable necessity for regulation, given other foci of intervention that may offer the chance of effective control. It *is* essential for reassurance, and also for centralized, standardized modes of regulation, but the point is to explore institutional options and implications. This question becomes especially germane as the accuracy to be expected of such data diminishes the closer one approaches it.

The attempt to classify hazardous wastes definitively at their point of generation is undermined by the fact that there are several technical options for classification, as analyzed in Chapter 5. It is also limited by industrial secrecy, and by the uncontrolled variability of the physical and chemical (and thus hazard) properties of many waste streams from day to day, even within accepted production parameters. However, even more significant than these difficulties of defining the "intrinsic" hazards of wastes at their point of production is the downstream indeterminacy of the life-cycle of any given waste. The physical and chemical fate of a particular waste is influenced by independent actors, rationalities, and interactions, some of which cross-cut the rationality of ideal central management. Through these situational changes, the "downstream" hazard of a waste may change, perhaps radically; for example, if a transporter dewaters a sludge to reduce transport costs and so inadvertently reduces the flashpoint to below a safe threshold. In other words, the variation of risk *within* the life-cycle of a waste may be greater than the distinctions of hazard between wastes *as generated.* This point is depicted schematically in *Figure 3.2.*

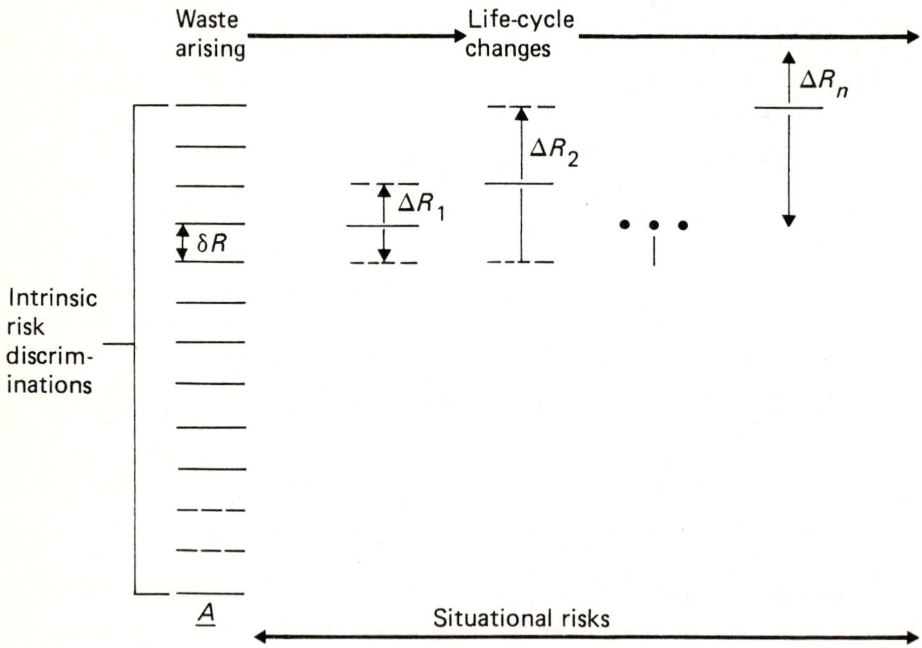

Figure 3.2. Possible risk variations for hazardous waste (schematic): "intrinsic" and "situational" risks.

We draw an important distinction between *intrinsic* risks, which are those associated with given wastes as produced – defined by toxicity of components, etc. – and *situational* risks, which are those created by the various situations, such as physical and chemical condition, environmental state, form of containment, etc., through which a waste will pass during its life-cycle.

Strictly speaking, all risks are situational; intrinsic risk measurements and specified tests still embody – usually implicitly – conditions and criteria of damage, exposure, etc., that imply a standard situation and act as a baseline for comparing measured risks reproducibly. For example, an intrinsic risk test that specifies a concentration threshold for an aqueous leachate assumes groundwater contamination from landfill as its critical scenario. Intrinsic risk approaches are equivalent to a heavy regulatory dependence upon the data and classification of hazardous waste *arisings.* Based on a standard scenario or situation, they imply a *behavioral* standardization in the risk-generating system that is equivalent to projecting the risk discriminations made at *A* (*Figure 3.2*) rigidly down the whole life-cycle.

Conventional approaches to environmental regulation extended to hazardous wastes may be unrealistic about the feasibility of this projection, because they normally have little or no such life-cycle dimension to reckon with.

This tension between managing intrinsic or situational risks is an important cross-roads of technical approaches and institutional structures in regulation, and is dealt with at length in Chapters 9 and 10. It is a special case of the more general point that the hazardous waste life-cycle allows different regulatory foci and approaches; these imply different institutional relationships, which in turn need different kinds of technical risk assessment. For example, an ideal system would be able to decentralize risk analysis and management to such a degree that an optimally efficient balance between costs and an accepted level of environmental protection could be achieved. If however, central institutions cannot trust local actors (regulators or regulated) to analyze and optimize in this way, they must impose more standardized risk frameworks, such as by banning landfill rather than saying that in some local circumstances (to be determined by local actors) it is acceptably safe. The technical risk frameworks in regulation embody assumptions and prescriptions about these *institutional* relationships, and a parallel balance between "intrinsic" (standardized) and situational (flexible) risk analysis and management. In Chapter 5 we outline technical methods of hazard classification of wastes. One of these – defining characteristics (e.g., toxicity tests) has potential flexibility toward unforeseen situational variations (e.g., by including analysis of *exposures*), whereas the others (e.g., defining constituents or waste streams) do not. But even the "characteristics" method embodies assumptions about risk *situations* that need strong local institutional frameworks for effective use.

3.3.4. Multiple possible decision levels

A direct consequence of the complex life-cycles of hazardous wastes is that there are several major, *different* points in the system where regulatory intervention can be made. This is shown in *Table 3.1* (see also *Figure 3.1*). The first point of direct control is where the wastes are actually produced. The aim of most existing regulatory frameworks is to control hazardous wastes "from cradle to grave", that is, all the way through their life-cycles to their final resting place.

Note that this can mean anything from merely trying to *track* wastes, so as to see where they go, to *controlling* their movements, e.g., by designating forms and operators of transport, specific T&D sites, or techniques, etc. The ultimate aim is the latter and, indeed, this happens for very special wastes, such as PCBs; but practical achievement is qualified by what is actually feasible. Where systems of regulatory authority do not have comprehensive data on industrial waste arisings, and have little or no

Table 3.1. Regulatory nodes, information, and instruments for hazardous waste management.

Regulatory node	*Monitoring or analysis*	*Information needed*	*Regulatory "action" possible*
Production process	Inputs, process design, waste streams, intermediates	Composition and interactions in process, to estimate waste streams e.g., waste coefficients of pdn	Dictate process choice and design tax/waste costs enough to influence process decisions
Waste arisings – factory gate	Volume, composition, place, form (e.g., packaging) of waste arisings – internal treatment – external treatment	Detailed knowledge of how producer defines and records wastes	Prohibit pdn of some wastes; waste tax; classify and tag some wastes to specific minimal levels of T&D; license generators
Transport systems and specific consignments	Monitor consignments, frequency volume, reported compostiton, and actual destinations	Actual vs reported final destinations; intermediate transfers, reloads, mixing, storage, etc.; prohibit some combinations	Regulator acts as transporter; license transporters; specify transport design standards and operating rules for given wastes, "trip-ticket" registration of consignments
Transfer or storage	Registration, inspection; license data on transfer	How many, where, what they receive, mix, repackage, and stations, etc.	License and inspect handlers Public ownership–management? Where send?
T&D plant receipts	Analyze waste receipts at T&D plant	Composition, volumes, origins physical state of wastes received	Licence plants to receive only specified wastes; analysis
T&D plant operation (+ design?)	Plant emissions – air, water (leachate, surface), soil, operators	Air emission rates (stack monitors); composition of leachate, surface water, operating environment	Specific plant performance design criteria, e.g., discharge limit + *prohibition* of some wastes
Environmental media ambient	Analyze most indicative media, monitor contamination	Media contamination levels, processes of environmental	Ambient concentration limits – air, water, soil, etc. movement and distribution
Specific environmental end-points e.g., aquifers, human doses and effects	Monitor potentially affected populations or critical groups	Full range of health effort possibilities	Calculate specific concentration limits, e.g., human dose limit equivalents + safety factor, derived working limits (DWLs), specific tissue and organ contamination limits.

sovereignty across the full waste life-cycle, tracking may be all that is feasible, and may be challenge enough. However, regulation may include *contracting* and radically simplifying the possible life-cycles to within the national system's own sovereignty, by imposing restrictions on export out of the waste production region. Further simplification may be effected by creating a public monopoly responsible for all wastes in a region. Trends have been toward more designation and restriction. For example, the original ambitions within Europe were confined to waste tracking, but FRG initiatives go beyond this, such as by requiring that for permitted exports (for which special exemption must be justified) the receiving T&D facilities must be at least as good as domestic ones.

A risk assessment scheme for chemical wastes could begin far "upstream" from normal regulatory points of application, at the stage of industrial process design (see *Figure 3.1*). This is recognized as the ultimate regulatory ideal, and is also accepted by some [37] as in industry's best interest, since it might prevent some hazardous materials (e.g., PCBs or 2,4,5–T) from ever becoming downstream "wastes", where they later engender regulatory reactions, which threaten established production commitments. In Hungary (see Chapter 8) the government regularly analyzes "internal" industrial data, but in less centrally planned economies industry strongly resists even external information gathering, let alone external direction of plant and process decisions.

The main problem with a regulatory strategy at this upstream point is that there is so much ignorance about the passage from conceptual design and early risk indicators to eventual waste situations and effects. The chances are, therefore, that only rather crude and extreme signals from downstream (e.g., threats of public opposition or steeply rising unavoidable costs) will appear back upstream, through all the intervening uncertainties, as sufficiently clear to warrant or enforce action. The credibility of technical analysis alone is likely to be very limited.

Quite apart from confidentiality problems, industrial processes usually produce multiple waste streams, and even within regular daily operating conditions the waste composition of a process will fluctuate quite significantly ("some waste streams go into and out of the hazardous category several times per shift" [38]). Therefore, the specificity and accountability of this level of analysis is always likely to be limited.

The usual starting point for regulation is waste arisings. The aim is to determine the "intrinsic" properties of the waste as it stands, and match its hazards to the cheapest effective technical option to reduce those hazards to acceptable levels. The situational variations downstream make risk analysis at this point uncertain, unless those degrees of freedom are reduced, not only by technical norms, but by institutional innovations too.

The US EPA regulation has attempted to develop an integrated risk-analytic model for this stage. It attempts to match and optimize combinations of waste–environment–technology (WET model) for different risk-cost objective functions using linear programming [39]. The developers recognize that in aggregating many factors (for example, to score all possible human health effects for an initial 140 different chemicals on a single scale), this risk assessment model is a very crude cut. Nevertheless it is intended that it will systematically identify combinations of wastes, environments, and technologies where regulation should increase or decrease in resolution and, at the limits, where total prohibitions or nonregulation should occur. As "a broad policy planning tool ... incapable of developing and revising specific regulations" [40], the WET model is essentially the first, crude step of a step-wise regulatory strategy to establish where analytical attention should be focused when regulatory attention and resources are far outstripped by the magnitude and complexity of the available problems.

In principle, WET matching should optimize the *situational* conditions (treatment, transformation and disposal, containment, isolation, etc.) and hazards from the *intrinsic* hazards of a given waste, thus refining the risk–cost trade off. Because many of the risk-varying factors are detailed (*ad hoc* situational realities) using the model as a surrogate for such local behavioral and scientific knowledge in a decision is impossible. Most regulatory systems have responded to inadequate implementation by attempting more intervention in choices of T&D for given wastes, via the use of such models or equivalent frameworks.

Whereas WET might be used, ideally, for planning a T&D infrastructure or at least issuing principles for T&D, a more realistic decision situation is one where the waste producer is faced with a range of facilities of different kinds at different distances and costs (including in-house treatment and disposal and export options). There may be some T&D facilities whose legal receipt of the waste in question is uncertain, as the license conditions and their enforcement are vague (see Chapter 7 for example). There are also often intermediate handlers who collect wastes from firms for a fee, then decide themselves what to do with them, so that the producers may not know what happens to them beyond the handler. Assuming that the producer or handler is a responsible one, there may still be large differences between legal disposal options, and transport to appropriate facilities also has to be considered. In the UK, for example, the closure of a high-temperature incinerator in Scotland left nearby producers of special wastes unfit for landfill with the option of transport to an incinerator in the south of England, at 1500 pounds sterling per wagon load. An in-house special incinerator is now being built by one such company instead.

It is generally true that decision makers in the waste life-cycle make *risk* decisions only indirectly. They make decisions based upon competing *costs*. These ought to reflect (*inter alia*) considered risk estimates, but the less direct this is, in a multiactor, badly defined life-cycle, the less controlled are the "risk estimates" that shape decisions. A combination of front-end waste arisings classifications, producer and handler commercial decisions, and T&D site-licensing restrictions creates certain broad networks with branching options, through which waste can move, be transformed, and eventually be disposed of. The overall balance of application and tightness of regulatory effort can vary greatly between systems, even when they may appear to have similar regulations.

Thus, regulatory approaches have attempted to establish accepted frameworks for evaluating wastes for their intrinsic hazard, then control their routing toward hazard-limiting T&D options. But this can be done in various ways, and to very different levels of determinism or specificity. The variables are not only technical, but also institutional. One option is to allow complete freedom of choice for the operator to dispose of his wastes once classified as hazardous, but only within a given menu of licensed handlers, transporters, T&D options, and operators. This approach would involve emphasis on risk control from the T&D site licensing end (the very back end), with a minimal "intrinsic" hazard classification of wastes from the generator end (the front end) and a large mediating middle ground of free enterprise. Insurance against the effects of front-end regulator ignorance of waste arisings is attempted by placing control emphasis at another, decentralized, downstream regulatory node. Risk analysis here is essentially a complex mix of *minimal conditions* at each end of the match – initial waste and final T&D – but allowing for normal commercial processes in between to optimize for all other factors that an ideal risk management decision would include. This is the least deterministic form of regulation. In putting regulatory emphasis on T&D plant licensing but not upon what wastes *arise*, it is also institutionally most distant from production process innovation, and least able to influence that phase. This kind of balance is, in principle, most nearly approximated by the UK, which has probably the least elaborated hazard classification for wastes, and the most strongly expressed dependence upon T&D site facility licensing, even if in reality this is highly variable in quality – see Chapter 7.

This balance can be changed by more elaborate front-end attempts to classify and direct hazardous wastes, as mentioned earlier. This is being tried in Austria and in the FRG. (In effect it is partly established already in Bavaria and Hessen.) A weaker version is being considered in the UK. It has also been introduced in the USA (1984) reauthorization and strengthening of the 1976 RCRA. Licensing waste producers and enforcement of full registration of all waste arisings, including that disposed of in-house, licencing intermediate waste handlers and transporters, and centralizing licensing of T&D facilities to be more specific in the wastes they are allowed to treat, are all supplementary methods of reducing behavioral uncertainty in waste life-cycles. The US EPA began in 1978 by considering that it would issue centralized criteria for T&D plant design and performance, but had to relax these in 1981 to "best engineering judgment" in a case-by-case manner, requiring *operators* to perform justifying risk assessments [41]. The uncertainty of this approach for operators caused the EPA to retreat to general standards, this time for *environmental* performance of T&D facilities. The reauthorization has returned to stricter controls.

Such regulatory moves push the front and back ends of waste life-cycles together, and structure the ill-defined middle. In theory this increases determinism, control, and effective regulation. In practice, however, if carried too far, it may *reduce* the coordination of the system, which may be only viable if a large degree of freedom, thus regulatory indeterminism and uncertainty, is allowed to private enterprise in matching specific wastes to specific end-points. The best balance may also depend upon the overall size of the economic–industrial network that is to be regulated.

In practice, countries differ widely in their choices with respect to these factors. Some require producers to register but not transporters, others require transporters to be licensed. Nearly all systems require registration of defined wastes "from cradle to grave" – the so-called trip-ticket notification system – but the definitions of waste differ and so do detailed ways of operating the registration procedures.

Some regulatory bodies, such as those in the UK, feel that hazard lists and classification schemes are *only* necessary for the transport phase – although they have not taken steps to license hazardous waste transport operators [42]. Trip-ticket notification systems for every waste consignment that leaves the factory gate are supposed to aid regulation, but it is widely accepted in practice that the sheer volume of paperwork and coordination necessary, the dislocations between brief, unit paper descriptions and real world transport arrangements (e.g., "season tickets" aggregating many loads into one recorded consignment), and the time lag between actual transport and receipt of all copies of the paperwork by an agency, all mean that trip-ticket systems can *at the very best* be only a retrospective infor mation gathering exercise, and not a form of real-time regulation (see Chapter 7).

The actual management of hazardous wastes goes through several distinct phases, the exact combination varying in each case. Each of these phases offers different opportunities for regulation, but regulates different actors, activities, and relationships, and requires very different kinds of risk analysis. For example, transport risks stress different criteria and define different materials as hazardous compared to, say, final disposal. Also, the degree of refinement needed (or possible) of risk analyses or regulatory instruments may differ at different points of the life-cycle. It may not matter to a waste producer or transporter, for instance, if the heavy metal concentration in an organic sludge is above a certain threshold. Yet this may make it unsatisfactory for high-temperature incineration, where aerial emission of heavy metals can be a hazard. All the five main phases listed in *Table 3.1* need some kind of classification of a hazardous waste in order to invoke controls. The question of how this classification is made, and also how it is *used* once defined, is very important, down to a much finer level of practice than normally recognized. The opportunities for both formally and informally inconsistent practical meanings and usages are endless as already intimated in this chapter. Appropriate technical methods can only be defined in relation to institutional parameters including their uncertainties.

3.3.5. Professionalism

The foregoing points about the fragmentation of hazardous waste life-cycles and the earlier contrast with nuclear waste throws into relief another significant aspect of the chemical waste issue, namely its historical lack of professionalism. This strongly affects the social structure and potential effectiveness of different regulatory strategies and arrangements. In the nuclear case, whether we are talking about radiation protection or the engineering safety of nuclear power plants, there has always been a well developed professional cadre of experts, highly funded and with a strong basis in fundamental research as well as more applied analysis. As Serwer has shown in the case of radiation protection [43], these cadres early on evolved a strong sense of professional *self*-regulation, even when this regulation was extended from radiology into industrial areas. This affected standards of vigilance, expertise, and quality control, which helped to maintain public credibility and acceptable regulatory standards, even when the "regulatory" experts were formally part of the industry.

Ensuing institutional structures of regulation were, at least, well developed and financed, coherent, and highly professionalized, with a strong ethos of enlightened self-interest (self-protection via preemption of public hostility through protecting public health). Furthermore, there was a continuous bridge – even an identity – between the government regulators and the regulated industry. The regulators were recruited from the industry and shared a common professional background, somewhat aloof from either the governmental or industrial context. This arrangement has drawbacks, but also important benefits. There is no doubt that it is far better than no regulation at all, which was the case for a long time with hazardous chemical waste. The point is that the risk source in the nuclear case was actually a prior professional activity (radiology), which generated professional bodies of regulation that later formed the basis of coherent regulatory bodies as the nuclear industry developed.

Similar kinds of highly professional, regulatory arrangements have developed in other fields. There is also a significant connection with the heterogeneity factor, because it is the relatively unitary nature of the nuclear industry as a later risk source that has allowed its regulatory cadres to be drawn from unitary technical disciplines (radiation biology and physics; nuclear engineering) and so enhance a technically strong (if institutionally sometimes "co-opted" professional identity.

The social structure of regulatory bodies is also an important factor in public perceptions and reactions. The hazards of nuclear processes and materials focus only upon radiation damage. Although this, in itself, involves an extensive differentiation of specialist topics within the field, it is nevertheless under a singular, professional discipline. The resultant *social* coherence of the expert community has meant that many questions about uncertainties and alternative techniques of analysis, etc., could, until recent years, be retained largely within the specialist community, away from the *public* regulatory agenda. In the case of hazardous wastes, with a more fragmentary array of technical issues and relevant specialties (even within toxicology alone), this expert privatization of uncertainty has been more difficult. Because of this lack of a coherent social containment insulating the public regulatory arena from esoteric specialist concerns, many technical uncertainties have been more easily drawn into the public policy/regulatory domain, leading to greater confusion in the setting and implementation of standards. The reasons why this has not yet led to public anxiety on the scale of the nuclear issue (except perhaps in the USA) lie elsewhere, in the perceived *familiarity* of chemical waste disposal institutions, and in different histories. The technical risk uncertainties of toxic chemicals overall dwarf those of radiation risk.

Biotechnology also provides an interesting comparison in relation to professionalism and self-regulation, because the emergence of regulation bears some hallmarks of the development of professional self-regulation of radiologists in the 1920s and 1930s and of its institutionalization in government regulatory arrangements. Thus, molecular biologists themselves effectively alerted public concern to the risk associated with their work in the famous Asilomar declaration of 1974 [44]. Although there was later a retreat from this position, the US National Institutes of Health Committee set up to "regulate" this field was not only *not* an executive agency, but was staffed by largely the same corpus of experts who had first expressed the need for systematic risk assessment and regulation of their own activities. Thus, the same kind of bridging, highly expert, cadre-structured regulatory developments occurred here as in the nuclear case. Professional self-regulation has been the norm in biotechnology.

Now, interestingly, the field of biotechnology, which started with scientists organizing self-regulation, has industrialized very rapidly under essentially the same regulatory arrangements, but in a diffuse, decentralized way very different from the nuclear case, so that this area of regulation bears some characteristics of both the nuclear and (in its industrial structure) the chemical waste regulation cases. Indeed, the industrial end falls under existing chemical waste arrangements, though with apparent confusion as to the nature of its wastes.

Because hazardous chemical wastes emerged as an issue from ordinary waste arrangements and perceptions, except in the largest chemical companies there was no professional status, no systematic attention, nor were resources given to deal with these wastes. Thus, many hazardous waste arrangements and regulatory personnel in governments and industries were originally of low qualification and status. Relevant personnel had no cohering sense of sound standards, professional identity, or reputation to defend. They were usually only organized, if at all, at diffuse and uncoordinated local plant and municipality levels. One of the repeated comments about the early days of the EPA Office of Solid Waste's dealing with RCRA, for example, concerned their lack of qualification for dealing with the complex problems of risk management and regulation that swept through and transformed their established, more mundane concerns about ordinary garbage [45]. Whereas the source of regulatory recruitment in the nuclear and biotechnology fields has been clear-cut and highly trained, recruitment to regulatory bodies for hazardous waste management seems to have been far more disparate and fragmentary, with corresponding difficulties in establishing an effective, professional identity that bridges regulators and all reaches of the industry, out to its smallest, most esoteric corners. A senior regulator in the UK, for example, could think of fewer than 12 people in the whole country qualified for the post of chief inspector in the Hazardous Wastes Inspectorate created in 1984. There are no higher degrees or chairs in waste management. Typical recruitment to hazardous waste regulatory positions appears to vary widely. In Britain, the government regulatory body is proud of the large number of man-years of industrial experience its staff has. In other countries recruitment seems to be mainly of engineers, ecologists, or chemists with no industrial experience [46].

The heterogeneity factor also enters here. The kind of regulation described above is inevitably highly interventionist, which is a major reason for the difficulties experienced in implementation. But it is intervention in a wide variety of different industries, plants, processes, and conditions. It implies a need for a very wide range of specific industrial expertises. Not surprisingly, compared with the experts in any given industry, the experts in such a regulatory body will probably be nonexperts, and conflict is highly likely, not just on specific points but, more corrosively, over competence and credibility in general.

Compare this situation with the nuclear case again. Here there is usually only one central decision-making frame [47], and the institutional structure of regulation outlined before means that the regulators are an integrated part of the industrial design, planning, and decision-making process anyway. Nuclear plant design, for example, is carried out in detailed consultation with regulatory experts to assess design features of requisite safety standards. Nuclear waste processing and transport are likewise designed and conducted with regulators as part of the decision-making process. The dividing line between regulation and industrial planning is nearly broken down altogether. Thus, the "interference" of regulators, due to the fact that nuclear waste is also packaged, is a nonissue because they are already integrated in the decision making anyway, and in a monopoly industry there is not the same extreme cutting edge of competition and ever-vulnerable profit margins that requires keeping regulators and their impact as distant as possible.

Another side effect of the same structural property is that, because the nuclear industry is virtually only one industry, there is no real ambiguity as to the "wastes–resources" boundary. There is, of course, one such conflict cleaving the whole area, namely the proper role of plutonium created in and extracted from spent thermal reactor fuels. The point is that for nuclear materials these conflicting options and definitions are dealt with by *whole policy systems*, and are uniformly applied, without ambiguity to the actors, within a given system once an option has been decided. There may be misleading signals given to the public, but that is a different issue. Unlike hazardous chemicals, these elements are not subject to variable definition and dynamic choice by a polycentric network of autonomous enterprises, interacting with diverse interests and perceptions within the regulatory system itself. Thus, one more major source of *intrinsic* confusion and limitation to effective implementation in hazardous waste regulation is essentially absent from the nuclear case.

This history of low status, institutional and technical fragmentation, lack of technical expertise and professionalism, and neglect of hazardous waste, combined with the extreme heterogeneity of the area, has given the field a very particular institutional structure of regulation. However, there has existed the requisite professional expertise, resources, and awareness for self-regulation in the specialist chemical companies, and trade associations have made important advances in levels of professionalism. Although there are national modifying factors, the overall ethos and implementation of hazardous waste regulation is deeply influenced by this general history of institutional development.

3.3.6. Inverse materials–cash exchange relations

Let us start by viewing hazardous waste T&D as just any other complex industry, regulated by conventional operating emission standards to air, water, and soil. The T&D industry has resources inputs – other industries' wastes – from whose industrial "conversion" it makes (or hopes to make) a profit. The only difference – and it is crucial – is that its resources inputs are of *negative value*. The T&D industry does not pay for its "resources"; it is *paid* specifically for taking them away and dealing with them, and it is paid before it converts its "resources" into final "products". This means that the T&D industry receives the money *as well as* the input materials.

Yet – unlike a conventional industry, which *pays* for its inputs – it is only really concerned to receive the former. Although a material goes in one direction, it is the (T&D) *service* that is central. This flows the other way, and is delivered to the *customer* once the material is removed from his site, *not* when (if) it is properly and finally disposed of. The item of *risk* interest is the material. Notice the extra complication that if the T&D service contractor sees the waste as recyclable, the whole framework is turned upside down! The relationships are represented in *Figure 3.3*.

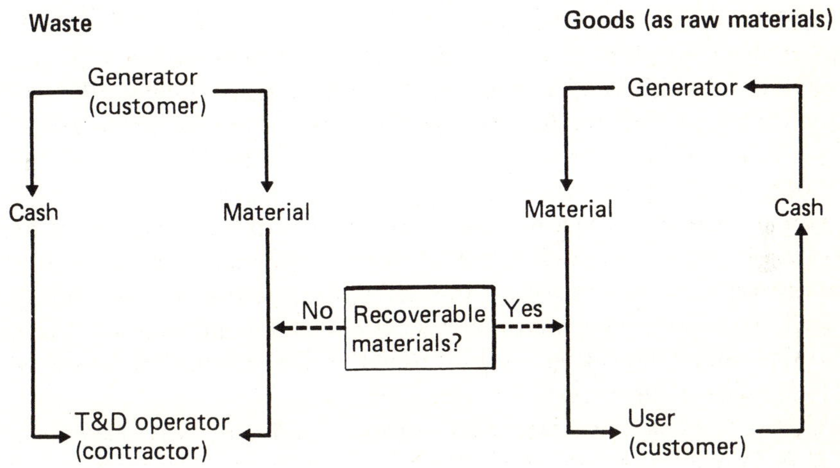

Figure 3.3. Materials-cash exchange relationships: hazardous waste and conventional materials. For waste, (*a*) the cash buys a service, not materials, and (*b*) the service is invisible to its buyers. (I am grateful to E. Finnecy for the original point from which this is developed.)

The regulatory situation is fundamentally affected by this structural property and can be summarized in four points:

(1) For waste, the cash buys a *service*, not materials. The service is the removal of materials, which is not identical to their proper disposal.

(2) The service is invisible to its buyer, who therefore does not know if it has been fully performed. Therefore, the servicer or carrier (T&D operator), who has *already* been paid, has no incentive to carry out the service (i.e., receive the hazardous wastes, or treat them properly once received) because this costs the servicer. This is a strong practical reason for avoiding arrangements where intermediate handlers are paid directly for collecting wastes.

(3) Because of the lack of value in the waste from the generator's point of view, it is not subject to any kind of "quality control", thus, its chemical composition, mixing, and even physical form may vary more or less uncontrollably. Yet this radically affects the feasibility of treatment, in which somebody has to invest resources. Such uncontrolled variation is a major problem for T&D operation and thus a key obstacle to more investment in T&D.

(4) For wastes containing valuable resources that could be extracted, the same variability problem arises. But, in principle, there may come a point where the value of extractable materials makes the "waste" worth paying for by the T&D industry rather than being paid to take it away. In terms of *Figure 3.3*, the value shifts from left to right, merely by fiat of the handler. However, this not only makes more incentive for quality control ("no quality, no pay" clauses), it also converts the "waste" into "not waste", and thus automatically exempts it from hazardous waste regulation. This is a major loophole because there are so many wastes, and the transition of each is open to diverse and fluctuating definition, as mentioned earlier.

3.4. Conclusions: Ignorance, Indeterminacy, and the Limitations of Risk Analysis

As suggested before, the hazardous waste problem is in important respects underripe for systematic risk management approaches, because the area is characterized by ignorance and indeterminacy in basic, elementary properties and parameters. Crucially, these are institutional or behavioral, of a different order to those definable technical uncertainties in damage-generating processes and substances. Thus, for example, although he felt it to be a flawed Act, the Dutch Environment Minister, L. Ginjaar, decided in 1979 to go ahead with the Chemical Waste Act because it, at least, contained registration requirements that would eventually produce information (at that time completely lacking) about the producers, volumes, compositions, and movements of the large variety of industrial wastes [48]; that is, about the *activities* of a whole network of agents concerned with hazardous wastes. In his view, only when this information was compiled and analyzed could a better Act be designed and proper research carried out to define risks and help choose better regulatory instruments and standards. Perhaps indicating the extent and intractability of this ignorance, although Ginjaar thought such revisions could be made in three years, the information was eventually published in 1985 and, although a T&D infrastructure is now at least planned, the envisaged tightening of the regulations is not yet in sight.

This situation, where regulation is nearly reduced to a passive information-gathering exercise, again derives from the issue-specific structural properties – the extremely poorly defined, dispersed, and behaviorally dependent nature of the regulatory issues.

Risk analysis of environmental exposure pathways, as opposed to risk analysis of release mechanisms, is recognized to be a field in its infancy compared with the questions confronting it. Hazardous wastes pose all of those release and environmental risk analysis problems in great diversity, but they pose the further problem of defining a dispersed and partly indeterminate behavioral–technical risk system. Implementation of regulation and supporting analysis are tangled together more closely, having to define risks of variable *situations* in a waste's life-cycle, and assume or negotiate what behavioral degrees of freedom exist in reality to influence the risks and the uncertainties. These degrees of freedom include the freedom to define the key terms, and hence the structure of the field, in different ways.

Practical management first requires the creation of an industrial infrastructure, which involves protection from close regulation, and relatively crude and flexible risk discriminations. To organize this industrial and trading activity in wastes into the appropriate T&D option is, as noted, inevitably more *directive* than is conventional environmental regulation. Since hazardous waste is concentrated and moved by human agents, it has a greater potential for dramatic impact on public perceptions. The practical influences that affect controls over waste production and movement are likely to be based as much upon unstable and badly understood processes of public perception and reaction as upon sophisticated risk analyses. The structural uncertainties outlined in the previous sections both reduce the relevance and meaningfulness of such analyses *and* increase the instability and effects of public perception processes.

The above observations of the underripeness of hazardous waste policy for formal risk assessment may be analytically valid; but the lack of structure in the risk-generating system is arguably a major reason for the fickleness of public reactions. Public reactions and pressures for justification are therefore creating an apparently irresistible demand for the issue to be defined more elaborately and formally, in terms of risk analysis. The gradual ordering of some of the extreme *behavioral* uncertainties in the system requires forms of risk analysis tailored to these poorly defined structural realities, and aimed at incremental situational improvements rather than idealized (e.g., standardized) situations. One of the observations of our research was that national systems vary in the extent of the distance they embody between the implicit standardized assumptions made about risk situations in the formal risk assessments and the variety of real situations of implementation that actually exist. Bringing the former nearer to the latter implies that stronger *institutional* resources are needed at local levels in order to *integrate* the local knowledge available only to them with the more general principles and system requirements. But the use of more centralized and standardized scientific language to reassure the public tends in the opposite direction. The way in which regulatory bodies treat intrinsic ignorance and institutional uncertainty in framing risk analysis will reflect a combination of universal pressures, such as the generally increased importance of *credibility*, the heterogeneity of waste life-cycles, and local pressures arising from the particular cultural context.

Also crucial will be the responses of policy to the conflict between the need to create and maintain a new industry as part of regulation, and the increasing need to subject it to formal regulatory controls before its *existence* is even secured. It seems increasingly dubious whether free market forces alone are adequate to cope with this dilemma, though significant elements of self-regulation have to be encouraged and allowed. In any case, it is clear that the risk management dimensions ultimately interconnect with basic institutional questions not only "internal" to hazardous wastes, but more broadly, of industrial investment, public financing, and economic management generally.

In Chapter 4 we describe a case study in which the conflicting overall ways of defining the hazardous waste issue, as outlined in this chapter, are evident, and in which local institutional realities exacerbated these intrinsic tensions in handling the issue. We also see from this case study how deep are the interactions between technical risk management frameworks and behavioral factors in the regulatory arena, an insight that is developed in later chapters.

Notes

[1] Resource Conservation and Recovery Act, US House of Representatives, Public Law 94–580, October 21, 1976 (Washington, DC).

[2] Levine, A.G. (1982), *Love Canal: Science, Politics and People* (Heath, Lexington, MA).

[3] Märka, E. (1984), Sonderabfall Beseitigung in der Schweiz, in W. Kemmenberg (Ed), *Wiener Miteilungen: Wasser, Abwasser, Gewasser, p.L-1* (Technische Universität Wien, Vienna); UK House of Lords Select Committee on Science and Technology Report (1981), *UK Hazardous Waste Management and Disposal,* Chairman Lord Gregson, Vol II, Appendix F, p. 210 (HMSO, London); Alberta Environment Report, No. 1184, *Hazardous Wastes in Alberta,* Reid, Crowther and Partners (Calgary, Alberta). The evidence of the Harwell Environmental Safety Group, UK, to the UK Royal Commission on Environmental Pollution, 11th Report, (1986) also makes this point.

[4] However, some systems (such as Austria) have begun to educate people about domestic wastes, and encourage the separation and reclamation of different elements.

[5] On radioactive waste management, see, e.g., Kasperson, R. (1980), The dark side of the radioactive waste problem, in T. O'Riordan and R.K. Turner (Eds), *Progress in Resource Management and Environmental Planning,* Vol. 2, pp. 133–164 (Wiley, Chichester, UK).

[6] Lee, G.F. and Jones, R.A. (1981), Application of site-specific hazard assessment testing to solid wastes, in R.A. Conway and B.C. Malloy (Eds), *Hazardous Solid Waste Testing: First Conference,* Special Technical Publication 760, American Society for Testing Materials, (Philadelphia) pp. 331–344.

[7] Kragg, B. (1983), *The Hazardous Waste Management Industry,* draft mss., Harvard University Business School, Cambridge, MA.

[8] International Solid Wastes and Public Cleansing Association and the Institute for Solid Wastes (1984), *Hazardous Waste Policy Issues: Multinational Perspectives*, proceedings of a symposium organized by the ISWPCA and ISW.

[9] For example, the study by EPA (1984), *National Hazardous Waste Generators and Treatment, Storage and Disposal Facilities Regulated under RCRA in 1981* (EPA, Washington, DC). See also Lavelle, R., p. 340 of US House of Representatives Hearings before the Subcommittee on Oversight and Investigations, of the Committee on Energy and Commerce, *PCB and Dioxin Cases,* 97th Congress, 2nd session, 19NN 1982, Serial 97–194 (US Government Printing Office, Washington, DC).

[10] Anon (1981), Allocating the costs of hazardous waste disposal, *Harvard Law Review*, **94,** 584–604.

[11] These figures are taken from the following sources: Crawford, P.J. (1983), International harmonisation of chemical control: Purpose and Prospects, paper presented at the conference on Chemical and Carcinogens Regulation, Bellagio, Italy; Nichols, J.K. and Crawford, P.J. (1983), *Managing Chemicals in the 1980s,* (OECD Environment Directorate, Paris); Conway, R.A. (Ed) (1982), *Environmental Risk Analysis for Chemicals* (van Nostrand Reinhold, New York, NY). The figure under item 4 (from Crawford) is open to some dispute. Majone (personal communication) related the same figure to US laboratory resources only. The difference may turn on what precisely is meant by testing. Crawford was referring to the level of testing indicated for carcinogenesis.

[12] Klein, R. (1981), *European Community Action and Orientation in the Field of Waste Management*, paper presented at the International Symposium on The Practical Implications of the Reuse of Solid Waste (Institution of Civil Engineers, London).

[13] Thomas, S. and Roberts, M. (1983), *Decision making in the US EPA on the Resources Conservation and Recovery Act*, draft mss., Harvard University, Graduate School of Public Health, Cambridge, MA.

[14] Waddel, T. (1981), *Integrated Waste Management in Europe: A Tentative Assessment,* Report to the German Marshall Fund and Institute for European Environmental Policy (Bonn).

[15] Burnabeau, C. (1984), US Federal Coordinator of programs relating to sulfur emissions, statement made at the IIASA International Forum on Science and Public Policy, Laxenburg, Austria.

[16] By "objective" here we do not mean unbiased in some ultimate sense, but that there are intrinsic dynamics that have a certain pattern and stability, which take over independently of specific interests and choices.

[17] This coincides with the widely attested rise of single-issue politics, in most liberal democratic societies at least. Pollak, M. (1983), The rationality of bureaucratic regulatory fragmentation, paper presented at IIASA Summer Study on Regulation, INS group (Laxenburg, Austria).

[18] Wolbeck, B. (1982), Political dimensions and implications of hazardous waste disposal, in J.P. Lehman (Ed), *Hazardous Waste Disposal,* p. 15 (Plenum, New York, NY).

[19] Department of the Environment (1985), *Report of a Review of the Control of Pollution (Special Waste) Regulations 1980* (DOE, London).

[20] *Op. cit.* [1], reauthorization 1984.

[21] Barnes, S.B. and Edge, D.O. (Eds) (1982), *Science in Context* (Open University Press, London). p. 290–302.

[22] Kleindorfer, P. and Kunreuther, H. (Eds) (1987), *Insuring and Managing Hazardous Risks: From Seveso to Bhopal and Beyond,* Proceedings of a Conference held at IIASA, Laxenburg, Austria, July 1985 (Springer, Berlin).

[23] Kragg, *Op. cit.* [7].

[24] Kragg, *op. cit.* [7], mentions that, for example, Rollins dropped out of the hazardous waste treatment field due to the brittleness that this factor created in the market. Another set of examples are the new incinerators built in several countries in the expectation of a large market, but which are in financial trouble because regulations have not generated such a market for them. Investors are indeed caught in a double kind of uncertainty, because from the opposite side, production innovations may unexpectedly remove a waste from generation and thus remove a large slice of a market. This happened to Rechem, UK, who built a new incinerator close by and specifically to service a Shell plant whose waste they had been treating much further away. Without telling Rechem, Shell changed their production process and Rechem was left stranded.

[25] Environmental Data Services Ltd. (1983), *ENDS Report 100*, pp. 14–16 May (Southwell Press, Surrey, UK).

[26] Bailey, G. and Hawkins, R. (1983), *The Future of Rubbish – Waste Management Options for the Environment Reviewed* (CPC, London).

[27] Thomas, S. and Roberts, M. *op. cit.* [13].

[28] Diver, C. (1980), A theory of regulatory enforcement, *Public Policy,* **28**, 257–299.

[29] *Ibid.*

[30] Kragg, *op. cit.* [7].

[31] Colen, H. and van Veen, F. (1983), *Hazardous Wastes, A Literature Review on Classification Systems*, Report No. 58070.01/R0.01, BV INFRA Consult (Deventer, Netherlands).

[32] Kasperson, *op. cit.* [5]; Surrey, J. (Ed) (1984), *The Urban Transportation of Irradiated Fuel* (Macmillan, London).

[33] Environmental Data Services Ltd. (1985), Trade Organics in British Aquifers: A Baseline Survey, *ENDS Report 129*, October, Southwell Press, Surrey, UK.

[34] Butlin, J. (1982), *Private Compliance Costs and Public Administration Costs in Hazardous Waste Management* (OECD Waste Management Policy Group, Paris); US EPA, Office of Solid Waste (1981), *Economic Impact Analysis of RCRA Interim Standards,* 2 Vols, Washington DC, November (US Government Printing Office, Washington, DC); EPA, OSW, *Characterization of Hazardous Waste Transportation and Economic Impact Assessment of Hazardous Waste Transportation Regulations,* (OSW-17c), March (US Government Printing Office, Washington, DC).

[35] See note 3

[36] Pollak, M. (1983), *Licensing procedures in comparative perspective*; paper presented to IIASA Summer Study on Regulation, INS group (Laxenburg, Austria).

[37] See, e.g., Jenkins, W. (1978), *Policy Analysis: An Organizational Approach* (Martin Robertson, London); Majone, G. (1985), *The Uses of Policy Analysis* (Yale University Press, New Haven, CT).

[38] Haymore, C. (1982), Incorporating risk assessment into the Resources Conservation and Recovery Act regulatory process, in F.A. Long and G.E. Schweizer (Eds), *Risk Assessment at Hazardous Waste Sites,* Symposium Series 204, pp. 111–124 (American Chemical Society, Washington, DC,). See also Chapter 10 in this volume.

[39] Haymore, *op. cit.* [38].

[40] US EPA (1980), *Federal Register*, 40 CFR 260-263, *45*, 19 May 1980, pp. 33060–33137, and 40 CFR 264-265, Hazardous waste management systems: standards for owners and operators of hazardous waste management facilities, pp. 33154–33258. Thomas, S. and Roberts, M. (1983), *op. cit.* [13]. Daniels, S. (1981), Development of realistic tests for effects and exposures of hazardous wastes, in R.A. Conway and B.C. Malloy (Eds), (1983), *Hazardous Solid Waste Testing: First Conference*, pp. 345–365 (American Society for the Testing of Materials, Philadelphia, PA).

[41] Haymore, *op. cit.* [38].

[42] UK Department of the Environment (1981), Evidence to the House of Lords Select Committee on Science and Technology, Lord Gregson, Chairman, *Hazardous Waste Management,* vol. 2, pp. 60–112 (HMSO, London).

[43] Serwer, D. (1978), *The Rise of Radiation Protection* (Brookhaven National Laboratories Report, Brookhaven, Long Island, USA).

[44] Eddy, H. (1983), *Regulation of Recombinant DNA Research: A Trinational Study* (Science Council of Canada, Ottawa); Krimsky, S. (1979), Regulating recombinant DNA research, in D. Nelkin (Ed), (1979) *Controversy; The Politics of Technical Decisions*, pp. 227–253 (Sage, London); Yoxen, E. (1983), *Who Should Control Genetic Engineering?* (Harvester Press, Hassocks, UK).

[45] This view has emerged from interviews with various US experts. See also Thomas and Roberts, *op.cit.* [27], for a concurring account.

[46] Interviews with different national regulatory bodies have consolidated this view, but it remains to be properly documented and perhaps quantified.

[47] Private initiatives in US nuclear waste processing have virtually collapsed.

[48] Interview with Dr. L. Ginjaar, February 1983. The first data from waste registration, analyzed by the Ministry of Housing, Planning and Public Health, were published in March 1985 (see Chapter 4).

CHAPTER 4

The Rationalities of Problem Definition: The Netherlands and Hazardous Waste Management

Brian Wynne and Dick Hortensius

4.1. Introduction

Any attempt to improve regulation must begin from an understanding of the problem structure that regulation confronts. As indicated in Chapter 3 there are certain properties of the hazardous waste issue that exist wherever it is regulated. One of these properties is the severely poorly structured and behavioral nature of the "system" of waste generation, handling, transformation, and eventual disposal. Within a single regulatory arena there are many interacting problem definitions being developed by actors with varying involvement and interest in wastes and their adequate management. The pattern of problem definition interactions varies according to the state of evolution of an issue, the particulars of specific subissues, and the institutional structures of the overall regulatory system. Whilst this pluralism and lack of clear structure is universal, its effect upon the optimality – even viability – of policies is not. The particular political culture and institutional milieu in which the issue is being managed affect the point in the waste system at which, and the degree to which, contradictory perceptions and interests confront one another or are "smoothed" so as to suspend or dissipate potential dislocation.

Plural problem definitions exist in any policy issue, even ones structured by strong technical content. A key point from the introductory chapters is that for hazardous wastes this pluralism extends to the technical heart of the issue, namely the terms hazard, waste, and risk; coherence will not be achieved by more intense technical effort to find precision, but from *institutional* mechanisms. In this chapter we focus upon a particular country, the Netherlands, whose institutional features have made the issue appear nearer to paralysis. This important case demonstrates the institutional complexities of achieving regulatory problem definitions that are sufficiently related to a broad enough set of key institutional actors to be practically viable. It is the minimum-feasible institutional consensus that defines the parameters of useable knowledge. This, in turn, suggests certain, more general practical conclusions about the best forms and focus of regulatory effort.

There are two interrelated points that are drawn from this analysis and developed in later chapters:

(1) The depth and substance of the diverse frameworks of rationality that thread through the overall hazardous waste issue – even in the formal institutional settings, let alone amongst less recognized actors – forces a reconsideration of approaches to risk management and regulation, which normally assume a singular problem definition and uniform frameworks of information. This, in turn, forces more careful consideration of the proper balance between centralized and decentralized structures of regulation in this issue.

(2) The Dutch use of precise and inflexible concentration thresholds for legally defining and ranking hazardous wastes is relatively unusual internationally. It is interesting because of the evident contradictions – more extreme in the Netherlands than elsewhere – between such an elaborate and precise normative technical framework of regulation, and the severe shortfall of adequate domestic industrial facilities to actually treat and dispose of wastes within this theoretical framework. Analysis of this discrepancy supports our general interpretation of the relationships between institutional uncertainties and technical frameworks. It also suggests the significance of indirect, symbolic roles of credibility and legitimation played by scientific language. A wider interpretation is offered, based on features of Dutch political culture and of processes of legitimation through symbolic action. It is a concrete example of the tension, outlined in Chapter 2 and elaborated upon later, between the direct and symbolic roles of risk analysis.

We treat problem definition as a "deep structure" underlying the specific interests, values, and arguments of relevant groups or organizations ("actors"). Their problem focus may not necessarily be primarily on hazardous wastes, yet their ensuing actions or perceptions may be relevant to its management.

It is also important to follow through from the early stages of issue emergence to formal legislation, and enactment in regulations and implementation and/or enforcement mechanisms. Some factors, such as the interests of certain actors, remain stable throughout this cycle; others, such as the patterns of interaction and the focus and language of action, may change. In this case we see the impetus for explicit regulation emerging from the local level, being adopted and shaped at national level, to return transformed in various ways, via the provinces, back to local levels. Tracing this process through its several incarnations is valuable in underlining the point made by, e.g., Lindblom [1], that policy problems are not natural or self-evident, but are socially shaped by the traditions, experiences, and values of their authors.

Here we focus on the *taken-for-granted* problem definitions of different actors: this involves more than identifying the diverse interacting *values* of those actors, because "values" suggests diverse *preferences* focusing on a common problem definition. The diversity of presumptions as to what is the problem being addressed is a more fundamental form of heterogeneity, yet it has some institutional pattern. Of central interest to this kind of analysis is the form of *interaction* between the physical–technical characteristics of the hazardous waste issue and the institutional context of policymaking and implementation. We offer a tentative analysis of the influence of Dutch political culture on the use of technical norms, and the ability of the system to implement regulatory policies which – in the nature of the issue – have to be coordinated and followed through across several widely separate points of application in the overall waste life-cycle.

4.2. Hazardous Waste Management – An Outline

A 1985 compilation from registration data by the Ministry of Environment [2] gave estimates (*Table 4.1*) for the total amounts of chemical wastes disposed of in the Netherlands in 1983. Before this survey the unknown volumes of waste disposed of privately on-site by generators had been the source of large discrepancies between estimates of waste generation arising and what was notified for regulated disposal. The main feature of the Dutch system has been the severe shortage of any significant treatment and disposal (T&D) infrastructure, despite early legislative action and relatively precise and rigorous formal regulations. The basis of the Chemical Waste Act's implementing regulations is the Ministry of Environment's Substances and Processes Decree (1981). In this, a list is given of 41 compound waste categories (e.g., "metal carbonyls") and nine composite industrial process categories (e.g., "production of caustic soda"). The compounds are ranked in classes A to D, each one given a precise concentration threshold which, if exceeded in a waste, automatically legally defines it as a chemical, i.e., regulated, waste. The thresholds are (mg/kg or parts per million): Class A, 50; Class B, 5000; Class C, 20000; and Class D, 50000.

Table 4.1 Reported chemical waste disposals in the Netherlands, 1983.

I	Disposed of on-site by generator (exempt from Chemical Waste Act)	530 000 tonnes
II	Notified and disposed of under Chemical Waste Act	251 400 tonnes
III	Disposed of otherwise by known methods (e.g., export)	130 000 tonnes
IV	Unknown disposal (rough estimates: 20% dumping, 10% recycling, 70% water discharge	100 000 tonnes

Various norms about weighting for specified toxic elements in compounds or mixtures are given to elaborate these rigid decision rules still further. For all their precision and inflexibility, however, and their sharp consequences, these norms are rather arbitrary, being related to attempts to identify "natural background" levels of the chemicals listed. They have the compensating advantage in an administratively fragmented system of at least being clear-cut – everyone knows where they stand (assuming, that is, that sampling and analysis are clear-cut, which they are not).

The norms have draconian consequences in principle. The slightest excess over a threshold and a waste cannot legally be landfilled in the Netherlands; it must be notified and subject to "controlled" T&D, or "disposed of" on site. In practice, given the dearth of actual domestic facilities, controlled T&D has often meant export (though a more recent requirement is that export will only be permitted to facilities which meet Dutch domestic T&D standards).

The various methods of disposal of notified wastes are given in *Table 4.2*. They show a large dependence upon foreign "facilities". If we include sea disposal, foreign disposal accounts for more than 60% of official Dutch hazardous waste management. This large and vulnerable dependence upon foreign receipt of wastes (and risks) should eventually be reduced by the enactment of a policy announced in March 1985, to have a joint government–industry investment in a secure, fully engineered containment landfill and in new high-temperature incinerators. The government will provide 100 million guilders. Investment in such an industrial infrastructure had previously been stalled for a decade, despite the recommendations of two government commissions, legislation to control wastes, private industry initiative, and various scandals highlighting the damaging lack of proper, legal T&D facilities. Government expectations that private industry would engage in commercially profitable T&D once controlled wastes were legislated into being have proved false. The earlier emergence of the issue onto the legislative agenda partly helps to explain this state of affairs.

Table 4.2. Reported disposal methods for notified wastes in the Netherlands, 1982–1983.

Method	*Total* ($\times 10^3$ tonnes)	*Treatment (% of total)*	
		National	*Foreign*
Overall	252	40	60
Distillation	10	27	73
Physical–chemical treatment	28	45	55
Dehydration	24	96	4
Incineration (land)	98	69	31
Incineration (sea)	18	81	19
Sea disposal	5	100	0
Dumping	68	18	82
Other	3	52	48

4.3. Hazardous Wastes – The Process of Issue Formation

As we have already noted, the informal issue agendas of different groups vary in policy, and the "official" agenda of policy issues, as measured, e.g., by legislative attention, is a complex, not very stable function of these interactions and conflicts. Not only is there interaction between competing problem definitions of the "same" problem, but parallel problems intersect and boundary definitions are also open to change.

Hazardous wastes became a public policy issue in the Netherlands before being a matter of widespread or urgent public concern. The latter has only arisen in the 1980s following the discovery of previous toxic waste dumps in residential areas such as Lekkerkerk, Gouderak, and Dordrecht. Whereas in other countries the issue has developed from municipal waste disposal, in the Netherlands its origins also lay in a long-standing national preoccupation with soil pollution. Legislation to regulate dangerous substances, such as explosives, has existed since the late nineteenth century. Other environmental risks have been regulated (in theory) under the Nuisance Act of 1875 (revised in 1952) [3]. Recognizing the hazards presented specifically by chemicals, the Dangerous Substances Act was passed in 1963 [4], which, in accordance with a recent EC directive, will be replaced by the Toxic Substances Act and the Transport of Dangerous Substances Act [5]. A general national law to prevent soil pollution does not yet exist, although a fragmented array of rules govern the use of soil and waste management on the national, provincial, and local levels. As early as 1904 the Inspector of Health reported several hundred cases of air, surface water, drinking water, and soil pollution; however, it was not until the 1970s that national regulation in these areas was established. Local provincial authorities could, and did, make their own statutory orders for environmental protection (VICAs), but without coordination with national criteria an effective, preventive policy on air, water, and soil pollution did not result.

This fragmentation of policy prompted a comprehensive national bill on soil pollution in 1971, which even today has still not passed Parliament. The speed of developments over hazardous waste overtook the original approach toward legislation on soil pollution. The complexities of the bill would have stalled any action with lengthy parliamentary debates, so the Government abandoned its plans for comprehensive soil pollution legislation and, instead, introduced four separate bills: The Ground Water Act [6], the Waste Act [7], the Chemical Waste Act [8], and the Soil Protection Act [9]. The latter two Acts, which govern chemical wastes and soil pollution, are the most relevant for our purposes.

Before turning to a discussion of the Chemical Waste Act, which was passed in 1976, it should be mentioned that once again a general law on soil protection was thwarted by events. The new Soil Protection Bill, introduced into Parliament in 1980, was already facing an uphill battle when the first major soil pollution scandal shook the Netherlands. A new housing development at Lekkerkerk was found to be built on seriously contaminated soil. Excavating under 270 houses, at a cost to the Government of 170 million guilders, became front-page international news. The magnitude of the national soil contamination problem became apparent following Lekkerkerk, when an urgent, rough inventory of suspected sites listed more than 4000, including even more serious cases than Lekkerkerk. The public anxiety and incredible expense of cleaning up these sites motivated political pressure groups and Parliament to force the Government to abandon the Soil Protection Bill and introduce the emergency Soil Clean-Up (Interim) Measures [10], preceding the Soil Clean-Up Act [11].

Although the two most important national laws governing chemical waste are thus now the Chemical Waste Act (1976) and the Soil Clean-Up Act (1983), an understanding of this legislation alone would not encompass the complex system of rules and laws at the national, provincial, and local levels that govern waste management generally, and chemical waste management in particular. The Nuisance Act still governs certain activities and the Chemical Waste Act does not include substances regulated by the long-standing Dangerous Substances Act, the Nuclear Power Act [12], and the Animal Waste and Pesticides Act [13]. Most provinces have separate rules governing soil and groundwater pollution, and many municipalities experiment with separate collection systems for specialized, and sometimes hazardous, wastes. There is, therefore, much scope for local variation in the standards of formal regulation (let alone enforcement).

The inventory of 4000 contaminated sites (later increased to 5000) following the 1980 Lekkerkerk events showed the need for a legal and financial framework to effect remedial action. The existing Soil Protection Bill (submitted in 1980) was inadequate for the emergency since it would take several years to pass Parliament and, anyway, hardly dealt with remedial actions. The Soil Clean-Up Act was therefore drafted and submitted in May 1982, passed in December 1982, and mostly enacted in January 1983. In anticipation of the Act, remedial measures were begun urgently in 1981 based on the systematic approach laid down in the bill. In 1981, 60 million guilders (roughly 15 million US dollars) were allocated for these measures, rising to 195 million in 1982. Under this so-called interim policy, the provinces drew up one-year programs for 1981 and 1982 and received financial support for projects, including orientation surveys, methodological investigations, clean-up operations, and inspection. To apply for central funds it was necessary to undertake extensive investigations into the nature and scale of the contamination and to ascertain the best clean-up methods.

Implementation of the emergency *interim measures* revealed a number of problems only some of which were solved by the full introduction of the Soil Clean-Up Act. Intractable issues included refusal on the part of the owner of the contaminated site to cooperate voluntarily in clean-up operations; finding temporary storage space for contaminated soil; adapting to new and uncertain clean-up techniques; and finding criteria to establish when "cleaned up" soil could be considered adequately clean. These implementation difficulties – related to technical uncertainties about "acceptable risk" – caused serious delays in spending the allocated amounts of money. The delays were exacerbated by institutional fragmentation between central and local government.

The Soil Clean-Up Act provided a framework for dealing in the short term with those cases of soil contamination where the soil pollution or potential pollution posed a *serious threat* to health or the environment. The Act was supposed to operate for five years, at which time it was assumed that the Soil Protection Act would be approved by Parliament, and the Soil Clean-Up Act would then be integrated into it. The Soil Protection Act would include preventive measures, and would regulate *all* necessary clean-up operations. It is indicative of unforeseen uncertainties that it is now planned to extend the so-called "interim" Soil Clean-Up Act to 1996.

The shaping of the hazardous waste issue as a policy problem cannot be understood without looking at institutional interactions at the time. Probably the first awareness of a problem of some kind developed amongst the very few existing provincial environmental inspectors and the municipal health authorities, who became concerned about municipal waste disposal, especially domestic landfill siting and safety. Fearing that chemical wastes were being dumped with domestic garbage, and in the absence of central government interest, some provinces in the early 1970s introduced local regulations, embodying a rudimentary waste notification and site-licensing system. These so-called VICAs were eventually elaborated and incorporated in the 1979 regulations enacting the Chemical Waste Act. However, these are now issued by the Ministry of the Environment.

Central government interest coincided with the creation of the new Ministry of Public Health and Environmental Protection in 1971. However, central–local fragmentation was exacerbated by ensuing processes, because the new ministry, in its need to create standing for itself in existing central government interactions, felt the need to take and control new initiatives. One of its first measures was to begin formulating the Chemical Waste Act, but, obsessed by its own standing in the adversarial context of central government, it did so with little consultation with local authorities. The ensuing lack of collaboration in shaping the Act may have contributed to later fragmentation between central and local government agencies over Induval and UNISER (see below) attempts to site T&D facilities, and other implementation problems.

The present system of legislation is, therefore, a mixture of frameworks, at times overlapping and competing, and with somewhat arbitrary boundaries defined as much by surprise events, evolving institutional relationships, and urgent reactions, as by any more consistent strategic principles. In addition, legislation in some "dispersive" areas (e.g., water and air purification) simply created more and more toxic filter sludges, etc., throwing the burden onto concentrated hazardous wastes instead. At the same time as the original provisional Chemical Waste Bill of 1973 was being reviewed, industry's options were being reduced by tightened controls on sea and land dumping. At this time (indicating the great uncertainties) the prevailing official estimates of the scale of the annual chemical waste disposal problem were about 2 million tonnes (mt). Only a few years later this estimate was reduced to 1 mt.

4.3.1. Local–central relations: the Induval plan

Unlike the branches of government, Dutch industry organized itself into a very coherent lobby on the chemical waste issue, and had its interests strongly represented by the Ministry of Economic Affairs. This resulted, for example, in forceful arguments demonstrating the extreme costs of tight regulation in this area and that industrial generators as commercial traders should be allowed to define what was waste. Seeing that legislation was inevitable, industry decided that the key need to comply happily with the emerging framework was a national, controlled dumping site. To achieve this aim, a number of large private chemical firms, such as Akzo, Unilever, and Dow created the coalition Induval foundation in 1973.

The initial plan contained a technical description of a controlled landfill and, by 1976, confidential investigation of four potential sites in the province of North Brabant had been conducted. The Ministries of Economic Affairs and Environment gave the go-ahead for the plan to be submitted to the local councils. At the same time, the plan's existence was leaked (by a government agency antagonistic toward cheap landfill disposal) to the national environmentalist group, Foundation Nature and Environment, which also strenuously opposed the idea of landfilling toxic wastes. They believed it would provide a cheap and convenient diversion from pressure that should be exerted on industry to seek cleaner production methods and better processing technologies.

The sudden exposure of the secret plan generated so much hostility in North Brabant that the local councils would no longer entertain Induval, despite pressure from the national government that the plan complied with the existing Chemical Waste Bill. Further tightening of the plan was conducted by Induval and the government, and in August 1977 a second version was presented with the express support of central government, but with no site named. By the spring of 1978, the provinces that had been approached – Overijssel, Gelderland, North Brabant, and Limburg – had all flatly rejected it. Their official reasons were: concern about leakage; that the actual wastes to be dumped were not specified accurately; and that it had not been shown whether such wastes could be processed by other, more acceptable means. By now the councils were also strongly influenced by local public hostility.

Following this emphatic provincial rebuff, Induval was abandoned. It is interesting to note that, at the time when local authorities began attempts to regulate via the VICA system, and even when the new Ministry of Health and Environmental Protection began to formulate a national Chemical Waste Act, the issue was not one of widespread public concern, and no environmentalist groups had taken it up. In this sense, the Induval plan's exposure and demise was a watershed. The whole problem of hazardous wastes was thereby broadened from a "management" problem under a relatively unified (though artificially restricted) set of values and interests, into one involving competing groups and conflicting problem perceptions and objectives, with the very authority of government thrown into question on this issue.

Amidst inter-Ministry recriminations over the failure of Induval, the Minister of Health and Environmental Protection set up the Hofman Commission [14] to review and establish policy on the storage and disposal of nonprocessable chemical wastes. The commission was composed of representatives of industry, and central and regional government, but not of the environmentalist movement until a second phase, when the proposals for permanent storage sites made in the first phase were being developed into detailed designs. Predictably enough, Foundation Nature and Environment rejected these designs in a minority report, because they involved nonretrievable disposal.

The lack of controlled domestic landfill facilities was a major bottleneck to chemical waste disposal, but it was exacerbated by problems elsewhere in the system. The only significant commercial waste incinerator

(AVR) nearly closed down in 1980 and had to be rescued by subsidies. Furthermore, the national-scale waste processing company, UNISER, was involved in a massive scandal, successfully prosecuted, and closed in 1981; this further undermined attempts to establish an industrial T&D infrastructure and to implement the 1979 Chemical Waste Act regulations.

The deepening paralysis was amplified by the decreasing willingness of local authorities to apply under the Chemical Waste Act for government dispensation for waste dumping, which they had previously condoned. Yet another commission was set up, the Kolfschoten Commission [15], with a more comprehensive brief than that of Hofman, to examine the best form of organization of a national waste disposal system with several integrated functions, not merely landfilling.

In retrospect, the Kolfschoten Commission can be seen to have marked a change in government thinking about the regulation problem. Up to about 1981 it had believed steadfastly that the problem was to create a "market" of wastes for controlled disposal, allowing free enterprise to see the opportunities for profit in *private* investment in T&D facilities. By the time Kolfschoten reported in April 1982, it was clear that the private route was failing and that government stimulation of joint investment, via a large leading stake in new T&D facilities, was on the agenda. Not only did such government capital investment encourage industry to participate, but also, if written off as proposed, it meant cheaper (and possibly internationally competitive) prices for T&D to waste generators. This last factor is important as a means of competing with the export of waste to countries with legal and cheap landfill, such as the UK. In fact, Foundation Nature and Environment and the government specialist Institute for Waste Research (IVA) separately proposed something akin to the "Bavarian solution" (see Chapter 6) for this problem of drainage of T&D business into cheap exports and away from domestic facilities. This would have involved introducing a legal obligation on waste generators to supply their wastes to the domestic facilities. Although this institutional mechanism has worked in Denmark, Bavaria, and Hessen, it has become problematic in the EC context due to the EC's commitment to free trade. In the Netherlands the industry lobby was far too strong for it to become a realistic policy option and, indeed, the government body whose scientists proposed it was vigorously taken to task by other, pro-industrial, branches of government for allegedly overstepping its allocated role.

The UNISER scandal was important as a further nail in the coffin of the original policy strategy of depending upon free enterprise to create a controlled T&D infrastructure. It also indicates dimensions of Dutch central–local authority relationships that are of more general interest. A brief account of that episode, and a parallel one involving the Booy Clean firm in Rotterdam harbor, is therefore in order.

4.3.2. EMK/UNISER and Booy Clean

The waste treatment firm Exploitatie Maatschappij Krimpen Ltd (EMK) operated from 1970 to 1980, when it was taken over by UNISER, the biggest waste treatment firm in the Netherlands.

In 1981 it was disclosed that UNISER was operating illegally: wastes supposedly for treatment were being discharged or dumped illegally throughout the Netherlands, and hazardous wastes were being sold as oil and, after mixing with coal, as solid fuel. This scandal eventually led to the prosecution of the top management of UNISER, in the biggest environmental court case in Dutch history.

In 1970 EMK settled on the former site of the Chemical Industry Uithoorn Ltd (CINDU) in Krimpen aan de Ijssel. CINDU had been processing tar and tar products under a 1965 Nuisance Act license. EMK described itself as a trading company for regenerating, reconditioning, and processing oils and fats. When EMK took over the site and premises, the CINDU Nuisance Act license was passed on and covered EMK activities provided the original terms were complied with.

From the outset EMK provoked complaints about severe smell and, after the Rijnmond Environmental Control Body, DCMR, had reported negatively on the situation, the Court of Mayor and Aldermen of Krimpen acted in August 1970. They requested that EMK apply for a new Nuisance Act license for their entire organization since, according to the DCMR, the old CINDU license did not cover all their activities. EMK repeatedly failed to provide a complete application. It took a number of deadlines from the municipality, as well as the threat of closure, before a complete application was finally submitted in August 1971. During this year there were continuing complaints. Despite the lack of a license for it, several shipments of the toxic compound Resinformer were landed at the EMK site, and in November 1970 a ship carrying about 530 tonnes of Resinformer sank at the EMK jetty. The municipality repeatedly threatened to close down (part of) EMK if the landings and related processing were not halted, but in September 1971 barrels of toxic chemicals, including Resinformer, were still being stored at the site.

From the summer of 1971 onward, several individuals and institutions reported to the Krimpen authorities on the deplorable situation at EMK and their possible infringement of numerous regulations. In a (then confidential) report from DCMR it was stated that:

> A vast part of the site is covered with a tar-like substance. A nearby ditch is filled with the same kind of substance. Since a pipe from one of the tanks empties itself in the ditch one gets the impression that the state of the ditch is not being improved.... One can state that here practically every regulation is being violated [16].

In spite of this devastating DCMR report, the Court of Mayor and Aldermen took no action, though they did increase inspection of EMK in close cooperation with the Control Body for Hazardous Substances and the Central Report and Adjustment (Regulation) Office, Rijnmond.

In November 1971, EMK was installing a number of storage tanks, although the municipality had forbidden their use without the necessary building and Nuisance Act licenses. The Krimpen municipality, with the help of the police, this time stopped EMK. In response an EMK director confidently announced that he would ask other government agencies to intervene. The same day officers from the Regional Environmental Inspectorate and the Rijnmond Authority contacted the municipality of Krimpen, asking whether such extreme measures were really necessary. The Inspectorate emphasized the importance of industrial removal, discharge, and processing of chemical waste and waste oils.

By this time (late 1971) the municipality had already been waiting several months for an answer from the national Ministry of Health and Environmental Protection to another problem, apparently created by the Ministry's allocation of part of EMK's site for the storage of barrels of hazardous waste awaiting ocean dumping. The lack of reply from the ministry, combined with pressure from the Inspectorate over EMK's tank installation, caused the Krimpen municipality to feel they had been abandoned by higher authorities.

The conflict between central and regional government and the Krimpen local authority reflected the central government's overriding concern to cultivate an industrial infrastructure for hazardous waste treatment. At that time it was preparing the Chemical Waste Act, a crucial point of which was to encourage and create such an industry. The already fragile attraction to private investments would be destroyed by too much regulatory zeal. The regional inspectorate was at that time concerned with the dumping of container rubbish, which frequently contained industrial and chemical waste, around Krimpen; however unsatisfactory, EMK at least *seemed* to offer a reduction in the openly illegal dumping of chemical waste.

After the storage tanks incident municipal control activities were intensified. Daily inspections were made of the EMK site, but they were not very effective because, as became evident later, the official in charge was unaware of the Nuisance Act license conditions. Nevertheless, substances stored on the EMK site were regularly sampled and analyzed.

At that time the engineering consultants, DHV, the municipality's adviser in the slow crawl toward a new Nuisance Act license, concluded that the EMK application still did not meet the requirements. The municipality decided not to grant a new license and instead sought an opportunity to shut down EMK. That opportunity occurred when analyses showed that EMK were storing substances not covered by their Nuisance Act license. In

May 1972 the municipality decided to order the immediate closure of EMK on the basis of severe health risks.

EMK lodged an appeal with the Crown and instituted a lawsuit against the municipality. This led to a settlement by agreement, suggested by the judge. The compromise was that the old Nuisance Act license from 1965 would remain valid with the addition of a number of new conditions that limited the kind of substances allowed to be stored and processed. It was also agreed that EMK would submit a new application for a license.

In subsequent years, the Papendrecht affair in early 1977 revealed that EMK had buried a number of barrels containing arsenic compounds on a dumping site in Papendrecht. For this the EMK director was prosecuted and sentenced to a fine and suspended imprisonment, which caused him to resign his membership of the national advisory committee on used oil, a statutory advisory body to the Ministry of Health and Environmental Protection. The fact that he had retained membership of the committee until that time can be seen as an indication of the importance attached by the Ministry to private initiative in treating chemical waste, without noticeable concern over the extensive local problems EMK had caused.

The granting of a new Nuisance Act license to EMK took a long time, partly because again the application still needed amplification. After the license was finally granted in August 1977 the Rijnmond Authority took over all tasks and responsibilities concerning the Nuisance Act in order to ensure better coordination of environmental protection at the regional level.

For EMK a large number of conditions in the new license were unacceptable so they again lodged an appeal with the Crown. By the end of 1977 EMK announced the transfer of its activities to the town of Moerdijk, so that the Rijnmond Authority agreed to a gradual renovation of EMK at Krimpen. Then EMK started the old game of moving the renovation plan deadlines, until they announced in early 1980 the closure of all activities in Krimpen, since it had become evident to them that doing business within the frame of the Nuisance Act was impossible. This had taken nearly ten years of local "control", undermined by central government.

In the meantime EMK had caused continuing complaints about the stench and numerous infringements of the Nuisance Act, all laid down in a so-called "black file" (complaint book) by the environmental protection agency of Rijnmond. When the final dismantling of the EMK site began, a true catastrophe was revealed. The ground turned out to be very heavily polluted to a great depth with oil, aromatic, and phenolic compounds. Enormous amounts of chemical waste were also found on the site in tanks and storage cellars, as well as in a moored boat. The first Dutch T&D facility turned out to be one of the worst pollution cases.

After leaving Krimpen, the illegal activities of EMK continued at Moerdijk, where they had already established the RTM (Recycling Terminal Moerdijk) in 1976 in a joint venture with Drisolco, a chemical trading and processing firm. EMK, RTM, and Drisolco were merged into one holding company, UNISER, at the end of 1977, which succeeded in suggesting that they were *the* waste processing company in the Netherlands.

The waste flow from numerous (chemical) companies to UNISER began to grow and in 1979 it received and processed some 100 000 tonnes of waste (probably over half the total "processed" in the Netherlands). In commercial publications UNISER advertised a division of labor between its companies: EMK took care of transport, Drisolco of storage, and RTM of processing waste oils and chemical wastes. But the actual situation was so different that the public prosecutor in the later court case against the UNISER management called its conduct of business "a great fake show".

The fake was exposed, however, when a suspicious leak from a storage tanker hired by Drisolco was discovered in early 1980, and judicial investigations led eventually to the arrest of UNISER managers in August 1981. The ensuing court case resulted in their imprisonment for periods of seven months to 2.5 years.

In Krimpen, all attention was given to the enforcement of the Nuisance Act, rather than other available regulations, such as the Surface Water Pollution Act [17] and the Provincial Regulations on Chemical Wastes. The aim of the Nuisance Act is to prevent "danger, damage or nuisance by appliances" to their surroundings. Dating from 1875, this Act is considered to be the oldest environmental law in the Netherlands, although it was not primarily intended for the environment. It is administered through licensing at the municipal level. Until 1979 the Nuisance Act contained only one administrative sanction: closure of the firm, preceded by a notice of proposed closure.

At first sight the Nuisance Act is a well prepared law providing for the necessary protection of surroundings against industrial activities, but in practice it is weak. Severe pollution could not be prevented in the Krimpen, nor indeed elsewhere. This was initially due to a weak attitude of the local authorities, but once the municipality did show a resolute attitude, it was undermined by strong pressure from central government to relax its stance.

After the Rijnmond authorities took over responsibility from the Krimpen municipality the Nuisance Act was enforced more rigorously. In fact, pollution did not end until EMK itself decided to take refuge at Moerdijk. The Nuisance Act was not very effective in controlling EMK; its enforcement needed central government support, which was lacking. One may wonder why the Surface Water Pollution Act was not invoked, because the damning DCMR report in 1970 gave reason enough to suspect EMK of severely polluting surface waters. But because the Act was rather new and the authorities had little experience in implementing it, it was too uncertain as an institutional resource.

It is clear from the EMK/UNISER case that different levels and agencies of government held fundamentally different perceptions of the main problem confronting them. This uncertainty and lack of coordination rendered formally available regulations and sanctions moribund. The case of Booy Clean underlines this picture.

The tanker cleaning firm Booy Clean operates in the Rotterdam harbor area, cleaning tanks and collecting cargo remnants and slobs (remnants mixed with washing water) from ships and fixed land installations. From 1970, local newspapers reported complaints of residents and environmentalist groups about the smell and illegal discharges. In 1973 an environmentalist group began a lawsuit against Booy Clean, in which a former employee declared that "poison, oil and chemicals were regularly discharged into the Oude Maas." But the judge acquitted Booy Clean because "there was no conclusive evidence that the declarations of the witnesses were based on facts observed by themselves" [18]. At that time, Booy Clean did not have a Nuisance Act license because it did not have to comply with the Nuisance Order, being bound instead by the Harbor Regulations [19]. Nevertheless, harbor inspectors had drawn up some 20 warrants against Booy Clean in the period 1971–1973 for infringement of the regulations [20], and in a 1983 report prepared by local authorities in Rotterdam it is stated that the river police alone issued no less than 60 warrants, among other things because of infringements of the Chemical Waste Act and the Pollution Act. Thus, civil servants in Rotterdam certainly knew of the pollution being caused by Booy Clean.

After the Surface Water Pollution Act was enacted, Booy Clean was a "fictive licensee" because it had discharged before the enforcement. Booy Clean had submitted an application for a discharge license in November 1979, which was granted in September 1980 under several conditions that restricted the substances to be discharged, plus a demand for a reconstruction plan involving a purification plant, within a year. The plan was submitted in March 1980 and deliberations with Rijkswaterstaat (the national water authority responsible for enforcing Booy Clean's compliance with the discharge license) were begun.

In August 1981 Rijkswaterstaat evaluated Booy Clean's observance of its license and concluded that the situation was intolerable. They initiated a coordinated action with the river police, harbor service, and the DCMR in order to force Booy Clean to observe the discharge license and to effect the reconstruction plan [21]. Booy Clean denied any of the infringements claimed by Rijkswaterstaat. In autumn 1981 the waste water of Booy Clean was frequently inspected and forbidden substances were again found in it. At that time it also became clear that the sludge in Geul harbor, where Booy Clean had operated since 1976, was very severely polluted with, amongst other substances, chlorinated hydrocarbons, discharge of which was forbidden.

By the end of 1981 a judicial inquiry was mounted into alleged environmental offences committed by Booy Clean, such as infringements of the Chemical Waste Act. In March 1983, the director and manager of Booy Clean were taken into temporary custody on suspicion of forgery and fraud. In early 1984 Booy Clean's discharge license was finally withdrawn by the Ministry, after several years of vain effort by local authorities to control the firm via various formally available instruments. The Rijkswaterstaat laid claim on the installations, buildings, and capital of Booy Clean to ensure that the firm would contribute to the cost of cleaning up Geul harbor.

In summer 1984 these measures were revoked by the State Council (the highest appeal court in the Netherlands) [22], according to whom there was lack of evidence that Booy Clean had caused the pollution in the Geul harbor. Shortly thereafter Booy Clean paid half a million guilders to escape further lawsuits and announced that it would make a comeback as a waste treatment firm in combination with other firms.

4.4. Chemical Waste – Problem Definitions and Institutional Fragmentation

As in the EMK/UNISER case, local regulatory authorities in the Booy Clean case (here, the DCMR again) found themselves unable to enforce the environmental control legislation for which they were supposed to be the statutory agent. In both cases legally adequate evidence proved extremely difficult to mount – in the Booy Clean case, impossible – despite endless reports, monitoring tests, etc., that indicated illegal discharges. Furthermore, the local regulators needed central government back-up in order to be effective, and central government was mainly busy trying to undermine them, because it had a fundamentally different view of the problem (the exception was the Rijkswaterstaat in the Booy Clean case). In the Booy Clean case, also, there was a specific and major local economic interest – the economic viability of the Port of Rotterdam – in maintaining the company, even if it was cutting corners. Thus, the Rotterdam Alderman van de Dunnen also reflected the view of central government when he observed that:

> Our harbor is visited by 40 000 vessels every year, and they bring in everything which God has forbidden. So there must be a firm to treat that rubbish. If not, ships with waste will take it to the open sea and discharge it anyway [23].

When Booy Clean was eventually prosecuted, even the public prosecutor was very restrained, observing that with some adjustments the company could fulfill an important function in the harbor.

The pragmatic restraint on the part of central government and other actors was part of an overall assumption that the central problem was to nurture the growth of a *free-enterprise* hazardous waste T&D industry. Ironically, "regulation" was turning into a form of determined "protection", even when that new industry was repeatedly contravening regulations. This general problem was outlined in Chapter 3. Furthermore, in the face of this *institutional* conflict of perspective, technical evidence was ineffective no matter how clear-cut and elaborate.

EMK/UNISER and Booy Clean were seen to be central to the creation of a new industrial sector of waste treatment and disposal, without which proper hazardous waste management would be impossible, no matter how elegant a framework of standards and registration was theoretically in place. The cases of these companies are also good examples of the administrative fragmentation that severely undermined effective enforcement of existing environmental legislation embodied in the Chemical Waste Act and other laws. In these cases, where local perceptions first provoked local regulatory reactions, it was mainly central government's divergent response that paralyzed effective action. Whilst one set of actors (central government) was urgently trying to nurture a whole new industrial sector into viable economic existence – and an inherently brittle one at that – local government and other parts of the center were trying to regulate this emergent industry in conventional, pollution risk control terms.

In the above cases, existing firms were the center of these contradictory forces. In the Induval case and in related attempts to establish new facilities, it was central government that was (along with industry) trying to take the initiative. Here, the fragmented structure of Dutch administration was even more prominent, in two respects:

(1) The adversarial nature of much of central government itself meant that interests within it who were hostile to landfill ensured that the secret Induval plan was leaked and, predictably, torpedoed.

(2) Local government had sufficient autonomous power to be able to refuse outright such facilities. With UNISER and Booy Clean, central government paralyzed local government initiatives; with Induval, local government paralyzed a central initiative. In other countries – the UK for example – they would have less power to refuse, *and* more responsibility for the provision of necessary facilities in their area in the first place. In both dimensions together – exclusion from policy responsibility, yet autonomous powers of refusal – the Dutch local authorities are formally structured to play an essentially negative, adversarial and interventionist role not wholly unlike that of a conventional "sectist" environmentalist group. Indeed, this loose analogy is underscored by the openness of Dutch local government (and, relative to other countries, central government) to public interest lobbying.

Overall it seems that the chemical waste issue was perceived to require central management and control, yet this management foundered on the fragmentary structure of government and its relative openness to plural external interests. There was little institutional "smoothing" of those interests into effective compromises. As a result, and its formal elegance and precision notwithstanding, the Chemical Waste Act was a weak one, passed largely only to gain information in the hope of supporting stronger legislation in the future. This observation is especially interesting in a comparative light because of the very similar formal structure of the Dutch legislation to that of other countries. The real driving forces lie not in the technical wording of formal programs, but in the informal infrastructures and institutions that underlie their formulation and implementation.

4.5. The Soil Clean-Up Issue

The Chemical Waste Act was developed as a legislative program before there was any widespread public concern. It was the exposure of the Induval plan – then a private part of the legislative agenda – that sparked off public attention and suspicion. Once in the public eye, it demanded – and received – more urgent attention than the slowly developing, all-embracing Soil Protection Bill, which was soon to be dismembered.

The Soil Clean-Up Act of 1983 was a very different matter. It cut sharply across and into all existing frameworks – including the Chemical Waste Act – and was hurriedly created in the crisis atmosphere following the 1980 Lekkerkerk episode, which precipitated the discovery of a major national problem of uncontrolled past dumps of toxic wastes, many of which, like Lekkerkerk, were situated in residential areas. Yet, although in its dramatic arrival as a policy problem it was very different from normal chemical waste problems, in other respects some identical processes were also at work:

(1) There was the establishment of a formal system of precise, technical regulatory norms before the practical realities of implementation (including even the existence of necessary techniques and storage or disposal sites) had been fully considered.

(2) There were severe ambiguities buried in the apparent precision of the technical norms. Technical uncertainties were brought into sharper profile by the need for (inflexible) precision to compensate for extreme *institutional* uncertainties in what was a dramatically new, unstructured, but urgent policy situation foisted onto an already fragmented political culture.

(3) There were fundamental structural differences in the perceptions and rationalities of different parts of government (though not perhaps so sharp as for the Chemical Waste Act), which strongly affected the technical interpretations and implementability of regulations.

The unexpectedness of the wide extent of contamination in past toxic waste dumps is indicated by the fact that neither the earlier Soil Protection Bill nor the Chemical Waste Act mentioned this possibility [24]. Yet only a few years later, a major part of the whole Dutch environment budget was devoted to the problem. In 1983 this was 115 million guilders, rising to 200 million guilders in 1984. Even so, the estimated scale of the required remedial action program is between the 1 to 2 billion guilders of government figures, and the 17 billion guilders of other political parties and environmentalist groups. Costs based on real experience have turned out to be far higher than these estimates, and the government figures include only the most serious threats to public health. What was intended as a five-year program has now been extended to 1996.

The Soil Clean-Up Act is devoted to remedial action to reduce the health and environmental risks from past dumps. This normally means cleaning up or removing the toxic soil and inhibiting migration of toxic materials. The Act's main lever is to finance any party – normally local government – required to undertake such action. There are several major areas where institutional uncertainties created by a totally unexpected need for interaction between different government levels and bodies, have amplified technical uncertainties, which in turn have paralyzed those institutional processes still further. The technical uncertainties included ignorance of the scale and distribution of the problem; unknown cost effectiveness and availability of different technical options for decontaminating sites; ambiguity as to what actually constitutes a site that needs cleaning; and at what levels of residual contamination can a site be regarded as being cleaned up?

We see again in this issue, as for the Chemical Waste Act, that rather precise technical regulatory norms were elaborated, effectively as if to try to fill an institutional void of fragmentation, unpredictability, and behavioral uncertainties. But in the absence of *social* consensus and informal trust, the technical ambiguities, e.g., as to what constitutes a "natural" background level of contamination, were even more sharply exposed.

4.5.1. The emergence of the issue

In 1979, soil contamination was discovered in Wierden, but it was seen as an incidental case and the minister did not become involved. In 1980, however, the whole country was shaken by the discovery of serious contamination at Lekkerkerk, which followed complaints of sickness from residents. This incident became headline news and the government, expecting that this was the worst of only a few isolated cases, decided to take immediate and strong action. A special law was passed enabling the government to buy all the affected property and 270 families were relocated in mobile homes in order to start clean-up activities. Not seeing its own commitments as the precedent for a large number of cases, the government assumed full responsibility and committed large sums to the sanitation effort. Lekkerkerk in the Netherlands, as did Love Canal in the USA, transformed the soil contamination problem into a highly visible and controversial public issue.

As a follow-up to Lekkerkerk, environmentalist groups (such as Poison-Free Netherlands) applied pressure on Minister Ginjaar to request that provincial and municipal authorities submit a list of confirmed or suspected contaminated sites. Within just three days the regional inspectorate of the province of South Holland located 141 suspected sites, and by the end of 1980 this first national inventory contained more than 4000 sites. Soil pollution gathered momentum as a public issue and a major political item; it became daily front-page news.

By late 1980 there was convincing evidence that soil pollution in the Netherlands was not restricted to a few incidental cases, but was a serious, widespread problem that demanded a large-scale approach. Far from being an isolated worst case, Lekkerkerk was beginning to look quite mild. With little political debate, all the parties involved – municipalities, provinces, the Inspectorate, national government, ministries, and environmentalist groups, as well as the public – agreed upon the need for a special bill to cover remedial actions as soon as possible and to implement an immediate interim policy. However, there was far less agreement with respect to actual procedures and the unprecedented nature of the problem bred further mutual uncertainty amongst the institutions involved.

Since enactment of the Chemical Waste Act, the general economic situation had declined dramatically by late 1980. The Ministry of Social Affairs was confronted with increasing unemployment, the Ministry of Economic Affairs with falling investment and economic growth, and the Ministry of Finance with an increased budget deficit. Municipalities saw national subventions as more difficult – but more valuable – to obtain for local services and jobs. The huge expenses that would be incurred in remedial actions would not be well received. The Ministries of Finance and of Economic Affairs, as well as industrial interests, thus questioned the size and the speed of a sanitation program. Scientific uncertainty with respect to the number, size, and risks of contaminated sites, as well as with respect to the development of cheaper sanitation techniques in the future, were arguments for reducing the amount of money allocated to the sanitation program from the 1 billion guilders announced by Minister Ginjaar. Environmentalists, however, argued that more detailed surveys would certainly increase the number of sites as well as the average size of a site's threat to health or the environment. This was largely borne out by later experience, but at this stage the perceived scale and severity of the problem oscillated wildly. For example, the Ministry of Health and Environmental Protection had first designated 350 sites for action from the mushrooming lists of over 4000 candidates discovered during 1980. Although this number soon grew to over 1000, the average costs of sanitation per site up to 1983 were, in fact, less than envisaged – about 1 million instead of 3 million guilders. This was because the worst (and thus more expensive) cases were ironically being delayed by further site investigation prior to action. Indeed, the percentage of sites needing further investigation before action (usually implying a worst case) turned out to be much higher than expected. This was partly because the nature of these sites changed, from actual dumps to sites of past industrial activity later re-used in ignorance of the toxic wastes left behind, but it was also due to the inherent delay built into the Act by the institutional processes that it required (see below).

Soil sanitation was a totally new activity for all the parties concerned. The ministry, as well as the provinces, had to reorganize in order to carry it out and the activities to be undertaken were also new. The proposed regulatory system was unusual in that it was rather centralized, and did not allow for public scrutiny. Neither, as became evident later, was it based on knowledge of the size and type of contamination, or on the possibilities, techniques, and total costs of future remedial actions. The first draft of the bill resulted in opposition from nearly all the parties involved. Under pressure from environmentalist groups, parliament decided to include a procedure for public scrutiny, despite objections of the ministry that this would extend procedures. Environmentalists also advocated that the program be financed by those responsible for the production of the wastes, i.e., mainly industrial firms. Municipalities disputed the financial provisions of the Soil Clean-Up Act, which were part of a broader context of contradictory organizational rationalities.

4.5.2. Divergent rationalities

The municipal contribution to the costs of cleaning up a designated site consisted of a basic amount, plus 10% of the remaining costs. The basic amount depended on the population of the municipality – 10 guilders per person, with a ceiling of 200 000 guilders. The government pays the rest, with a (remote) theoretical possibility of recovering the money from the original polluters, if they could be identified and successfully prosecuted in the courts.

Under this arrangement, municipalities were in a quandary. On the one hand, they were pressed directly by resident and environmentalist groups to find, report, and have sites cleaned up. On the other hand, they could incur large (and open-ended) financial burdens by too zealously reporting sites for government designation. This specific financial strain was multiplied by the institutional structure of local government, as explained below.

The procedures envisaged that local authorities would identify contaminated sites, and then the provinces would rank them for their risk characteristics, and thus for priority in the overstretched program budget. This identification and ranking was to be performed, in theory, according to precise technical criteria (see below). The ranking was then to be given, via provincial governments, to central government, who would decide which sites would actually be funded for remedial sanitation. In formal terms, therefore, local government played a very subordinate role, and it was effectively left to central (and, partly, provincial) government to determine which sites would suffer disruption and what financial burdens local authorities would incur. In practice, there was some room for municipal authorities to maneuver informally in reinterpreting the centrally imposed site ranking criteria – and there was more reason for them to want to do so.

The specific financial relationship between central and local government embodied in the soil clean-up program was complicated by the general financial system. At any given time, a single municipal authority might be receiving over 100 various subsidies from central government for various local activities, such as building programs for schools, leisure facilities, houses, hospitals, sewers, roads, etc. Many of these projects are site-specific and would be suspended indefinitely if a nearby (or even, as happened, the same) site were recognized to be contaminated and subjected to the major physical disruption of a clean-up program. In this event, the facility, the local jobs in its construction development or repair, and the local injection of central finance, all disappeared. Most suspected sites were much less clearly risky than Dordrecht, Gouderak, or Lekkerkerk (and even this was later claimed to have been exaggerated), so that there was ample opportunity as well as incentive for municipal authorities to avoid identifying sites that, if designated for clean-up, might lose them substantial existing revenues in addition to incurring their share of the (open-ended) clean-up costs.

The formal criteria that a local authority was supposed to use in identifying and ranking a site were, on the surface, quite precise and technical:

(1) The nature and degree of contamination (as measured by the concentrations of about 50 toxic chemicals and families of chemicals similar to those on the Chemical Waste Act hazard list) is evaluated by a "test framework" for distinct soil phases, solid and fluid, at three levels, A, B, and C. Level A concentrations are supposed to reflect the natural occurrence in soils, or the detection limit. Levels B and C were apparently derived from level A by a multiplication factor based mainly upon the scientific judgment of government experts concerned with drafting the act. Level B acts as a threshold for priority site investigation. Exceeding level C triggers immediate investigation of the best remedial techniques and urgent execution of sanitation. Level

A then acts as the criterion for determining whether sanitation has been adequately performed.

The "test framework" is acknowledged to be somewhat tentative, yet it is very precisely formulated. As such its role is evidently to reduce the severe institutional uncertainties experienced between the various agencies and branches of government involved in the identification of risky sites, financial allocations, technical clean-up choices, and triggering the full gamut of ramifying social effects created by such decisions.

(2) The local contamination situation is then assessed in terms of the possibilities for diffusion of the hazardous materials, in nearby watercourses, soil types, aerosols, etc. Conventional environmental transfer models could be developed and used to assess this dimension.

(3) The use and function of the affected soil is then evaluated, e.g., the proximity or exposure of human populations in residential sites, schools, and hospitals. Recreational areas (e.g., parks), water supplies, and building sites should be given priority.

These criteria are quite elaborate, yet there are ambiguities even before one considers the relation of the local authorities to other policy actors. Leaving aside the scientific *origins* of the standards, the accuracy of measurement of contamination levels (especially in bulk soil in heterogeneous sites) is notoriously very low, and the precise sampling of a site has a crucial effect upon its measured level of contamination.

In practice it has been found that the scientific uncertainties concerning the validity of the actual A, B, and C norms return to influence the judgment as to how contaminated a site is. In other words, the administrative aim has been to categorize in precise, inflexible numerical standards the broader policy *risk evaluation* (acceptable risk) problem from the supposedly more objective and confined *local implementation* problem of measurement of contamination levels, allowing the latter to determine day-to-day decisions on clean-up action or inaction. In practice, the measurement allows evaluative stances to be embedded within it. This embodiment of institutional interests in "objective" technical knowledge undermines the attempted *institutional* division of control designed into the technical framework of test criteria. This problem was also observed in the cases studied by van Eijndhoven [25].

However, the most significant difficulty arose through the different objective situations in which local and central government agencies found themselves. Faced with a major crisis of public confidence, central government had to create a regulatory framework that was centrally controlled, technically precise, uniform, and apparently coherent, with clear decision rules and action thresholds. Inevitably, this implied reducing the scope of local authority discretion.

To local authorities, on the other hand, the clean-up procedures and formal site ranking criteria meant that if a suspected site were near, say, a building site or a school development, it should be given priority. Yet the real effect of this would be to lose the extra jobs, services, and central funds for those activities. Thus, at the local level, the rationale was almost directly opposite to that of the central government regulatory criteria, especially given that the health effects were often nebulous. The municipalities had problems, pressures, and reference groups to consider that were not recognized by central government, whose precise, formal technical framework of norms were artificially stripped of any such social context. This local divergence from central expectations was exacerbated by a further requirement of the soils clean-up policy.

The Act required that provincial authorities clean up sites, yet the techniques to do this were simply not available. Thus, the only action possible in many cases would be to excavate the polluted soil and store or dump it somewhere else. Some soil was incinerated, but the large volumes involved, the lack of facilities, and the mixed results made this a relatively minor option. Thus, the very problem that beset the Chemical Waste Act – safe landfill sites – returned to haunt municipalities and provinces in the form of a proposed remedy for past, unsafe toxic dumps. These sites were vigorously resisted nearly everywhere. In some cases, local residents alleged that municipalities were dumping excavated soil that would have been defined as hazardous under the Chemical Waste Act, and which would thus be legally banned from landfill, on municipal refuse tips!

These difficulties for local authorities were made worse by the long delays in the decision procedure, exacerbated by lengthy site investigations. The passage of proposed sites up through provincial to central government for review meant a concentration and bottleneck at central government level, often leaving two or more years for local uncertainty and conflict to proliferate at the expense of the local authorities' credibility. This unusual degree of centralization was brought about by the need for a symbolic indication to a restive public of the seriousness of the national government's response, but it was especially inappropriate for the large number of relatively small cases that dominated the decisions numerically. Provinces argued that they should be allowed to make their own decisions within certain central guidelines and advice, financed from a central budget levied from industry. After a delay they partly succeeded, and now have their own budget (centrally provided) of 10 million guilders for cleaning up smaller cases.

4.5.3. Risk management in and out of context

Overall, local authorities operated under pressures and interests that were not just very different from those of central government, but were also largely unrecognized by central government. Local government has had to shape its responses in *ad hoc* fashion, so as to retain credibility with local residents and other interests, to avoid unknown financial obligations, and, as far as possible, to reduce central government control, decision delays, and associated uncertainties. The cases given by van Eijndhoven *et al.* show the very different pressures and rationalities at local level [26]. Despite the tight technical framework, therefore, local authorities, who in this issue were the *targets* of regulation, were forced to exercise a cross-cutting rationality in accordance with these local realities; their interpretations of the technical norms were thus very different from those the center may have envisaged. As an overall result, the main *practical* thresholds that determine whether or not a particular soil pollution case will be taken up are:

(1) Whether "techniques" exist (i.e., usually, publicly acceptable landfill sites) for the site clean-up.
(2) Whether substantial, indirect local finances and other activities are not too badly undercut by the direct site disruption and decision time involved.
(3) Whether the site contamination is small enough not to threaten a major, open-ended local expense, and small enough to avoid the need for further research and the associated long delays and uncertainties.
(4) In addition, of course, the environmental context as outlined in the test framework is relevant (e.g., the proximity of housing), but even here it is evident that those areas in which strong institutional arrangements exist effectively impose the criteria of evaluation. Thus, since the water authorities are very powerful and established across the central–local divide, potential water contamination is often the strongest trigger to action. This was the case with Lekkerkerk, where in May 1980 the pollution of an emergency drinking water supply precipitated the regional public health inspector to recommend evacuation; residence on the site and health complaints alone were insufficient to force evacuation.

The local authorities exercised "risk management", in the full sense, of optimizing their position on a broad and shifting range of agendas cutting through and incorporating the soil clean-up issue. The specific result of this kind of *contextual* rationality overall was to list sites with a different profile from that envisaged in central government's "optimal" risk management and rationality (which reflected its own institutional context).

4.5.4. Summary – soil clean-up

To summarize the soil clean-up aspect of the Dutch hazardous waste issue, therefore:

(1) The Dutch regulatory system was already fragmented institutionally and under pressure from existing chemical waste problems.
(2) Into this milieu burst the sudden crisis of the discovery of numerous past toxic dumps, of unknown risks, often in the heart of urban areas.
(3) The extreme institutional insecurity created by these two factors combined, and the crisis of national government credibility, engendered an elaborate, centrally controlled, and apparently technically precise decision framework. This was necessary to define the peripheral behavioral roles and criteria in a highly uncertain context, and to demonstrate symbolically that the problem was being taken very seriously.
(4) Technical uncertainties in establishing risk decision norms to identify and analyze situational risks have been made to appear even *greater* by institutional uncertainties, which have created a need for compensating (artificial and symbolic) technical "certainties
(5) Because of the widely different, often unrecognized, institutional realities of peripheral actors (local authorities), the technical test framework was creatively reshaped in the local process of identifying the sites that required remedial action.
(6) The overall result has been not only a very different framework of actual risk management from that envisaged, but has also created delays that have tended to multiply uncertainties, which in this context produced further delays, tending toward a self-confirming spiral.

4.6. Conclusions

From the many wider issues generated by the empirical research of this chapter, we emphasize those that coincide with the themes of other chapters.

The first point to be underlined is just how historically accidental are the frameworks of legislation, institutions, and technical approaches that end up as the regulatory imprint upon a given issue, shaping interrelations with overlapping issues, specific problem definitions, options, and instruments. Thus, superbly rational approaches to localized implementation decisions may be undermined because the definition of the original decision problem itself may not have coalesced sufficient institutional interests to carry implementation through the recurrence of underlying value conflicts. Hazardous waste legislation must be seen in a broader context, especially because it came after, and was thus shaped by, previous environmental control commitments. The general rise of environmental concern in the 1960s exposed the limitations of old, diffuse laws like the Nuisance Act, and also enforced a greater national government role, extending to full crisis management in the 1980s. A rational approach originally appeared to be to focus control legislation according to environmental media – water, workplaces, air, and soil – with their different properties. However, institutional realities, especially the existing levels of competence, interests, and reference groups of the various ministries, were the most powerful influences on the basic shaping of the legislative and regulatory frameworks. Agriculture took the Pesticides and Fertilizers Acts; Transport and Water Management the Water Pollution Act; and Economic Affairs the Nuclear Energy Act. For similar reasons the (concentrated) waste issue was split into two separate Acts, leaving chemical waste bereft of any practical disposal implementation infrastructure (which the other half of the Waste Act did have) reaching into local government.

Arguably, hazardous wastes were treated separately because a new ministry needed a specific crusade, and anyway Dutch political culture, being more fragmented and openly adversarial, tends to be more "reactive" than most of its European counterparts (see Chapter 12). Notice the stark contrast with the UK, where hazardous waste is legislatively and institutionally managed within the local government framework of domestic waste disposal. Thus, the Chemical Waste Act's elegant, formal regulatory structure was divorced of any practical means of implementation – and, possibly, also of its main avenue of legitimation.

This interpretation is strengthened by the later observations of Environment Minister Ginjaar, who was responsible for the 1979 regulations. He noted that he was faced with the dilemma of either waiting indefinitely whilst the fragmented interests of government were negotiated into some rough approach to coalition on the issue, or of doing something immediately, as was being vociferously demanded. Since he was more or less forced to do the latter, the regulations were precise, but also inevitably weak and expressly provisional (though they have still not been revised). The notification and licensing system was a compromise between the industry's intransigent belief in the freedom to make its own decisions and environmentalist demands that government should intervene to encourage less waste-producing technologies. The substances and processes decree – the hazardous waste classification system – for all its precision, was scientifically arbitrary and limited in scope. The provinces and environmentalist groups saw the ensuing framework as a formalized *restriction* of the authority of government on hazardous wastes.

Thus, a second point emphasized by the Dutch case is the relative unimportance of refined technical risk assessment as a regulatory resource, at least until certain more basic institutional requirements have been established. The technical frameworks were elaborated as a surrogate for the lacking sense of institutional order and predictability. It was not that prior technical uncertainty created *institutional* uncertainty, but that institutional uncertainty created a need for a high degree of technical certainty.

The development in the Netherlands of an industrial waste T&D sector under a private market strategy was paralyzed by various fragmentary properties of the government administrative structure (though it has to be admitted that free market strategies are also in trouble elsewhere – see later), leaving a largely symbolic framework of precise regulatory standards and definitions, which in practice could deal with only about 10% of the total chemical wastes being generated.

A further aspect of the progression from issue emergence, to its formal definition in legislation, to implementing regulations, to enforcement, is that the underlying structure of conflicting interests and perceptions hardly alters through the various phases, even though the principal actors and terms evolve. Various political and other demands combine to generate a need for legislation, and from this certain technical standards (in the Dutch case, very precise and, on the face of it, inflexible ones) are defined. However, even at the "back-end" implementation stage of the policy cycle, the political interests are able to pursue their conflicts right into the heart of the technical domain. They can exploit intractable ambiguities in hazard classification, and even in chemical sampling and analysis, to defend their own interests and perceptions.

Even the technical domain is thus to be seen as yet another arena where divergent economic, administrative, and policy concerns are pursued, albeit within the confines of technical language. This way of perceiving a symbiotic relationship between institutional contexts and technical knowledge is of importance throughout this book. In this chapter the relationship is developed by the argument that a culture of adversarial competition and fragmentation dominates government institutions in the Netherlands – the so-called "pillared" political culture – and, as a result, understandings and approaches to issues cannot be left implicit, informal, and flexible to further negotiation or discretion. Parties need to have more explicit, formal, "legalistic" statements of norms and mutual responsibilities in order to reduce institutional uncertainties if possible. Precise scientific statements appear to give such guarantees. This atmosphere naturally pervades regulatory relationships with industry and other organizations and is especially acute in new issues such as the soil clean-up program – and where, as a result, no stable relationships and patterns of expected behavior between parties have emerged. Technical "certainty" is required in proportion to the degree of institutional *un*certainty. In this respect the contrast with the UK is striking (see Chapter 7). There the same basic relationship is demonstrated, but at the opposite end of the scale. Interinstitutional confidence is manifested in a lack of concern for precise, technical norms and a trust in the flexible use of discretionary judgment. Furthermore, in the UK there was no felt need to symbolically isolate "hazardous" wastes as a supposedly clear population of nasty materials, and thus established institutional mechanisms for "ordinary" wastes down to local levels could be used and developed.

In this chapter on the Netherlands, therefore, we have seen some important elements of the general relationship between technical regulatory instruments and their institutional settings. The form of such technical instruments is, indeed, a coded statement of behavioral relationships between key parties, blending both future-oriented, normative requirements and a memory of past institutional experience.

In Chapter 5 we focus directly upon what is usually regarded as the major regulatory instrument and a major problem in harmonizing regulatory approaches – hazard classification of wastes. By examining the development of such classification schemes and their variations in several countries, we will be in a better position to analyze the options and limitations in international standardization, and the implications of this for national systems themselves.

Notes

[1] Lindblom, C. (1959), The science of muddling through, *Public Administration Review*, **19**, 79–88.

[2] The Ministry was originally set up in 1971 as the Ministry of Environment and Public Health. In 1982 it was rationalized into the Ministry of Housing, Physical Planning, and Environment. We use throughout the simplified title.

[3] Nuisance Act (1875, revised 1952) (Government Printing Office, The Hague) (in Dutch).

[4] Dangerous Substances Act (1963) (Government Printing Office, The Hague) (in Dutch).

[5] EEC Council Directive (1984), The transfrontier movement of toxic and dangerous wastes, 84/631/EEC, *Official Journal*, L-326 (EEC, Brussels).

[6] Ground Water Act (1981) (Government Printing Office, The Hague) (in Dutch).

[7] Waste Act (1977) (Government Printing Office, The Hague) (in Dutch).

[8] Chemical Waste Act (1976) Government Printing Office, The Hague) (in Dutch).

[9] Soil Protection Act (1980) (Government Printing Office, The Hague) (in Dutch).

[10] Soil Clean-Up (Interim) Measures (1981) (Government Printing Office, The Hague) (in Dutch).

[11] Soil Clean-Up Act (1983) (Government Printing Office, The Hague) (in Dutch).

[12] Nuclear Power Act (1963) (Government Printing Office, The Hague) (in Dutch).

[13] Pesticides Act (1962) (Government Printing Office, The Hague) (in Dutch).

[14] The Hofman Commission (1980), *Interim Advice Related to Dumping Non-treatable Chemical Wastes* (Government Printing Office, The Hague) (in Dutch).

[15] The Kolfschoten Commission, (1982) *Strategy for the Layout of a Factory Increasing Disposal Capacity of Chemical Wastes in the Netherlands* (Government Printing Office, The Hague) (in Dutch).

[16] DCMR (1983), Some remarks referring to occurrences around and visits to EMK, 1971, in *Report of the UNISER Commission*, Annex 5 (Government Printing Office, The Hague) (in Dutch).

[17] Surface Water Pollution Act (1969) (Government Printing Office, The Hague) (in Dutch).

[18] *Algemeen Dagblad* (1983), 12 March (in Dutch).

[19] Commission on Harbour, Economic Development and Environmental Afairs of Rotterdam City Council (1983), *Report on Booy Clean* (Rotterdam) (in Dutch).

[20] Gosewehr, D.N.R. and Maas, W.P.M. (1984), Booy Clean – Research into Administrative and Juridical Environmental Complications around the Rotterdam Cleaning Firm, *Mededelingen* **45**, p. 26 (Faculty of General Political and Social Sciences, University of Amsterdam) (in Dutch).

[21] *Op. cit.* [20], p. 26.

[22] *Op. cit.* [20], p. VII..

[23] *Algemeen Dagblad* (1983), 16 March (in Dutch).

[24] In fact, the Soil Pollution Bill had provisions for uncontrolled dumps, but only possible future ones.

[25] van Eijndhoven, J., Hortensius, D., Nauta, C., Nieuwdorp, G.H.E., and Worrell, C.W. (1985), *Hazardous Waste in the Netherlands: Dutch Policies from a Local Perspective,* Collaborative Paper CP-85-37 (International Institute for Applied Systems Analysis, Laxenburg, Austria)

[26] *Op. cit.* [25].

CHAPTER 5

The Listing and Classifying of Hazardous Wastes

Michael Dowling and Joanne Linnerooth

5.1. Introduction

Central to any regulatory program designed to manage the disposal of hazardous wastes is the development of a system for defining and distinguishing hazardous wastes from the plethora of nonhazardous wastes. As discussed in Chapter 3, a definitive discrimination between hazardous and nonhazardous wastes at the point of waste *production* is strictly impossible, because situational changes in waste life-cycles change their hazardousness. Nevertheless, some kind of initial classification would seem to be a crucial prerequisite of any attempt to control them into an acceptable life-cycle and "grave". In fact, as we shall see, one of the reasons why no single universally true hazard classification scheme can be constructed is because classification requires assumptions about what downstream freedoms are allowed, and this legitimately varies with institutional setting.

A classification system involves both defining the concept of waste in general and distinguishing the subset of wastes that is considered "hazardous" and thus requires special control. This process would seem to be straightforward, involving the appropriate scientific expertise to identify which wastes are potentially harmful to health or the environment, and listing and classifying them in some logical way. Economic efficiency arguments would suggest that this process would include developing a "degree-of-hazard" system, i.e., identifying those wastes that are most hazardous, less hazardous, etc., so that resources could be efficiently allocated to control the most hazardous wastes more stringently.

In reality, this process has not been straightforward at all, for reasons which have already been explained in previous chapters. In the variety of industrialized countries that have attempted to deal with the problem, attempts to define and classify hazardous wastes have been characterized by

(1) Multiplicity of technical options for classification of chemical wastes.
(2) Lack of information
(3) Scientific uncertainty.
(4) Differences in legal versus administrative or scientific definitions.
(5) Trade-offs between comprehensiveness, precision, and implementability.
(6) Considerations of environmental protection and costs to industry.
(7) Problems in balancing between standardization and the need for flexibility.
(8) Wide differences in the political cultures that affect responses to all of these difficulties.

As a result, definition and classification schemes vary significantly between countries and even between states or regions within countries. They are designed to meet different administrative purposes and institutional constraints. Indeed, even within a single regulatory system, different classification schemes exist for different parts of the waste life-cycle, creating potential confusion and loopholes. Although strong pressure toward international harmonization has resulted in the development of a cross-referencing system for the purpose of eventually harmonizing national lists to control cross-border shipments of hazardous wastes in OECD countries [1], coordinating bodies have tried but failed to develop an internationally accepted, practically usable list. An example of the curious results of multiple classification schemes is the current version of the EC directive on the transfrontier movement of toxic and dangerous wastes as operated in the UK. Thc UK Special Wastes regulations to control hazardous waste shipments from UK generators include waste solvents. The EC transfrontier directive excludes them (a separate directive is planned for them eventually) so that two loads of exactly the same waste, (say, of benzene or trichloroethylene) could be traveling around and being traded in the UK, one regulated the other not. In theory, both would still be controlled by yet another, different but overlapping classification system which is supposed to come into play in disposal site licensing. In reality, this site licensing classification system fragments into many different ones according to the way license conditions are written and used in daily practice.

In 1981, the late Dr. Berndt Wolbeck, one of the FRG's leading hazardous waste experts remarked:

> During the past ten years, the political and regulatory discussion has intensively focused on the question of what constitutes a hazardous waste. Despite these efforts, an international consensus could not be achieved on the issue. One of the primary reasons for this seems to me that the question has often been posed without indicating clearly enough the legal requirements which the definition and classification were to satisfy [2].

A British hazardous waste expert has observed that, in addition to formal legal requirements, hazardous waste classification schemes depend upon a complex structure of regulatory and administrative purposes that may vary significantly between countries [3]. We would go even further, to say that elements of local political culture determine central features of hazard classification systems and technical definitions.

Understanding how these variations arose and how explicit or implicit criteria for listing and classifying wastes were chosen reveals much about the general interaction between institutional constraints and science in developing environmental control regulations. From this understanding, two things of particular importance emerge.

First, there is a heavy reliance upon, and sometimes a wholesale transfer from, existing regulatory "knowledge" in other areas of environmental concern and in other countries. Second, defining and listing hazardous wastes is inevitably a political process since choices of method and criteria depend on local economic, administrative, and institutional needs, as well as upon scientific and technical factors. This means that any attempts to standardize definitions and lists of hazardous wastes must *inter alia* confront the same social, political, and legal interests that shaped the national lists in the first place.

In this chapter, we examine these forces that shape the hazardous waste classification systems in two key countries, the USA and the FRG . These countries were two of the first to develop hazardous waste regulation systems, both of which have served as models for other countries. For comparative insights, we also briefly describe the classification schemes of Austria, where the extent of government–industry cooperation make the procedure leading to a listing scheme somewhat unique, and the UK, where the main listing and classification scheme itself is of comparative interest. Finally, we discuss the general issues that emerge from our case studies, showing the scientific and institutional uncertainties that surround definitions and classification systems, and how the outcomes of an apparently technical problem ultimately reflect social, political, and legal factors.

5.2. Types of Hazardous Waste Classification System

Before turning to our case studies that show the history of hazard classification in specific countries, it is necessary to outline the various kinds of systems that can be used to define hazardous waste. Essentially, it is possible to designate a waste as hazardous by describing its general characteristics, such as flammability, toxicity, etc., or by noting the presence of designated hazardous constituents, or by simply determining what specific waste types are hazardous and listing them [4].

5.2.1. Hazardous waste characteristics

According to what has become known as the "characteristics approach", general hazardous waste properties, such as flammability, corrosivity, toxicity, etc., are identified and test procedures are specified to determine whether a waste exhibits any of these characteristics. For example, with regard to flammability a threshold flashpoint temperature is given, below which a waste is considered hazardous. Notice that, in principle, if test requirements were fulfilled at every stage, this approach may be sensitive to situational changes in a waste's condition, which offers a more refined risk-analytic framework. However, this could be *too* refined (e.g., a drummed volatile waste may have zero vapor pressure, but who would assume the drum is unbreachable?), and anyway implies a daunting frequency of testing.

5.2.2. Hazardous constituents

The "hazardous constituents approach" identifies the presence of specific toxic chemical constituents, possibly above a certain concentration, which then define a waste as hazardous. Wastes must be tested for the presence of such constituents. These constituents may also be listed (for example, the UK list consists primarily of chemical constituents), but the listing may be elaborated into individual compounds (e.g., "mercuric chloride, mercuric nitrate", etc.) or in constituent families (e.g., "mercury, mercury-containing compounds").

5.2.3. Hazardous waste lists

The most common approach to waste classification is the listing of specific (category) wastes identified as hazardous, for which no testing (except for its presence) is required. If the waste is on the list, it is automatically legally regulated. Several types of waste list can be used, which is rather imprecise, but has great advantages of simplicity.

(1) Lists of generic hazardous wastes, i.e., wastes arising in many different industries or from many sources. Examples are "waste lubricating oils" and "halogenated organic solvents".
(2) Lists of industry-specific wastes, such as pickling liquor from steel manufacturing.
(3) Lists of specific chemical products, which, if discarded, are to be considered hazardous, such as DDT or chlordane.

If the lists, whether of wastes or constituents, include the materials to be controlled, it is an inclusive list – anything not listed is unregulated (but see below). Another approach, which was adopted by the UK in its initial hazardous waste regulations, is the exclusive list. An exclusive list describes wastes that are *not* hazardous, and therefore not regulated. Note that an exclusive list "fails safe" in the event of incompleteness. Another "fail-safe" approach is to back up a waste and/or constituents-inclusive list with a requirement to test all un-listed wastes for defined characteristics. In the cases that follow, we examine how classification schemes using different combinations of these methods were developed and used in various countries. Notice also that even the same method can be used differently (in terms of what it contains, whether it uses concentration limits, etc.), and can be applied to different parts of the waste life-cycle. Different levels of discretionary freedom to interpret rules are also allowed.

5.3. The USA

The Resource Conservation and Recovery Act (RCRA) of 1976, reauthorized to strengthen it in 1983, is the formal legislation for controlling hazardous wastes in the USA. This legislation replaced the previous Solid Waste Disposal Act of 1965 and placed more emphasis on "resource recovery" or the recycling of waste components, such as glass, metal, aluminum, etc. Subtitle C of RCRA focuses on the control of hazardous waste, which at the time of its passage was viewed as the least significant section of the law. Coming nearly two years before the widely publicized Love Canal episode, government, industry, and environmentalist groups generally did not view the problem of hazardous wastes as a separate and distinct issue from the more general problem of handling and recycling solid industrial and municipal waste. Therefore, the Act was passed virtually unopposed by industry and unnoticed by environmentalist groups. Its passage, to a large extent, resulted from the efforts of a few individual Congressmen who felt that a law concerning solid wastes was the one unfinished piece of environmental protection legislation, after the passage of the Clean Air and Clean Water Acts in the early 1970s. Ironically, the Environmental Protection Agency (EPA) had little involvement in the preparation and passage of the RCRA legislation, yet it was later charged with the task of developing and implementing the elaborate regulatory system outlined in it [5].

Subtitle C of RCRA lays out a broad framework for the comprehensive control of hazardous wastes with five major elements:

(1) A federal classification system.
(2) A trip-ticket control system.
(3) Federal standards for generators, transporters, and disposal facilities.
(4) A permitting program.
(5) The authorization of state programs as substitutes for the federal program [6].

The details of the framework were developed within the EPA and promulgated as regulations on May 19, 1980. The 1983 reauthorization extended the controls, especially under 1, 3, and 4 above, and introduced greater precision, for example, by laying down concentration thresholds for permissible landfilling of toxic wastes.

5.3.1. The US definition of hazardous waste

Material cannot be classified as "hazardous waste" unless it first qualifies as a "waste" or, more specifically, as a "solid waste", defined in Section 1004(27) of RCRA as:

> ... any garbage, refuse sludge ... and other material including solid, liquid, semisolid, or contained gaseous material resulting from industrial, commercial, mining and agricultural operations, and from community activities [7].

Interestingly, the US definition of *solid* waste includes *liquid* and *gaseous* wastes, but by *solid* the act means "packageable wastes" as opposed to those discharged by *dispersal* into the air and water, which are regulated under the Clean Air and Water Acts. Hazardous waste is defined in Section 1004(5) of RCRA as:

> ... a solid waste, or combinations of solid wastes, which because of its quantity, concentration, or physical, chemical, or infectious attributes may:

(a) cause, or significantly contribute to an increase in mortality or an increase in serious irreversible or incapacitating illness, or
(b) pose a substantial present or potential hazard to human health or the environment when improperly treated, stored, transported or disposed of, or otherwise managed [8].

Section 3001 of RCRA directed the EPA to develop criteria for identifying the characteristics of hazardous waste and criteria for their listing. In adopting these criteria, Section 3001 required the EPA to take account of "toxicity, persistence and degradability in nature, potential for accumulation in tissue, and other related factors such as flammability, corrosiveness, and other hazardous characteristics." In response, the EPA specified a two-tiered system: a waste generator determines whether a waste is hazardous by checking if the waste appears on one of the EPA's published lists or, if the waste is not listed, by testing the waste for four designated characteristics of hazardousness. We discuss each of these tiers in turn.

5.3.2. The first tier: The hazardous waste lists

The EPA selected two sets of criteria for listing hazardous waste: criteria for wastes that are acutely hazardous and criteria for other toxic wastes. The criteria for listing acutely hazardous wastes were intended by the EPA to meet Part A of the statutory definition, i.e., to identify wastes that may "cause or significantly contribute to an increase in serious irreversible or incapacitating illness." The EPA defined these wastes to be either:

(1) Fatal to humans in low doses.
(2) Have animal toxicities of oral LD50 (lethal dose for 50% of the animals tested) of less than 50 mg/kg in rats, inhalation LD50 of less than 2000 mg/m^3 in rats, or dermal LD50 of less than 200 mg/kg in rabbits.

These animal toxicities were equivalent to US Department of Transportation, Consumer Product Safety Commission, and National Academy of Sciences definitions for poisonous or acutely toxic substances [9].

Other wastes that were not acutely toxic were to be listed if they were carcinogenic, mutagenic, teratogenic, phytotoxic, or toxic to aquatic species. Using these characteristics, the EPA identified over 380 chemical substances that, if present in a waste, would make it hazardous, unless it could be shown that the waste was not hazardous after consideration of 11 factors. The factors included, for instance, the nature of the toxicity of the constituent, the concentration of the constituent in the waste, the quantity of waste generated, and "such other factors as may be appropriate" [10]. As a result, formally speaking at least (i.e., discounting the constricting effects of litigation, lobbying etc), the Administrator of the EPA had a large measure of discretion in deciding whether a waste must be listed or not. Although any of these 11 factors could be used to justify *not* listing a waste, the actual process was described by the EPA as follows. If a waste contained one of the 380 constituents identified as hazardous, it was then analyzed to see if the constituent or constituents were present in significant concentrations. If so, the waste was most likely listed, unless there was "a very strong likelihood that the constituents were incapable of migrating even if improperly managed" [11]. The burden of proof for delisting lies on the waste producer.

The actual compilation of the hazardous waste list thus began with the identification of toxic constituents. However, toxicity testing, especially for carcinogens, is a complicated, costly procedure. Tests must be carried out under carefully controlled laboratory conditions and at best produce results that usually require judgment as to their significance. Fortunately for the EPA, other environmental legislation had already required that many toxic constituents be identified. The EPA relied almost exclusively on these lists developed for other environmental regulations to compile its list of 380 toxic constituents for RCRA. Specifically, it took approximately 300 entries from the list prepared for the Clean Water Program, six or so from the Clean Air Program, approximately 20 from the EPA List of Toxic Substances, and approximately 20 from those identified by the EPA Cancer Assessment Group [12].

Since data from the Clean Water Program were used so extensively, it is interesting to note how this list was compiled. Its history can be traced back to 1974 when environmentalist groups sued the EPA for not implementing Section 307 of the Clean Water Act, which required the EPA to identify and regulate specific toxic water contaminants. In reaching a compromise with environmentalist groups, the EPA hastily compiled an initial list from reviews of the scientific literature. One source provided the bulk of the information, a book entitled *Water Quality Criteria*, edited by McKee and Wolf. It was first published in 1952 and has been repeatedly revised up to its last edition in 1971. The book contains a survey of potential toxic contaminants of water with reference to the US and foreign literature, giving general information on the effects on aquatic life (not humans or other mammals) [13].

With the help of these identified hazardous constituents, the EPA developed its list of hazardous wastes. In this list, the EPA described wastes in two ways. First, it listed a large number of wastes from standard manufacturing or industrial processing operations known to contain toxic constituents. Second, the EPA also listed hazardous commercial products that became wastes when discarded. These lists were developed by examining some 200 studies of industrial wastes that had been compiled for other pollution control at the EPA prior to the RCRA legislation. From these studies approximately 125 wastes were identified as hazardous [14]. However, the EPA estimated that there were over 10000 major industrial waste processes, of which only 200 had been examined; therefore, the identified wastes did not begin to encompass the full gamut of hazardous wastes. Recognizing the severely limited scope of the hazardous waste lists, the EPA decided to require generators to test their wastes if they were not listed, a procedure that is unique to the US system. For this purpose, the EPA developed testing procedures for four designated characteristics of hazardousness (see below).

Industrial studies are continuing at the EPA and since the promulgation of the 1980 regulations some 10–12 additional wastes had been identified by 1985. According to one EPA staff member, however, the program can never hope to cover the dynamic universe of wastes produced. Indeed, some EPA staff members feel that the lists should be dispensed with and that the regulations should rely exclusively on the characteristic approach [15], which can be effected independently of knowledge about what wastes are produced.

The EPA lists included all three types of inclusive list explained in Section 5.2, i.e., generic lists of wastes arising in many industries, industry-specific waste streams, and a list of chemical products to be considered hazardous if discarded. Specifically, the lists comprise 13 hazardous wastes resulting from nonspecific sources (generic wastes) including various spent solvents, sludges, and similar material; 76 hazardous waste streams from specific industrial sources, such as wastewater treatment of sludge from the production of chrome yellow and orange pigments (K003); and more than 400 hazardous chemical products, such as acetaldehyde (P023). Such chemical products become controllable wastes only if discarded. Of these products, 196 were listed as acutely hazardous and over 200 classified as simply hazardous.

The acutely hazardous wastes are subject to much tighter controls based on a quantity exclusion level, which was a singular and limited attempt by the EPA to impose a degree-of-hazard system. Under the RCRA regulations, a hazardous waste was to be regulated only if a generator produces more than 1000 kg of the waste per month. This was reduced to 100 kg in the 1983 reauthorization. However, those wastes listed as acutely hazardous (Section 261.33e) must be regulated if more than *1 kg* per month is produced. By introducing this distinction the EPA recognized that some wastes are more dangerous than others and should be subject to tighter control. The EPA also considered more detailed degree of hazard classification systems including:

(1) Using a threshold quantity for hazardous constituents of a waste to determine whether regulation is required, i.e., a concentration level for particular constituents that must be exceeded before the waste is considered hazardous.

(2) Developing degree-of-hazard categories for particular wastes based on their characteristics.

The EPA rejected these and other suggestions for degree-of-hazard systems, explaining that:

> ... given current knowledge and information these assessments cannot be made for most wastes with sufficient precision to determine the specific quantities which represent a threshold for finding a waste hazardous. The Agency must, therefore, consider all quantities of any waste listed as hazardous [16].

This introduced something of an imbalance between listed wastes and those incorporated as hazardous by characteristics test, since the central "elutriate procedure" (EP) tests do embody a concentration threshold. (In fact, the 1983 reauthorization did introduce threshold concentrations of constituents to define wastes banned from landfill. This limited use of thresholds borrowed the already established Californian values.)

The detailed justification for listing each waste in the regulations was contained in background documents. The documents included:

(1) A summary of the administrator's basis for listing each waste.
(2) A brief description of the specific industry.
(3) A description of the manufacturing process.
(4) An identification of waste composition, constituent concentration, and annual quantity generated.
(5) A discussion of the basis for listing each waste stream.
(6) A summary of the diverse health effects of each of the constituents of concern [17].

Despite this elaborate justification, the EPA admitted that decisions to list a waste were often based on qualitative judgments, generally involving expert assumptions rather than precise scientific knowledge or field measurement [18].

The original waste list was subjected first to internal EPA review and then, according to US administrative law, was published for public comment in 1978. Although this procedure generated many comments from industry, the content of the list was not substantially revised and only six or seven wastes were removed as a result [19].

Finally, the regulations also provided a procedure by which any person can petition the EPA to have a listed waste "delisted". This challenge can be based on:

(1) The contention that the EPA reached an erroneous conclusion in its evaluation of the scientific grounds for listing.

(2) The assertion that the waste of the petitioner differs from the material on which the EPA focused in listing the waste as hazardous [20].

The first approach is significantly more difficult since it requires the petitioner to present scientific evidence to rebut the conclusions reached by the EPA [21].

5.3.3. The second tier: The characteristics approach

The EPA felt that its listing procedure would not cover the entire range of hazardous wastes. To fill this gap, it required generators to test those wastes that did not appear on the lists to determine whether they exhibited certain hazardous characteristics. The USA is the only country to have adopted this characteristics approach as a supplement to more administratively straightforward lists of wastes, waste processes, etc. The draft regulations originally proposed eight characteristics that required testing, but these were reduced to four in the final regulations. They include:

(1) Ignitability – liquids with a flash point of less than 60°C and flammable solids or semi-solids.
(2) Corrosivity – wastes that have a pH of 2 or less, or of 12.5 or more; or wastes that corrode steel at a rate greater than 6.35 mm per year.
(3) Toxicity – wastes that cause acute or chronic adverse health effects in persons exposed.
(4) Reactivity – wastes that react when compounds/solutions are mixed or initially interact.

The toxicity characteristic was by far the most controversial, mainly because the EPA encountered great difficulty in trying to develop testing procedures to measure toxicity [22]. The EPA finally promulgated an Elutriate Procedure Test that specified the laboratory steps to be followed in analyzing representative samples of each waste for 14 contaminants listed in the US National Drinking Water Standards. If these contaminants were present in an aqueous leachate extracted from the waste at levels 100 times or greater than the concentrations allowed in drinking water, then the waste is considered hazardous. The factor of 100 is regarded as an "attenuation coefficient" to take into account the fact that pure leachate from a landfill will be attenuated and diluted in reality before it reaches groundwater, and further diluted before it reaches water supplies. This factor was originally only 10 (meaning a 10-times stricter criterion), until it was discovered that nearly all sewage would qualify as hazardous waste on that basis. This test and the "100 times" standard have been subjected to heavy criticism due to the large scientific uncertainty involved [23].

5.3.4. Summary

The regulatory system developed by the EPA to implement the RCRA legislation represents, in the words of a former deputy administrator of the EPA, "one of the most intricate and bewildering regulatory frameworks that exist in any area of federal law" [24]. It is an attempt to devise a complete control system and in developing its multilevel classification and listing scheme, the EPA tried to cover the universe of hazardous wastes being produced throughout the USA. The system included all three types of classification schemes mentioned earlier: waste lists, characteristics, and constituents lists. There was no direct mention, however, of specific concentration levels. Given its complexity, multiplicity, and attempted comprehensiveness, this approach presented formidable implementation challenges not only to federal authorities, but also to state authorities if they chose to take over the management of hazardous wastes in their states. The 1983 reauthorization is effectively recognition of the intrinsic infeasibility of such an ambitious technical control framework, after repeated failure to meet Congressional deadlines and reportedly widespread lapses of implementation, even of regulations that had been established.

5.4. The FRG

The FRG, one of the first countries in Europe to deal with the hazardous waste problem, was also one of the first to grapple with classifying and listing hazardous wastes. The history of hazardous waste classification and listing begins with the Waste Act of 1972, amended in 1976 and 1984 (for a more detailed description see Chapter 6). The Federal Waste Act was designed to deal with the disposal of all wastes. After its enactment a series of scandals occurred concerning industrial hazardous wastes, and the *Länder* responded by pressuring the federal government to change the Act so that it would deal more specifically with industrial wastes. On June 21, 1976, the Act was amended with a revised definition of hazardous wastes and the introduction of specific control measures, including rules for siting treatment and disposal facilities; requirements for disposal facilities to appoint plant waste supervisors; and a strengthening of penalties for violations [25].

5.4.1. Two definitions

In the FRG (as in the UK, though with different meaning) the term "special waste" is used rather than "hazardous waste". In the amendments to the Waste Act, two related but somewhat different definitions of special wastes were given. The first definition was a technical one defining certain wastes from industrial sources that:

> ... because of their nature, composition, or quantity are especially dangerous to health, air, or water quality, are explosive, flammable or could promote infectious diseases, and therefore special requirements for their control are necessary [26].

The second definition was a more administrative one, stating that certain wastes:

> ... because of their nature and quantity should be excluded from disposal with household wastes [27].

Negotiations between the federal government and the *Länder* took place prior to the passage of the 1976 amendments, during which the *Länder* expressed concern that the term "hazardous" might cause unnecessary public alarm. The term "special" did not include the wastes covered by the second definition, which are really "excludable" wastes. In the end, both the words "special" and "hazardous" were left out of the Act. The term "special" is still used, but its meaning is somewhat ambiguous. In general, "special wastes" are those referred to by the aforementioned first definition found in Section 2, Paragraph 2 of the Waste Act [28]. However, the term special is still used by some authorities to mean any waste (e.g., snow from road clearing) excluded from disposal with household wastes. This confusion has yet to be cleared up, though efforts to do so are underway and are discussed below.

5.4.2. LAGA and the waste catalog

Following passage of the 1972 Waste Act requiring the *Länder* to develop waste disposal plans [29], the state governments cooperated to develop a nomenclature system for all wastes from which a list of special wastes requiring more stringent controls could be derived [30]. The forum for this cooperation was an organization called the *Länderarbeitsgemeinschaft Abfall* (LAGA, the State Working Group on Wastes). LAGA is an organization of the various environmental ministries responsible for waste disposal from the 11 *Länder* within the FRG and also includes representatives from the federal government. It has no legislative or executive functions, but serves mainly as an advisory body to the federal government, providing an opportunity for the *Länder* governments to meet and establish common positions on issues regarding waste law and policy. LAGA meets twice a year in plenary session, but often forms smaller working committees on particular issues [31].

In 1974, LAGA formed a working group for special wastes. The committee consisted of approximately 12 representatives from the *Länder*, the Ministry of Interior, and the Federal Environment Agency (Umweltbundesamt, UBA). The committee was charged with making the concept of "special waste" more concrete. It thus first organized categories of wastes into a coherent and comprehensive system so that the subset of special wastes could be more easily defined. As the chairman of the committee put it, "The catalog was really a by-product of our main objective; however, it has proven to be the more lasting contribution" [32].

The basis of the catalog was a comprehensive survey of waste producers in one of the *Länder*, Baden-Württemburg, conducted from 1970–1973. This survey collected detailed data on waste types, amounts, and disposal practices. The LAGA committee took the typology developed in the survey and refined it with the help of the Federal Statistics Agency. Wastes were divided into five major groups divided into subgroups, and finally listed as individual waste types identified with a five-digit number. The nomenclature system was a mixture of chemical compounds and generic industrial process descriptions. For example:

(1) Waste Number 55315: Methanol
(2) Waste Number 31103: Oven ash from metal processing

The five major categories were:

(1) Plant and animal waste.
(2) Mineral waste.
(3) Chemical waste from processed and synthetic products.
(4) Radioactive wastes.
(5) Municipal wastes.

Over 600 individual waste types were listed [33].

From this list of wastes, the LAGA group developed a smaller inclusive list of "special wastes" which, in their opinion, required special handling and disposal. This special waste list served as the basis for individual *Länder* regulations. These efforts were soon overshadowed, however, by a federal initiative to adopt a list of special wastes on the basis of the 1976 amendments to the Waste Act. The federal initiative related to the development of proposals for an EC directive on hazardous wastes (the eventual 1978 directive).

5.4.3. The federal regulations

In the 1976 amendments to the 1972 Waste Act, the definition of special wastes (without calling them that) in Section 2, Paragraph 2, also authorized the federal government to issue regulations that listed wastes to be included in the trip-ticket control system. These regulations, listing some 86 waste types, were promulgated on May 24, 1977.

The preparatory work for these regulations began at the UBA in Berlin, when the Minister of the Interior asked the UBA to develop a set of draft regulations. The in-house experts at the UBA were forced from the beginning to choose between two strategies. They could either develop an inclusive waste list based on the LAGA catalog, or put together a list of toxic constituents that, if present in a waste, would subject the waste to regulation. They also considered fixing concentration levels for these hazardous constituents in order for a waste to be controlled by the system. In the end, however, they decided to adopt the system developed by LAGA. There were a number of reasons for this: sampling and analysis are time-consuming and expensive, and scientific analysis was not advanced enough to make the testing of concentration levels very practical; it would be too easy to circumvent the regulations by simply diluting or mixing wastes; the *Länder* had already developed regulations based on the LAGA catalog and a new system would have been very difficult to implement.

The UBA scientists chose wastes for their list based on the following criteria:

(1) Source of the waste.
(2) Composition.
(3) Amounts produced.
(4) Disposal practices.
(5) Environmental dangers.

These criteria were not weighted in any quantified manner, but one of the compilers of the list felt that emphasis was placed on the environmental dangers of the substances in the wastes [34]. The UBA considered the lists of other countries (the Netherlands, Denmark, and Canada) and the list proposed for the promulgated EC directive on hazardous wastes. This process led to a rather long list of wastes that was then shown to *Länder* authorities and industrial associations for comment. The *Länder* protested against the length of the list and felt that the federal regulations should only include those wastes that were produced in *all* of the *Länder* [35].

Finally, a much smaller list was submitted by the UBA and adopted by the Ministry of the Interior. This list as it emerged was not meant to be a comprehensive description of the most problematic wastes that should be controlled by all of the *Länder* [36]. There was no procedure given for adding or subtracting wastes from this list, but *Länder* governments had the authority to require trip-ticket regulation for other wastes and some did increase the number of regulated wastes substantially (see the discussion of Bavaria and Hessen below). The federal regulations contained no explicit production quantity cut-offs, as did the US regulations for the wastes listed.

More subtly, the FRG chose to embody quantity considerations in the listed description of the source of the waste. Whereas the USA separated the hazard definition from questions of quantity produced, the FRG combined their judgments of the latter in their hazard definitions. This was done as shown in *Table 5.1*. Column 1 of the list gives the generic name of the waste, column 2 gives the chemical name and the waste number, and column 3 lists the industrial source of the waste.

Table 5.1. Example of FRG hazard definitions.

Column 1	Column 2	Column 3
Halogen-containing organic solvents	Chlorobenzene 55202 Chloroform 55203	From the chemical industry, gas works, pharmaceuticals

In listing the sources in column 3, the UBA listed only those industries or production processes that they considered as producing significant quantities of the listed waste. So, a waste comes under the regulations if it is listed in columns 1 and 2, *and* column 3, i.e., is produced by one of the industries or as a result of one of the processes listed in column 3. The same waste produced by an industry not given in column 3 *is not* covered by the federal regulations [37].

In contrast with the EPA in the USA, the UBA relied only on the listing method for classifying hazardous wastes and did not adopt the characteristics approach for identifying hazardous wastes not covered by the lists. The scientific analyses required for testing waste characteristics, it was felt, were both difficult to standardize and would place unacceptable cost burdens on industry. It is interesting that this attitude prevailed *before* formal consultation with industry, suggesting that the regulators already had a fairly well developed sense of industry's perspectives and practicalities. Notice that the FRG method also requires institutional confidence that consultation with *Länder* and industry will identify all significant wastes and sources.

5.4.4. *Länder* lists – Bavaria and Hessen

In addition to the 86 wastes that must be controlled under the federal regulations, environmental *Länder* authorities are free to require consignment notes and other controls for additional wastes. The two *Länder* that became most actively involved in hazardous waste regulation were Bavaria and Hessen. It is interesting to compare their waste lists with the federal one.

Bavaria had developed its own waste list before the federal regulation. This list was also based on the LAGA catalog, but included quantity cutoffs for four classes of waste. For each category, there was a minimum amount that had to be produced in order to trigger regulation (*Table 5.2*).

Table 5.2. Triggers for Bavaria's initial waste categories

Waste class	*Cut-off* (tonnes/year)
Class I	1
Class II	10
Class III	100
Class IV	1000

This was, in effect, a degree-of-hazard system with the wastes in class I being the most hazardous, those in class II less so, etc. It was put together by scientists within the Bavarian Ministry of Land Development and Environment with some consultation with industry experts. Since Bavaria had financed and constructed treatment and disposal facilities very early on (see Chapter 6), it had gained a good idea of the amounts and nature of wastes being produced in the *Land.* Its regulatory system had to be abandoned, however, when the federal regulations were adopted, since it did not conform. Bavarian authorities expressed frustration at having developed a control system too early and then having to change it at considerable cost [38].

Bavaria finally promulgated new regulations with a hazardous waste list that incorporated the federal one plus those wastes that it wanted to regulate in addition. They printed the entire LAGA catalog in their regulations and distinguished four types of wastes (*Table 5.3*). In addition to the 86 wastes covered by federal law, Bavaria added 153 wastes to its list for a total of 239 wastes under its regulatory system [39].

Table 5.3. Bavaria's four waste types.

Marking	**Explanation**
X	Wastes that should generally not, or only under certain conditions, be disposed of with household wastes, but do not require a trip-ticket
N	Wastes that require a trip-ticket and special disposal
Bold print	Wastes that under federal regulations require a trip-ticket and special disposal
No marking	Wastes that can be disposed of with household wastes

Hessen also began to regulate hazardous wastes early, a result of some highly publicized scandals in the early 1970s, which led to public pressure for more active government control. The organization of the central Hessen authority for hazardous waste disposal, HIM (see Chapter 6), and *Land* regulations resulted from this public concern. The waste list developed by Hessen was also based on the LAGA catalog. Again, in-house scientists advised the Environment Ministry in Hessen as to which wastes to regulate. Hessen divided the wastes listed in the LAGA catalog into three categories, based not on quantity cut-offs, but on the kind of disposal best suited to a particular waste. The three categories are:

(1) *Category I*: Wastes that generally can be disposed of with household wastes.
(2) *Category II*: Industrial wastes that cannot be disposed of with household wastes and must be disposed of in special waste facilities, such as secure landfills.
(3) *Category III*: Industrial wastes that are especially hazardous and require special disposal techniques in special waste facilities, such as incineration or deposition [40].

These categories, in effect, produced a degree of hazard classification system based not on quantity cut-offs, as in the USA, but on guidelines on how to dispose of particular wastes. Such a system reflects the degree of hazard if one assumes that treatment categories are ranked according to risk. There are exceptions; for example, heavy metal wastes might be disposed of more safely in a secure landfill (category II) than by incineration (category III). The Hessen classification system reflected the existence of HIM, which could give direct effect to the distinctions in treatment and disposal (T&D) methods designated.

In developing this system, the authorities in Hessen consulted with industrial trade associations on the make-up of the list. At first, industry resisted the proposed system and many joint meetings and conferences were held by the State Environment Ministry. In the end, the authorities persuaded industrial representatives that the system would be simpler, reducing both analysis costs and the time required to process wastes [41].

The final Hessen list included 312 waste types in categories II and III, including the 86 wastes in the federal regulations. Hessen, therefore, listed over 70 more wastes to be regulated than did Bavaria. Ironically, these two *Länder*, who had pushed for a short federal list, went on to develop considerably longer lists on their own, and significantly different systems of classification. These differences led to problems in tracking cross-border shipments of hazardous wastes, and has led to pressure for the development of a more comprehensive federal system [42].

5.4.5. Proposed changes in the federal regulation.

The difficulties presented by the differing waste lists of the *Länder*, and local public opposition to *Länder* facilities, have inspired efforts to develop a more comprehensive, unified federal list for all of the states. The work is being conducted by the LAGA, again by a smaller working group.

This group is currently revising the LAGA waste catalog with the intent of publishing a "special waste" catalog, i.e., listing only those wastes that should *not* be disposed of with household wastes. (This was also the goal of the earlier LAGA working group.) This new catalog would categorize special wastes by the preferred disposal method, similar to the Hessen three-category system, but also reflecting some refinements made by Austria (see below) in developing its hazardous waste regulations. The LAGA hopes that this special waste catalog will be promulgated by federal government as binding for all the *Länder*, greatly enlarging the current federal waste list [43]. Since LAGA enjoys representation of all the *Länder*, and is a consensual negotiating body, the acceptance of the enlarged list should be uncomplicated once promulgated. Again, consultation with industry is planned before a final list is developed. Industrial trade associations will not participate directly in development of the list, but LAGA plans to discuss all drafts of the list with industry. The LAGA working group also plans to conduct discussions with some environmentalist groups in the FRG, who are beginning to take an active interest in hazardous waste management.

The FRG's somewhat pioneering work on the listing of hazardous wastes has served as a model for other countries in Europe. For example, Hungary, in its hazardous waste legislation, adopted the catalog originally published by LAGA and refined by Hessen. Austria, which is discussed below, has also made extensive use of the FRG catalog in developing its own hazardous waste regulations.

5.5. Austria

Austria began relatively late to develop a hazardous waste control system. As in other industrialized countries, several well publicized scandals concerning illegal dumping and poor management of wastes spurred regulatory action. Austria passed its Special Waste Act to deal with the problem in March 1983, and it went into effect in January 1984. In identifying the universe of wastes to be controlled, the Austrian authorities took advantage of their late start to observe the experiences of other countries. The procedures by which the authorities then developed their waste list are unique in the extent of cooperation between industry and government, and the merging of technical and social criteria.

The task of developing a waste list in Austria was given to the Österreichishes Normungsinstitut, the Austrian Standards Institute, a nongovernmental, technical standard-setting body that advises the government on a variety of standards – from fire extinguishers to the size of trash cans. Its work is conducted by a number of Fachnormenausschusse (FNA), Expert Standard Committees. These committees are divided into smaller working groups on particular subjects. The membership of the FNAs includes representatives from federal, state, and local governments, scientific experts from the academic community, and industrial representatives, usually from industrial trade associations and trade unions. When the FNAs divide into working groups they also may invite experts from outside the FNA to assist them [44].

The Standards Institute's FNA-157 for Waste Disposal was given responsibility for developing a waste list by the federal Ministry for Health and Environment. As a first step the FNA defined the term special wastes for Austria:

> Special wastes are solid or liquid wastes which because of their composition or quantity cannot be disposed of with municipal waste, ie., household garbage, without special preparation [45].

This was an administrative definition similar to that used in the FRG. On the basis of this definition the FNA developed a special waste catalog that was patterned after the FRG catalog and used the same five-digit numbering system. But it contained only special – not all – wastes, listing over 400 waste types. And rather than specifying the source of the listed waste, the catalog specified the treatment methods best suited for disposal of each of the wastes, in a similar manner to the system developed in Hessen. The methods included:

(1) Municipal landfill.
(2) Special waste landfill.
(3) Municipal incinerator.
(4) Special waste incinerator.
(5) Composting.

For each waste the recommended treatment method was indicated by a numbering system that indicated whether the method was:

(1) Suitable.
(2) Suitable under certain conditions.
(3) Not suitable.

This system effectively resulted in a degree-of-hazard classification system. For example, waste number 55315, methanol, was listed as not suitable for a municipal landfill, conditionally suitable for a municipal incinerator, and suitable for either a special waste landfill or special incinerator [46].

The development of the catalog was relatively uncontroversial initially, since it was then legally nonbinding, making only recommendations. Controversy arose, however, when the Ministry of Environment asked the FNA to develop from the catalog a list of hazardous special wastes, to be regulated under the Austrian trip-ticket system required by its new legislation. The ministry agreed to accept the list of the FNA with compromises with the industries it involved. However, if the FNA failed to find a compromise list, the ministry threatened that it would develop its own, perhaps stricter list.

The anticipated perils in putting together a list of hazardous special wastes prompted the special waste working group of FNA-157 to expand its membership to a record 120. The majority of the members were industry representatives from trade associations and firms, but also included federal, state, and local government officials, and a few scientific experts from universities.

Because of the unique interest in the issue, the FNA decided in this case to accept any industry representative who requested participation (usually only selected representatives are chosen by the FNA). However, an environmentalist group called Critical Chemistry was denied participation on the stated grounds that "there were already enough chemists on the committee" [47], which implied that they were not an interest group, only supplementary neutral expertise. The FNA changed its rules again and allowed the entire working group and the members of the FNA to vote on the final list of wastes. The working group formed a smaller subgroup of 15 members to make an initial selection of wastes for the list, which began by listing those wastes in the special waste catalog that were designated as requiring the most stringent disposal methods. This subgroup also examined the hazardous wastes lists of other countries, most notably the FRG and Switzerland. Their draft list contained some 160 wastes, which was then reduced in a negotiation process within the full working group.

The rules of the Standards Institute require that working groups reach decisions *unanimously*. A potential deadlock in reaching consensus on the list could have resulted by representatives from particular industries vetoing inclusion of their wastes. But there was strong pressure to compromise, since the Ministry of Health and Environment had threatened to produce its own list if agreement was not reached. One representative claimed that members were most often persuaded by health effects arguments made by the scientific experts. In addition, the earlier nonbinding classification system developed for the special waste catalog, which was put together with the help of industry, gave clear guidelines for which wastes to include. It was very difficult for industry representatives, who had accepted that catalog, to argue for changes in its principles [48].

A compromise list of 148 "special wastes requiring supervision" was agreed upon and published by the Standards Institute for public comment. This terminology was used instead of "hazardous special wastes", supposedly to avoid public alarm. About 25 objections were submitted, reviewed by the FNA, and rejected. The list was published in final form on December 1, 1983 [49], and several months later the Ministry of Health and Environment issued regulations that declared the list binding and introduced a manifest system for all listed wastes. The ministry also introduced quantity production cut-off levels similar to those in the USA. Nine waste types are regulated if more than 20 kg/year are produced; all other wastes listed are regulated only if more than 200 kg/year are produced [50].

In summary, the Austrian system, like that of the FRG, relies on an inclusive hazardous waste list and does not consider hazardous waste characteristics or constituents. Quantity cut-offs are set, introducing an informal degree-of-hazard system as in the USA, and recommendations are given concerning the appropriate disposal method, similar to Hessen's three-category list. The most interesting and unique aspect of Austria's list is the process by which it was developed, especially the degree of industry's early participation in the process. In comparison, especially with the USA and, to a lesser extent, with the FRG, the Austrians relied very little on scientists and scientific justification for the entries on the list, but relied heavily on industry's expertise and cooperation, building such values into the hazard classification itself, rather than having these exercised in its separate practical operation.

5.6. The UK

This section gives a summary of the more detailed information analyzed in Chapter 7. In the area of hazardous waste management the British are in many respects nonconformers. Not only is the practice of co-disposal, or disposing of hazardous wastes in landfill with municipal household garbage, widely practiced and endorsed by the authorities, but also the definition and classification of hazardous wastes has been, in many respects, unique. When other nations were deciding what is hazardous, the British were deciding what is not hazardous; when most nations were defining hazard in such a way that their control system could be standardized and streamlined from "cradle-to-grave" control, the British were decentralizing, giving more discretion to local authorities and facility operators. In so doing the UK devised two sets of definitions, one for the road "to the grave", and the other "at the grave"; when some authorities were setting testing procedures to catch those wastes that were not explicitly listed but nonetheless could be hazardous, the UK authorities were setting testing procedures to catch those wastes that were listed but possibly not hazardous. These and other anomalies of the British system render it an intriguing and challenging source for comparison with the USA, the FRG, and Austria.

The 1972 UK legislation, the Deposit of Poisonous Wastes Act, mandated a regulatory regime that was especially comprehensive regarding the scope of what was to be considered hazardous for the purpose of regulation. A list was drawn up and all wastes that were *not* on the list were to be considered hazardous, i.e., the *exclusive* list system. In other words, this was a list of safe wastes, and if a waste generator did not find his waste on the published list, then his waste was legally hazardous and subject to the regulations.

This system, it seems, could only err on the side of safety, but, as we have pointed out in the previous case studies, the definition and classification of hazard serves not only the (artificial) purpose of differentiating the safe wastes from the hazardous ones, which would be a singularly difficult scientific exercise, but also serves the purpose of bounding or limiting the scope of the regulatory problem to dimensions that are institutionally and economically manageable. In the case of the UK's exclusive list, the large numbers of wastes that were thereby reported to the authorities overwhelmed the regulatory capacity of the small staffs of the local authorities.

In 1974, the Control of Pollution Act was passed. Under section 17 of this Act the Special Waste Regulations were drawn up in 1980 and came into force in March 1981. The purpose of the new regulations was to narrow and clarify the scope of control, to apply it to a less diffuse category of risk, and to simplify the notification system. Accordingly, a far narrower concept of hazardous waste was developed in an inclusive list which was very similar to that of the 1978 EC Directive, to replace the exclusive list, which had operated since 1972. Indeed, the new system was presented as implementation of the EC Directive.

5.6.1. UK definitions of "hazardous" waste

"Hazardous waste" is not a legally used term in the UK. According to the Control of Pollution (Special Waste) Regulations 1980, a *special waste* is defined as any controlled waste that:

> Consists of or contains any of 31 presently listed substances *and* by reason of the presence of such substance,
>
> (i) is dangerous to life, or

(ii) has a flashpoint of 21 degrees Celsius or less, or
(iii) is a prescription, medicinal product.

The *and* (italicized above) is highly significant, as it means criteria (i)–(iii) are *delisting* properties, which the producer can interpret, leaving the local authority to prove otherwise. A waste is to be regarded as dangerous to life for the purposes of the regulations if:

(1) A single dose of not more than 5 cm^3 would be likely to cause death or serious tissue damage if ingested by a child of body weight 20 kg.
(b) Exposure to it for 15 minutes or less would be likely to cause serious damage to human tissue by inhalation, skin contact, or eye contact.

This definition, like the US one, is a two-tiered system of evaluation, but is entirely the opposite way round. Whereas the US characteristics *supplement* their constituents list, the UK defined characteristics *subtract* from the UK list. (These tests are given in *Figure 5.1*). Another important point of contrast is that the UK toxicity characteristic is based upon the "worst-case" scenario of a child eating waste, or skin contact on an uncontrolled landfill (it was such an event that sparked off the early legislation). The US toxicity test (the EP test) models potential human exposure via leaching and contamination of drinking water. The 1978 EC Directive defined "toxic and dangerous waste" to mean any waste containing or contaminated by one or more of the 27 categories of waste constituents listed in the annex to the Directive, in such quantities or concentrations as to present a risk to human health and the environment [51]. The Directive did not specify any concentration limits. In the UK, the authorities adopted the EC list of 27 constituent categories with the addition of four categories. In fact the UK list is a hybrid of chemical constituents and industry-specific wastes.

According to the listing procedure set out above no specific concentration limits are given for determining whether the hazardous constituents contained in the waste are sufficiently concentrated to qualify the waste for regulation, except for the 1% carcinogen limit. Rather than checking the concentration, the generator can check whether the waste exhibits any one or more of four characteristics, as shown in *Table 7.4*: ignitability, carcinogenity, corrosivity, and toxicity.

The UK Government Waste Management Paper No. 23 [52] gives rather detailed descriptions of testing procedures for determining the four characteristics used in the UK – ignitability, carcinogenity, corrosivity, and toxicity. In addition, this technical guide – advisory only – expands the generic lists of 31 constituents to include over 1500 specific constituents, with data on the characteristics of each. Choosing one entry, for example acrylic acid, we find that it has a flash point greater than 21° C, that the LD50 in rats is 340 mg/kg^{-1}, that it is not a known carcinogen, and that 15 minutes

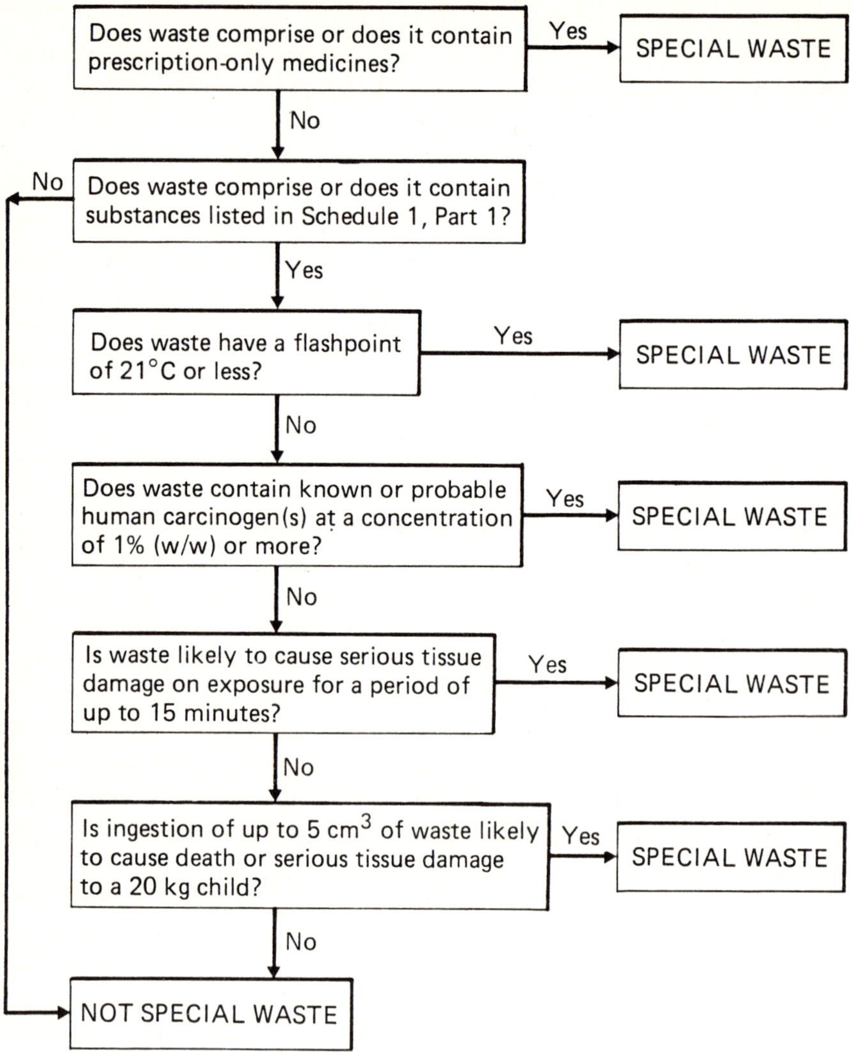

Figure 5.1. UK Assessment procedure for Special Wastes. This represents a logical approach to the assessment of a waste's status according to the defined criteria. The order in which the questions appear does not imply any grading of importance, but will usually provide the quickest reasoned decision.

of human skin contact will have a tissue damage of 10%. If a waste contains large amounts of this acid, its corrosivity would qualify it as special; if the concentrations are low, then the generator must test the waste to determine its overall corrosivity. There is some substantial room for discretion, depending on how the waste is sampled for testing and on how scientific data and tests are interpreted. In fact, there is plenty of scope for discretion and disagreement with regard to each testing procedure, even if the test models an appropriate real-world exposure scenario – an equally open question (see Chapter 7).

Ironically, a major concern with the new UK regulations and definitions of special wastes, which were meant to limit the scope of regulatory control, is that the scope is now too narrow to protect the environment adequately. The local authority associations estimate that about two thirds of the substances controlled under the 1972 Act are now outside the scope of the present regulations. The resulting loss of notification of waste transfers presents a significant handicap in their view in managing wastes liable to cause environmental harm other than danger to human life, a concern that is not explicitly addressed in the definition of special wastes [53]. Indeed, they and others also argue that the criteria used to define "danger to human life" are too limited *and* too vague to be legally enforceable.

This rather complicated classification system thus allows the generators considerable discretion in judging whether their wastes contain the right quantities and mixes of hazardous constituents to qualify them as special. The system was designed within traditional UK assumptions of informal collaboration, of regulation as a private partnership between industry and government at all levels. Hence regulations and definitions are imprecise, and often advisory only.

It is of considerable interest that this classification system is relevant only for the notification required for transporting special wastes from the producers to the handlers, or "up to the grave". How wastes are controlled "at the grave" is a separate matter for which a different system of classification is relevant, as described below. The main system of control as mandated by the Control of Pollution Regulations is the licensing of treatment and disposal facilities and, formally speaking, this part of the regulatory regime makes no use of the classification scheme described above.

5.6.2. UK "at-the-grave" definition of hazardous waste

The 1974 Control of Pollution Act provides for a system whereby all waste disposal facilities must be licensed by the relevant local Waste Disposal Authorities (WDAs). Basically, the WDAs, in conjunction with the Water Authorities and River Purification Boards, have to decide which wastes can safely be land disposed or treated at a given facility, and which conditions are appropriate.

Nowhere in the system of regulations are there standard rules prescribing how specific wastes must be handled, for example, deposited in a secure landfill, incinerated, etc. (specific guidelines are being considered). Essentially, the facility operator and the local WDA have full discretion and responsibility for these prescriptions based upon certain conditions suggested by the government in Waste Management Paper No. 4, *The Licensing of Waste Disposal Sites* [54].

The decisive question is whether a hazard could be caused by the treatment or deposit of the waste. The list of constituents, as discussed above, does not comprehend wastes that may cause environmental damage, only danger to life, and is thus not an appropriate guide for licensing conditions. For land disposal, which is the predominant handling method in the UK, along with a "dilute and disperse" philosophy, a variety of factors are important to consider, including the physical form of the waste, its concentration and amount, and also the circumstances of the land disposal site. Properties that are likely to influence water pollution include solubility in water; acidity or alkalinity; chemical reactivity; and capacity for absorption, adsorption, biological breakdown, and natural ion-exchange processes that may occur in the various materials encountered down the water table or surface water sources.

A WDA may wish to impose conditions relating to the records kept of disposal, treatment, or transfer operation at a licensed site. For hazardous wastes, possible additional types of information may be required, such as the disposal contractor involved, the physical form of the waste, the final location of the waste on a landfill site, etc. Officially, only for purposes of record keeping, therefore, there is a need to define and classify hazardous wastes so as to distinguish them from nonhazardous wastes. For this purpose, Waste Management Paper No. 4 contains in Appendix 4 a list of wastes and industry-specific wastes that are called "difficult". On this second, "difficult" list, which is made up of waste streams and not waste constituents, there are around 150 entries classified into some 19 categories. In general, difficult wastes are those that were considered notifiable by the previous (1972) DOPW system.

5.7. The Institutional Framing of Hazard

The case studies of the development of hazardous waste classification systems in the USA, the FRG, Austria, and the UK illustrate how both science and institutional needs shaped the regulatory programs that were constructed. In developing classification systems for hazardous wastes, administrators were confronted with highly uncertain knowledge regarding the various actual situations in which wastes occur, and regarding the toxic effect of chemicals even when tested in highly controlled laboratory situations. In many countries the regulatory agency relied heavily on expertise from other countries. Yet the rules adopted in each country, and therefore the outcomes of an apparently technical problem, ultimately reflected the political and legal frameworks in that country. The systems adopted were shaped to a large extent by social and administrative factors, even down to fine levels of technical detail, and this explains the difficulties encountered by international organizations in attempting to develop standardized classification systems.

5.7.1. Dealing with scientific uncertainty

One of the most important constraints in hazardous waste control concerns the scientific information regarding the hazards actually associated with the wastes, which is either uncertain, unknown, or unknowable (see Chapters 9 and 10). The uncertain role of science is illustrated by the deliberations over using degree-of-hazard systems for classifying wastes. Both the FRG and the USA have seriously considered, but decided against, developing such systems for their regulatory control programs, including establishing concentration levels for hazardous constituents in a waste to serve as cut-off points for control.

Few dispute the theoretical advantages of classifying wastes by different levels of risks, so that the control systems could be designed and resources allocated accordingly. Unfortunately, the information needed for such classification is difficult, if not impossible, to obtain. The "hazard" of a waste depends on a variety of factors, including physical form, composition, concentration of constituents, toxicity of constituents, method of disposal, etc. All of these factors are impossible to measure rigorously and to standardize in some form of classification system. As discussed in Chapters 3 and 10, they are not only imprecise, but *indeterminate*. Toxicity testing, for example, is filled with uncertainty concerning dose–response relationships, extrapolating animal models to humans, and the lack of standardized testing procedures. The fate of a waste can vary, depending on how it is traded, the properties of the waste, i.e., solubility, volatility, pH, etc., and the conditions of disposal, i.e., soil conditions, geology, etc. [55]. Such uncertainties led the EPA to conclude:

> The agency does not believe any of the degree of hazard systems suggested by commentators (or any the Agency could itself conceive) are capable of actually distinguishing different degrees of hazard among the myriad hazardous wastes and also relating management standards to the degrees in a technically and legally defensible way.

Yet the EPA did respond to political and economic pressure by introducing a limited degree-of-hazard system with its distinction of small quantity cut-offs for some acutely hazardous wastes. Federal authorities in the FRG also rejected the idea of introducing a formal degree-of-hazard system in their regulations, but the *Länder* have not been so reluctant. Hessen, with its three categories of wastes, has used a simple degree-of-hazard system. Also, in the USA, several states have adopted degree-of-hazard systems in developing their own regulations to implement the RCRA program [56]. Given the life-cycle (including human–behavioral) determinants of hazard (see Chapter 3), it may be more feasible to design degree-of-hazard frameworks at local regulatory levels. However, in the larger system this creates inconsistencies that could only be avoided by reducing waste movements between local regulatory areas.

An interesting comparative example is the Netherlands. As we saw in Chapter 4, under its Chemical Waste Act the Netherlands differentiated between four classes of hazardous wastes and determined differential concentration limits of chemical constituents for wastes to be considered hazardous [57]. These were precise, but rather arbitrarily fixed by negotiation between the Ministry of Economic Affairs, representing industrial interests, and government waste disposal experts. Other countries have expressed an interest in, and industries have lobbied heavily for, using such concentration limits in hazardous waste regulation, but outside the Netherlands this has been rejected as impractical for several reasons. First, it is very difficult and time consuming to sample and test wastes to determine the concentrations of particular substances. Second, as the UK demonstrates, the spirit of the regulations can easily be circumvented simply by diluting or mixing wastes so that hazardous constituent concentrations are below the cut-off point.

The notion of classifying hazardous wastes on the basis of degree of hazard is so compelling that pressure for the adoption of such systems will undoubtedly increase. But regulators faced with developing such systems, and limited by scarce resources, will have to carefully consider whether complicated, precise classifications are possible given the fundamental uncertainty of the scientific information on hazardous wastes. For example, as Majone has pointed out:

> ... even in the United States with its enormous scientific, technical, and financial resources – no more than 500 chemicals can be tested each year because of the limited availability of trained toxicologists, laboratory facilities, and test animals. This is barely sufficient to keep up with the flow of new chemicals, let alone to investigate the existing stock of well over 50 000 chemicals in commercial use. International cooperation in toxicological testing would have obvious benefits; but serious (if ill-understood) differences in methodology, risk philosophies, and regulatory approaches make cooperation difficult, and even reduce the value of the limited amount of information that is available [58].

It should be mentioned that the EEC, the USA and the OECD have been relatively successful in developing international protocols for testing toxic *goods.* Wastes, however, are much more difficult to test because of their heterogeneous nature and complicated life-cycles. Hazardous goods are often tested by agreed laboratory protocols, which ensure scientific quality control, but there is little incentive for such testing of wastes.

5.7.2. Political objectives

It is tempting to say that the observed variation in hazardous waste classification is a result of the large scientific uncertainty, which allows local political goals and administrative factors to "bias" the ensuing technical framework. However, as discussed in Chapters 9 and 10, this betrays a false assumption that "the" regulatory problem for which the classification system is developed presents itself in the same *institutional* form in the different countries. The technical hazard classification system is only a prism that refracts behavioral prescriptions between parties and relevant factors from the institutional system. For example, in the UK elaborate legal specifications for every chemical compound that could be toxic, and inflexible, precise hazard criteria, are felt to contradict the social relationships in regulation already established between industry and government. In the USA the opposite is true. This difference is not the result of "scientific uncertainty", but of different administrative and regulatory styles or political cultures (see Chapters 2 and 12). Of course, scientific uncertainties are very real and aggravate the institutional uncertainties and dislocations embedded within the "technical" hazard classification frameworks of different regulatory systems. However, the primary factor is that the basic problem is not a scientific one, but an integrated behavioral–technical one, and it therefore arises in different forms in different regulatory jurisdictions that have characteristic decision-making styles and institutional structures.

In the USA, the original goals of the RCRA program were to control all hazardous wastes generated in the USA. This comprehensive system was clearly mandated by Congress because it felt that a uniform system was necessary in order to ensure that some states would not become "dumping grounds" for others [59]. Later, as the EPA began to develop its regulations, the legal, economic, and logistical constraints of a total control system began to be recognized and *ad hoc* compromises in the philosophy of total control were introduced. These included a reduction in the number of hazardous characteristics to be tested, the introduction of a procedure for delisting wastes, the exclusion of small-quantity generators from the regulations, narrower definitions of treatment, storage, and disposal centers, etc.

In the FRG, the purpose of federal regulation never was to develop national all-inclusive standards. As with other types of regulation, the responsibility for implementing hazardous waste control was handed to the *Länder*. Federal regulations identified only those hazardous wastes that were common to all *Länder*, and the *Länder* were allowed to develop and greatly expand their own classification systems. The decision not to adopt a constituent list approach in the federal regulations was also based, in part, on the fact that *Land* governments had already developed their systems from the LAGA catalog. This deference to *Land* authority and the traditional role of the UBA as a technical advisor, not as regulator and supervisor (the role of the EPA in the USA), is reflected in the hazardous waste control system that was developed. The disadvantage of the system, namely that decentralized control can lead to wide disparities between the different *Land* regulations and make cross-border transport more difficult, has now led to pressure for a more centralized classification system and standards.

In Austria, the implementation of the system was thought to be impossible without the consent and cooperation of industry. The traditional spirit of social partnership holds that all social and economic problems can be dealt with through negotiation and cooperation between government, industry, and labor unions [60]. This belief led to a process that, from the very beginning, involved both industry and government in developing a compromise list of wastes, with the more or less explicit recognition that the definition of hazard was politically constructed. Inclusions on the waste list were voted on by representatives from industry, government, and academia.

Finally, how classification systems are shaped and influenced by political and institutional needs is demonstrated most clearly by the UK. The original "exclusive" list, which was reportedly constructed to encompass all wastes that might be hazardous, recognizing the uncertainty of the scientific knowledge, swamped the institutional capacity of the enforcement agents and placed allegedly large overheads on industry. The "scientific" pendulum then swung in the other direction, and hazard (for the purposes of transport) was defined very narrowly, creating a regulatory problem of more manageable dimensions. The bounds of the "scientific" problem were thus changed radically to meet political and institutional needs. More basically still, the dual system of hazard definitions, including a second one for disposal-site licensing, reflected the long-standing UK dependence on planning consents and licensing as an all-round means of industrial development, control, and regulation.

5.7.3. Scientific justification

In addition to the different political goals and attitudes that influenced the processes of hazard classification in the countries studied, there were differences in the amount and kind of scientific evidence presented to support the inclusion of certain wastes on the lists. In the USA, each listed waste was supported by a background document giving detailed justification. Such detail is supplied, in part, as a precaution against future legal challenges. As Brickman *et al.* [61] have pointed out in their comparison of carcinogen regulations in four countries, the courts in the USA play a much broader role in reviewing administrative action than those in Europe. For example, they point out that public interest groups and industry enjoy a much more liberal access to the courts in the USA than in the

UK, FRG, or Austria, where rules governing access to courts are much more restrictive. In addition, US law offers comparatively clearer definitions of agency responsibilities and statutory deadlines, which have allowed environmental groups and industry over the last 15 years to challenge (or compel) agency actions. Finally, US courts have gone beyond procedural scrutiny to review the substantive basis for agency decisions, while European courts have been much more deferential to agency findings of "fact".

More active judicial review in the USA, in part, led the EPA to provide elaborate justifications for listing wastes, while no such detailed justifications were deemed necessary in the FRG, Austria, or the UK. But, in addition to such legal differences, European approaches to regulation reflect long-standing traditions in most of Europe (though see chapters 3 and 12) that rigid, statutory regulations are less effective than standards and codes developed in cooperation with industry [62]. In contrast, US approaches to environmental regulation have been more adversarial in nature, where industry is generally forced to comply with rules that protect the public health and the environment. Even if agencies are reluctant to take on such adversarial roles, they are often compelled to do so by pressure from public interest groups and/or judicial action.

5.7.4. Political legitimacy

These differences in the handling of scientific uncertainty, political goals, and scientific justification all reflect cultural differences inherent in the approaches of the different regulatory agencies in legitimating their actions. In the USA, the regulatory philosophy that guided the development of the hazardous waste regulations was adversarial and characterized by centralized expertise and control. Interest groups were only allowed to comment on published draft regulations, not participate in their development. In the FRG, expertise and control were more decentralized, with regulatory responsibility in the *Länder* and the federal agencies mainly playing an advisory role. Consensus on regulatory objectives was sought – indeed, expected by the force of tradition – through the LAGA, between levels of government, not between government and interest groups. Discussions with interest groups were held, often after regulatory strategies had been chosen, but before the regulations were finalized and published. In the UK, informal consultation with interested parties is always sought before legislation or the issue of regulations, but the lobbying of industry is normally far stronger, more organized, and at least partly embedded in the administrative organs of government by strong social networks and even personnel exchanges. The distinction between scientific and other judgments is blurred. Austria explicitly involved government and industry at an early stage (although it excluded environmentalists) in a process that reflected traditional values of cooperation through social partnership. Although this process ensured acceptability among the industrial interest groups, it did leave the government open to harsh criticism from environmentalists and the media [63]. These countries represent a continuum of adversarial versus cooperative philosophies of regulation. Similar differences have been observed in the area of carcinogen regulations and the setting of occupational health standards [64].

This diversity in regulatory practice explains, to a large extent, the difficulties in developing international hazardous waste classification systems. However, some limited consensus has been achieved. The EC has issued a *Directive on Toxic and Dangerous Wastes* that lists 27 waste categories to be regulated by the member nations [65]. A NATO study identified 38 waste types that are regulated in three or more NATO countries [66]. The OECD has developed a means of cross-referencing a waste listed in one country to the list of another [67]. This scientific consensus has not, however, led to standardization. Achieving any degree of harmony will undoubtedly be frustrated by the fact that the many varied national lists reflect not only differences in the interpretation of scientific data, but, to a far greater extent, differences in political and institutional needs, which the lists were developed to meet.

5.8. Conclusions

The case studies clearly show that what appears a straightforwardly scientific exercise, namely the definition and classification of hazardous wastes, is not primarily technical. Rather, the definition of hazards ultimately depends on the political and institutional frameworks and objectives for which the definition is being developed. Scientific information, characterized by large uncertainties, can legitimately be molded to fit other needs. This analysis suggests that more attention must be given to defining realistic goals for hazardous waste definition and classification in the light of institutional and political constraints, so that these goals can have a reasonable chance of being achieved.

The widespread assumption is that the main problem is to define "hazardous waste" more rigorously, which will automatically lead to precise standardized definitions. This will then, it is believed, lead to closure of regulatory loopholes. In this chapter we demonstrate that this set of beliefs, enshrined in policy and analysis, is founded on false premises about the basic nature of regulatory problems in this area and about the role and nature of technical knowledge in relation to institutional factors. It is salutory to see that, even when hazard classification schemes are identical, or nearly so, they are different in practice, so their regulatory *use* must be examined as well. These differences are significant, and belie the apparent similarity of technical frameworks. This underlines the importance of our chosen analytical focus in the other case study chapters – implementation. In the next three chapters we look at widely different institutional settings before returning to develop this point more generally.

Notes

[1] Yakowitz, H. (1985), Harmonization of specific descriptions of special wastes subject to national controls for eleven OECD countries, *Transfrontier Movement of Hazardous Wastes* (OECD, Paris).

[2] Wolbeck, B. (1983), Political dimensions and implications of hazardous waste disposal, in Lehman, J.P. (Ed), *Hazardous Waste Disposal*, p. 8 (Plenum, New York, NY).

[3] Interview with Ted Finnecy, Harwell Laboratory, UK.

[4] Lehman, J.P. (1983), Hazardous waste definition and recommended procedures, in Lehman, *op cit.* [2], p. 45.

[5] Epstein, S. and Brown, L.O. (1982), *Hazardous Wastes in America*, p. 3 (Sierra Club, San Francisco, CA).

[6] Quarles, J. (1982), *Federal Regulation of Hazardous Wastes: A Guide to RCRA*, p. 3 (Environmental Law Institute, Washington, DC).

[7] Resource Conservation and Recovery Act, Public Law 94-580, October 1976.

[8] *Op. cit.* [7].

[9] *Federal Register* (1980), **45** (98), May 19, p. 33106.

[10] *Ibid.*, p. 33121.

[11] *Ibid.*, p. 33113.

[12] Interview with Gary Dietrich, Former Head, Office of Solid Waste, Environmental Protection Agency, Washington, DC.

[13] McKee, J. and Wolff, H.W. (Eds) (1971), *Water Quality Criteria* (State Water Resources Control Board, Sacramento, CA).

[14] *Op. cit.* [9].

[15] Interview with Matthew Strauss, Environmental Protection Agency, Washington, DC.

[16] *Op. cit.* [9], p. 33112.

[17] *Ibid.*, p. 33113.

[18] *Ibid.*, p. 33114.

[19] *Op. cit.* [9].

[20] *Op. cit.* [6], p. 56.

[21] *Ibid.*, p. 57.

[22] *Ibid.*, p. 54.

[23] *Op. cit.* [9], p. 33112.

[24] Op. cit. [6], p. 3.

[25] Kuntze, R. (1977), Das Abfallbeseitigungsgesetz, *Der Landkreis*, **8–9**, 282.

[26] Federal Waste Act, Section 2, Paragraph 2 (FRG).

[27] *Ibid.*, Section 3, Paragraph 3.

[28] Hosel, G. and von Lersner, H. *Recht der Abfallbeseitigung*, pp. 1120–1122, (E. Schmidt, Berlin).

[29] Federal Waste Act, Section 6 (FRG).
[30] Merkel, E. (1976), Abfall: Übersicht und Bewertung, *Wasser und Boden*, **8**, p. 207.
[31] Correspondence with Minister of Nutrition, Agriculture, and Forests, Silesia-Holstein, FRG, current chairman of LAGA.
[32] Interview with Barthl Fürmaier, Bavarian Ministry for Land Development and Environment, FRG.
[33] *Op. cit.* [29], p. 209.
[34] Offhaus, E. (1977), Überwachung der Beseitigung bestimmter gewerblichen Abfälle, *Der Landkreis*, **8–9**, p. 299.
[35] Interview with Axel Szelinski, UBA, Berlin, FRG.
[36] Weinheimer, T.F. (1977) Besonders Überwachungsbedurftige Abfälle, *Umwelt*, **3**, 223.
[37] *Ibid.*, p. 227.
[38] *Op. cit.* [31].
[39] Bavarian Waste Regulations from August 10, 1978.
[40] Waste Catalog of Hessen, October 1981.
[41] Interview with Carl O. Zubiller, Ministry of the Environment, Hessen, FRG.
[42] *Op. cit.* [34].
[43] *Op. cit.* [40].
[44] Information on the process of listing hazardous wastes in Austria was given in an interview with Dr. Gerhard Vogel, a professor at Vienna's Business University and a member of the Institute Expert Standard Committee on Waste Disposal.
[45] O-Norm S-2000, Abfall Begriff, Osterreichisches Normungsinstitut.
[46] O-Norm S-2100 Sonderabfallkatalog.
[47] Votzi, J. (1984), Steyrer's Müll Abfuhr, *Profil*, March 5, p. 21.
[48] *Ibid.*
[49] O-Norm S-2101 Uberwachtungs bedürftrage Abfülle.
[50] Regulations for the Control of Special Wastes, the Austrian Federal Minister for Health and Environment, January 19, 1984.
[51] EC (1978), Directive on Toxic and dangerous wastes, *Official Journal*, **L84**, 1978/319/EEC, Brussels.
[52] Department of the Environment (1981), *Special Wastes: A Technical Memorandum Providing Guidance on Their Definitions*, Waste Management Paper No. 23 (HMSO, London).
[53] Department of the Environment (1985), *Report of a Review of the Control of Pollution (Special Waste) Regulations 1980* (HMSO, London).
[54] Department of the Environment (1976), *The Licensing of Waste Disposal Sites*, Waste Management Paper No. 4, (HMSO, London).
[55] Office of Technology Assessment (1983), *Technologies and Management Strategies for Hazardous Waste Control*, p. 229 (OTA, Washington, DC).
[56] *Op. cit.* [9], p. 33104.
[57] Dirven, J. (1983), *Aspects of Hazardous Waste Policy in the Netherlands*, Mimeo (International Institute for Applied Systems Analysis, Laxenburg, Austria).
[58] G. Majone, (1982), Prevention and health standards: American, Soviet and European models, *Journal of Health, Politics and Law* **7**, 630.
[59] *Op. cit.* [7], p. 269.

[60] Matzner, E. Social Partnership, (1977) in Fischer, H. (Ed), *The Political System of Austria*, p. 432. (Europa, Vienna).
[61] Brickman, R. *et al.* (1985), *Controlling Chemicals: A Cross National Study of Policy and Politics* (Cornell University, Ithaca, NY).
[62] *Ibid.*
[63] *Ibid.*
[64] Op. cit. [46], p. 22.
[65] *Op. cit.* [46], p. 22.
[66] *Op. cit.* [51].
[67] Lehman, *op. cit.* [2].
[68] Lehman, *op. cit.* [2].

CHAPTER 6

Government Responsibility for Risk: The Bavarian and Hessian Hazardous Waste Disposal Systems

Joanne Linnerooth and Gary Davis

6.1. Introduction

Most Western, industrialized nations have passed legislation to identify hazardous waste streams, track their transport, and control and monitor their disposal. Despite the ambition of this legislation and its implementation, there is concern on the part of government officials, industry, and the public that severe problems continue to exist in controlling the hazardous waste life-cycle from its "cradle" to its "grave". In other chapters, a serious gap is identified between formal regulatory systems and the general compliance with these systems; in many countries, this gap can be attributed, in part, to the lack of a comprehensive industrial and economic infrastructure for handling the large volumes of hazardous wastes generated, even before considering whether wastes are processed through this infrastructure.

A whole spectrum of options exist for the management of hazardous wastes, including: comprehensive, integrated facilities that offer a full range of disposal and treatment methods or smaller, specialized operations; different combinations of government and private ownership of the waste handling facilities; public and private financing with or without subsidized prices to the users; and other pricing schemes coupled with incentives to encourage compliance.

The type of system that develops will depend ultimately on the political and regulatory culture of the country, but will also hinge on perceptions of hazardous waste as an industrial risk problem, or the ways in which the problem is defined and redefined by the many actors involved. For instance, the issue may be framed by some groups as a problem of the illegal dumping of wastes, in which case government ownership and subsidization of the facilities may encourage legal disposal by lowering disposal costs to the generators. Alternatively, for other groups the issue may emerge from environmental concerns over the large and increasing volumes of hazardous wastes being generated, in which case an aggressive policy of high prices will encourage waste reduction – as well as illegal dumping. This tension between encouraging the legal disposal of wastes and promoting the reduction of wastes is one of many issues to consider in making investment and financial choices for waste handling facilities. In analyzing these choices, it is instructive to examine the accumulated experiences of those countries with established and functioning infrastructures for the management of their hazardous wastes.

Two of the *Länder* (states) in the FRG, Hessen and Bavaria, along with the Scandinavian countries, have developed a unique system of public and industrial financing and ownership of integrated hazardous waste management facilities, which has effectively condensed the diffuse, multiactor hazardous waste life-cycle into a relatively comprehensive regulatory–management organization. These *Länder* have large, integrated facilities for the storage, treatment, and incineration of wastes, which are for the most part equipped with up-to-date environmental technology; furthermore, the facility managers have control over the waste from the factory gate. As a result, compared with other countries relatively little of the hazardous waste generated in Hessen and Bavaria is deposited directly in landfills; rather it is first pretreated or incinerated at high temperatures.

In this chapter we describe the history and operation of the Hessian and Bavarian systems, contrasting these systems with the more market-oriented approaches found, for example, in neighboring North Rhine-Westfalia, as well as in the USA. Our analysis of these contrasting management styles forms the basis for a discussion of the merits and drawbacks of public ownership, integrated facilities, monopoly markets, and subsidized pricing schemes. We also contrast how organizations have dealt with the inherent conflicts between promoting a market for capital-intensive facilities and motivating generators to reduce their wastes.

Since the management systems in Bavaria and Hessen cannot be directly transplanted to other countries, we view these experiences within the general political, institutional, and economic context in which they have evolved and are expected to work. Even if no "transplant" is envisaged, much of value to other systems can be learned from this kind of contextual, institutional analysis of risk management. We suggest in this chapter that a major, and hitherto unrecognized, benefit of the public, integrated management strategy is its reduction of the behavioral and physical uncertainties and indeterminacies throughout waste life-cycles.

6.2. Technical Options for Handling Hazardous Wastes

The predominant method for dealing with hazardous wastes in most industrialized countries remains land disposal. This includes landfill, where hazardous wastes are placed in or on the ground; surface impoundment or ponding, where liquid wastes are contained in natural or man-made depressions; land spreading or farming, where wastes are tilled into the topsoil for biological degradation; and deep well injection, where liquid wastes are pumped below groundwater in formations thought to be sealed off from above. In the USA, for example, deep well injection, surface impoundments, and landfills account for an estimated 90% of the disposal of industrial hazardous wastes [1]. Many other industrial countries report roughly similar figures; for example, in Europe the estimate is almost 70%. The land disposal of hazardous waste may be as a "secure" concentrate and in contained landfills that are designed to prevent contamination of groundwater, such as those with a natural or synthetic liner, and with a leachate collection system, with groundwater monitoring and a cover to prevent infiltration. Alternatively, landfill may involve "dispersal and dilution", aided by natural transformation processes in the ground, which can be beneficial. The trend has been toward containment landfill where no alternatives exist, though the UK continues to maintain that "dilute and disperse", landfill actively reduces the hazards over time, whereas containment leaves the same concentrated hazard, which will possibly have to be dealt with in the future.

Despite the engineered features, containment landfills can present two main types of environmental problems: contamination of surface- and ground-waters, and fires and explosions producing air pollution. There is a growing consensus that even state-of-the-art landfills are not appropriate for certain wastes (e.g., liquid organic solvents and inorganic acids), and generally too little is understood about the synergistic effects of chemical mixing or the reliability of natural or synthetic liners to guarantee proper long-term containment of hazardous substances [2].

Increased documentation of the risks from land disposal has led to a growing interest in the development and promotion of the numerous technological alternatives available. Based on evidence of the risks to the environment, the following technological hierarchy has been proposed as desirable management strategy [3]:

(1) Waste reduction or recycling: Preempting the generation of hazardous wastes by process changes and reuse of valuable chemicals.

(2) Physical, chemical, and biological treatment: These include physical processes, such as mechanical filtering, chemical processes, by which the molecular structure of the waste is changed, and biological processes that rely on microorganisms to treat organic materials.

These treatment techniques can render wastes innocuous or reduce their toxicity.

(3) Thermal destruction: Controlled thermal treatment, such as high-temperature incineration, destroys or renders organic wastes less hazardous and can be used to recover energy. High-temperature incineration is generally considered one of the safest methods for treating organic wastes. Recent concern about incomplete destruction of halogenated organic compounds has led to processes in which pyrolysis precedes incineration.

(4) Solidification and/or stabilization of remaining residuals before landfill: This method consists of various techniques to "solidify" or encapsulate wastes to make them less likely to migrate when placed in landfills.

Any management hierarchy for hazardous wastes is necessarily based on generalizations about the relative risks, which can be inadequate and uncertain in specific cases. For instance, high-temperature incineration is widely considered as a relatively safe method of destroying organic wastes, yet there is concern about the emission of dioxins when chlorinated organics are burned. Little is known about the chemistry of incineration, especially what happens with particular mixtures of hazardous wastes, even for those cases in which the breakdown characteristics are understood.

Furthermore, management hierarchies, such as that listed above, follow primarily from technical considerations of the environmental risks and are based to a far lesser extent on institutional considerations involved in implementation. This is true for landfill, which relies on the site operator's diligence and knowledge of mixing wastes (e.g., the UK defends "sensible" landfill codisposal, but unfortunately has not carefully analyzed what is practically involved and whether it is widely feasible and enforceable). The same is also true for high-temperature incineration, which relies on fairly narrow optimal conditions to ensure the complete combustion of wastes and the minimization of toxic emissions. Incineration of some wastes may be the best risk option in theory, but worse than landfill if performed at less than optimal conditions. Given the manifold technical and operational uncertainties, it is important to develop a management system, including the institutional base for implementing it, that is resilient to human factors and to changes in scientific knowledge.

Hessen and Bavaria stand out as having been apparently successful in implementing this preferred technical management hierarchy, reportedly placing few hazardous wastes directly into landfills. This stands in contrast to other nations, such as the USA, the Netherlands, where an estimated 88% of hazardous wastes is sent to domestic and foreign landfills [4], and the UK, which disposes 75% of its hazardous wastes in landfills [5]. Differences in definitions and the reporting of hazardous wastes make it difficult to compare these figures across countries. It is reported, however, that in the FRG as a whole there is more reliance on direct land disposal for hazardous waste management than in Bavaria and Hessen [6].

6.3. Hazardous Waste Legislation in the FRG

6.3.1. Background

The constitution (*Grundgesetz*) of 1949 established the FRG as a federation of what are currently 11 autonomous states, called *Länder*. Most regulatory power is distributed between the federal government and the *Länder*; however, in contrast to the USA, the FRG constitution vests in the *Länder* the primary responsibility for implementing and enforcing laws enacted by the federal parliament. The *Länder*, therefore, enjoy a more powerful position in relation to the federal government than do their US counterparts.

Political decision making in the FRG cannot be understood adequately unless one also takes into account attributes of the parliamentary system. In contrast to the US system, which fosters a unique competition between the executive and legislative branches (often represented by different parties), in most European countries the legislative majority party and the executive are the same, and thus the the powerful ties of party membership greatly reduce the likelihood of discord. In chapters 3 and 12 Wynne analyzes the implications of this difference of political cultures for environmental policy analysis; in addition, from an analysis of recent chemical control legislation, Brickman *et al.* have also concluded that many of the systematic divergences between US and European environmental laws can be attributed to this difference in executive–legislative relations and to the different underlying political cultures [7].

In the FRG, the federal ministries and *Land* representatives generally prepare the legislative details of a bill before its formal consideration in parliament. This institutionalized cooperation mirrors a more general cooperative ideology between the federal government and the *Länder*. For example, there is no federal environmental regulatory agency in the FRG and, in addition to direct regulatory responsibilities, *Land* agencies also cooperate in the development of standards that may ultimately be adopted by the federal government. The practices of each of the *Länder*, however, vary considerably [8].

Implementation of environmental legislation is the full responsibility of the *Länder*; it is usual for government authorities to work closely with industry and possibly other interested or influential parties in formulating policy. At this level, *Land* and local officials generally have significant discretion in developing "workable" regulations and standards with industry. This historical partnership between government and industry, in addition to the power that the *Länder* enjoy in implementing federal legislation, forms the basis for the close government and industry cooperation in the management of hazardous wastes in Hessen and Bavaria.

6.3.2. The Federal Waste Disposal Act

Hazardous waste in the FRG is regulated under the Federal Waste Disposal Act (*Abfallbeseitigungsgesetz*) of 1972, as amended in 1976, 1980, 1982, and 1985. As in other countries, the legislation was designed to cope with waste in general, not only hazardous waste. Certain types of waste are regulated separately and excluded from the Waste Disposal Act, such as waste oil, nuclear waste, waste water, military wastes, and wastes from mining. Dumping at sea is regulated by the Dumping at Sea Act of 1977.

The Waste Disposal Act and its accompanying Administrative Orders were formulated at the federal level and implemented by the *Länder* in cooperation with counties (*Kreise*) and municipalities (*Gemeinde*). Consistent with FRG regulatory tradition, this legislation lays out a framework which the *Länder* are obliged to follow, but which is general enough to allow them a great deal of discretion in choosing how the statutory goals will be met.

The federal legislation requires each generator to cede his wastes to the competent county or municipal (*Kreise* or *Kreisfreie Städte*) authorities for treatment or disposal. At the same time, it assigns responsibility to the local authorities by obliging them to handle the wastes generated in their region by providing adequate facilities. With the consent of the respective *Land* authorities, these local bodies can be relieved of their obligation to handle hazardous wastes, however, and most local bodies have followed this course.

In an Administrative Order of 1977, 86 specific waste types were listed as hazardous ("special") wastes (see Chapter 5). No procedure was specified for adding to or subtracting from this list; however, the *Länder* have the right to supplement it in their own waste plans.

The Waste Disposal Act also lays out a "cradle-to-grave" control system, where a mandatory trip-ticket procedure traces the path of each hazardous waste from its generation to its place of disposal. In addition, operators of certain facilities that handle hazardous wastes are required to appoint a waste disposal agent, protected from dismissal by law, who monitors the production, transport, and ultimate disposal of wastes.

A waste disposal plan must be drawn up by each *Land*, and is binding for local and district authorities. In principle, wastes may only be treated, stored, or deposited in approved installations and transported by certified operators. Waste disposal plants are generally licensed via regional planning permission proceedings, involving public comment and public hearings.

From this brief outline the FRG appears to have a relatively comprehensive legislative system, at least formally, for regulating hazardous wastes. Commentators and regulators, however, point to a number of weak points in the system. First, the effectiveness of the regulatory framework to "protect the well-being of the general public" depends crucially on whether all the wastes that are, in fact, hazardous are included in the universe of regulated wastes or the 86 wastes defined (a decade or more ago) as potentially dangerous. Some *Länder*, notably Hessen, have greatly expanded this list. Ways of legally evading the system also exist. A serious loophole appears to be the exemption of used oil, since used oil can legally contain a certain amount of hazardous substances, and has sometimes been sold as heating oil for apartment buildings and houses. A second, equally important loophole is the universal one concerning the recycling of waste; generators may declare their waste as an economic good if they can claim a willing buyer, and in so doing are not required to enter it in the notification system. A general clause allows the Minister of Defense to exempt military wastes from the Waste Disposal Act and the corresponding administrative orders. In addition, in the federal framework the legal possibility exists for generators to export their wastes to other *Länder* or countries with less stringent requirements. Bavaria and Hessen, however, have imposed their own export restrictions in order to protect the market of wastes for their own public facilities. This principle may be extended to all *Länder*.

These loopholes are considered to be serious, and attempts to close them by amending the Waste Disposal Act are currently underway. A 1984 amendment to the Act was intended to discourage the export of hazardous wastes from the FRG and to control imports. Discussions are also underway on how to deal with contaminated waste oil; it is likely that a testing requirement will be imposed and that certain waste oil will be regulated as hazardous waste.

A main concern in this chapter is the strategic and organizational difference between those states that have publicly owned or publicly supported integrated facilities and those that rely on a larger number of privately owned and decentralized facilities. *Figure 6.1* [9] shows how Hessen and Bavaria have organized waste disposal around a few comprehensive facilities. These are partially or fully owned by the local or *Land* authorities, compared, for example, with North Rhine-Westfalia, which operates with numerous private waste enterprises and a recently built (but underutilized) public facility. These differences go beyond organization, since the outcomes of the systems, themselves, are very different. As mentioned above, Hessen and Bavaria place a relatively small proportion of their wastes directly in landfills in comparison with other *Länder*. In the next two sections, we describe in some detail the Hessian and Bavarian waste management systems, including the ways in which these *Länder* have defined hazardous wastes, the technologies employed, and their economic, legal, and institutional context.

6.4 Hazardous Waste Practices in Bavaria

Bavaria is considered by many waste authorities in industrial countries to have one of the more successful systems for managing hazardous wastes. This is due principally to its modern integrated treatment, storage, and disposal facilities, as well as to its general performance record with respect to the way in which wastes are managed.

Bavaria is the largest of the FRG *Länder*, with an area of 70,500 km^2 and nearly 11 million inhabitants. Yet it is not highly industrialized, producing less hazardous waste than, for example, Hessen or North Rhine-Westfalia. Currently, a total of 239 wastes requiring regulation are listed in Bavaria (for more detail, see Chapter 5). Around 417,000 tons of industrial and commercial waste were produced in 1983, 118,000 of which were considered hazardous wastes as defined by the list. It is therefore a relatively dispersed system of waste production, with approximately 6 000 hazardous waste generators and around 120 000 shipments of hazardous waste per year [10].

Bavaria is of special interest because it recognized the hazardous waste problem very early on and became a forerunner in hazardous waste practices, even shaping the federal legislation. As early as 1966, the district of Mittelfranken founded a municipal cooperative [Mittelfranken Cooperative for Special Waste Management (ZVSMM)] responsible for the disposal of special wastes, and in 1970 a semi-public organization [The Association for the Management of Special Wastes in Bavaria (GSB)] was created to handle special wastes for the rest of Bavaria. In both instances, central facilities were built for treating, depositing, and incinerating hazardous wastes.

In 1983, 228 000 metric tons and 117 000 metric tons of industrial and commercial wastes were managed by the GSB and the ZVSMM, respectively. The proportion of these wastes that were incinerated, treated by chemical or physical methods, and deposited in secure landfills (mostly treatment residues) is shown in *Table 6.1*.

Table 6.1. Waste disposal in Bavaria [11].

	GSB		*ZVSMM*[a]	
Incineration	53 000 t	29%	18 000 t	14%
Chemical/physical treatment	97 000 t	45%	45 000 t	34%
Landfill	55 000 t	26%	70 000 t	53%

[a]The relatively large amount of wastes sent to landfills is principally due to one generator who disposes of large quantities of wastes containing vanadium and chromium, which cannot be treated profitably.

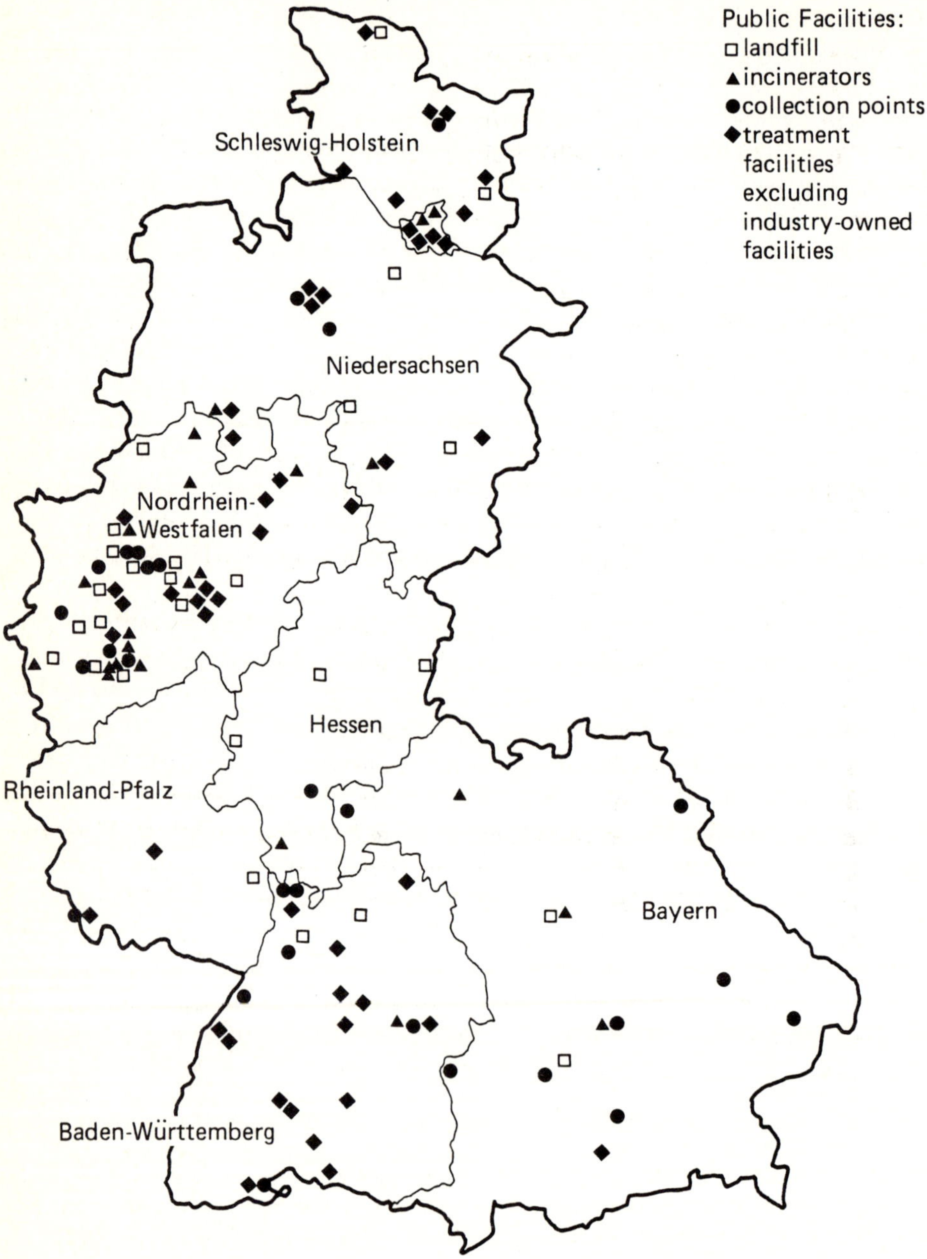

Figure 6.1. Location of hazardous waste treatment and disposal facilities in the Federal Republic of Germany.

In addition, unknown amounts of hazardous waste are managed at approximately 20 industrial facilities where the generators have been given permission to handle their own wastes.

The Bavarian law echoes the framework legislation of the Federal Waste Disposal Act of 1972. More detailed provisions that govern special waste management are contained in the Bavarian Waste Plan, adopted in 1977. Waste generators must notify the Bureau of Environmental Protection (*Landesamt für Umweltschutz*) of the Bavarian Ministry for Land Development and Environment of the types and quantities of wastes they produce. The Bureau then decides whether these wastes should be managed as hazardous wastes and in which category they belong.

The export of hazardous wastes from Bavaria is prohibited without permission from the Bureau (exemptions from this prohibition can be secured for those wastes that cannot be burned, treated, or deposited safely in Bavaria; or, permission to export may be justified by the ability of a large company to treat wastes at its own facility in another state), and generators of hazardous wastes must obtain permission from the Bureau to manage their wastes on-site. Very large chemical companies, such as Hoechst, burn their own wastes. This practice, however, is generally discouraged and often stricter environmental standards are applied to on-site facilities than to the GSB and ZVSMM plants. In 1983, approximately 9200 tonnes of industrial wastes were exported to other *Länder* in the FRG [12].

6.4.1. ZVSMM

Concern about groundwater contamination from the dumping of hazardous wastes in municipal waste dump sites inspired politicians in the district of Mittelfranken to form the ZVSMM in 1966, at which time there was no effective legislation in the FRG for dealing with the disposal of hazardous wastes. In 1968, the organization designed and constructed what was probably the first "secure" landfill, with a clay liner and leachate collection system. The management soon recognized, however, that not all hazardous wastes were suitable for landfilling and so began construction of treatment facilities in 1969 [13].

The ZVSMM is of special interest in that it is a fully public enterprise, in contrast to the joint government–industry owned facilities in Hessen and the rest of Bavaria. The operation of the ZVSMM is the responsibility of six different state, county, and municipal authorities [14].

Today, the ZVSMM owns and operates the Schwabach facility, which consists of a rotary-kiln incinerator for organic wastes, a physical and chemical treatment plant, a wastewater purification plant, and a clay-lined hazardous waste landfill (which includes leachate collection and treatment). ZVSMM has also recently opened another landfill site at Raindorf, approximately 20 km from Schwabach. These facilities serve, in addition to Mittelfranken, other parts of Bavaria and Baden-Württemberg, a total area of 21 000 km^2 (about 10% of the area of the FRG), 3.7 million inhabitants, and approximately 4000 industrial companies.

The Schwabach facilities are financed by the ZVSMM with assistance from the Bavarian state government. As of 1985, a total of DM 44.5 million had been invested in the facilities. The members of ZVSMM – five large towns, seven county districts, and seven small towns – raised DM 400 000 of the original capital investment of DM 5 million. The Bavarian government added approximately DM 2 million in the form of a grant, and the remaining funds came from loans (including a DM 1 million low-interest loan from the Marshall Fund). Subsequent construction has received a 30–50% subsidy from the Bavarian government, which will not have to be repaid [15].

The state government, therefore, subsidizes the ZVSMM facilities directly by contributing to the capital investment. The fees charged to generators for the management of wastes at Schwabach cover the remaining costs, including the unsubsidized capital costs, operating expenses, and interest payments. ZVSMM operates on a nonprofit basis, and prices are set at the end of the year to reflect the anticipated costs for the next year. Prices for 1985 ranged as in *Table 6.2.*

Table 6.2. ZVSMM prices for waste management in 1985 [16].

	Price	
Method	*DM per tonne*	*$ per tonne*
Landfill	65–195	22–65
Treatment	70–550	23–183
Incineration	80–580	27–193

At the direction of the district of Mittelfranken, ZVSMM is currently installing a flue gas scrubber for the incinerator, which will result in a price increase of approximately DM 100/tonne for incineration.

6.4.2. The GSB

The GSB manages hazardous wastes in the remainder of Bavaria. All special wastes other than those handled by the ZVSMM must be delivered to the GSB, unless permission is granted for on-site disposal or export. The GSB has four facilities and seven transfer stations in Bavaria, which consolidate wastes for shipment to these facilities. These transfer stations also dewater sludges and oil–water emulsions and perform some acid–base neutralization to reduce volumes before shipment.

The Ebenhausen plant, about 50 km from Munich, includes two rotary-kiln incinerators, a physical and chemical treatment plant, and a wastewater treatment plant. The Schweinfurt disposal plant consists of an incinerator for certain industrial wastes, such as paper contaminated with oil, that can be incinerated with domestic refuse. GSB also operates a solvent recycling facility near Munich and a large landfill at Gallenbach. These facilities have been described in detail elsewhere [17]. For our purposes, it is important to note that most hazardous wastes containing organic compounds are incinerated, and most toxic inorganic wastes are treated to reduce toxicity and mobility. The landfill sites do not generally accept liquid wastes or any wastes that have not been pretreated.

The original capital stock for the GSB facilities of DM 1 million was raised by 76 hazardous waste generators (30%), the Bavarian government (40%), and member communities (30%). The original DM 1 million of GSB member stocks had risen to a total of DM 21 million as of 1980. Bavaria now has an interest of 78%, industry 14%, and the municipalities 8%. The remaining outlays have been financed by a combination of direct government subsidies, indirect subsidies of low-interest government loans, and user fees. GSB prices for hazardous waste management in 1985 ranged as in *Table 6.3*.

Table 6.3. GSB prices for waste management in 1985 [18].

	Price	
Method	*DM per tonne*	$ per tonne
Landfill	64–195	(28–84)
Treatment	70–490	(30–211)
Incineration	5–620	(50–267)

6.5. Hazardous Waste Practices in Hessen

Hessen is also considered to have a successful hazardous waste management system. Emphasis is put on waste reduction; most hazardous wastes are incinerated or treated, and few hazardous wastes are deposited in landfills. According to the Hessian Ministry for Land Planning, Environment and Forestry, approximately 300 000 tonnes per year of hazardous waste are currently generated in Hessen and sent off-site for management [19].

Hessen has devised a degree-of-hazard system for the management of all the wastes in its waste catalog and has greatly increased the number of waste types requiring treatment as hazardous wastes. Wastes are divided into three categories, each with a specified management method:

I. Industrial wastes that are similar to municipal garbage and can generally be treated as such.
II. Industrial wastes that are hazardous and cannot be disposed of with household wastes, and thus require special handling, such as treatment, incineration, or deposition in a secure landfill.
III. Industrial wastes that are especially hazardous and require treatment under "special technical conditions", i.e., salt mine deposition or high-temperature incineration.

Categories II and III can be regarded as hazardous wastes, where category III wastes are special priority or especially hazardous [20]. From the 563 waste types listed in the LAGA catalog (see Chapter 5), 262 are identified as category II, and 37 as category III. This is compared with 86 hazardous wastes on the federal list.

For the approximately 300 000 tons of hazardous wastes in categories II and III, the breakdown in *Table 6.4.* shows how they were managed in 1983.

Table 6.4. Management of waste categories II and III [21].

Method	*Amount in tonnes*	*Percentage*
Incineration	43 000	(15%)
Chemical–physical treatment	100 000	(34%)
Landfill	136 000	(47%)
Underground deposit	13 000	(4%)

Much of the waste deposited in secure landfills was comprised of residues from incineration and physical–chemical treatment, and a large portion of the wastes were sent to other *Länder* for disposal, since Hessen lacks landfill capacity.

6.5.1. The legal and institutional framework

The organization of hazardous waste management in Hessen is legally based on the Federal Waste Disposal Act and Special Waste Order, the Hessian Waste Law (1978), and the Hessian Waste Management Plan, particularly the partial plan "Special Wastes from Industry and Firms" (1976). This plan assigns full responsibility for special wastes to the Hessian Industriemüll GmbH (HIM), which was established in 1974. The HIM is currently jointly owned by the Hessian government (26%) and a consortium of 25 Hessian waste-producing industries (74%). The Hessian government has three votes on the board of HIM, compared with industry's eight. Originally, the HIM was a private waste management company financed by hazardous waste generators, but when it encountered financial difficulties the Hessian government rescued and assumed joint responsibility for the facilities. Along with this public responsibility, the government also established central control of wastes by restricting export and competition in order to protect the economic viability of the management infrastructure.

Government authorities in Hessen, thus, exert a direct influence over hazardous waste management through their part ownership of HIM, and an indirect control through regulation. The responsible agency is the Hessian Ministry for Development, Environment, Agriculture, and Forestry, which categorizes wastes, specifies how they will be managed, and promulgates standards for emissions from industrial facilities. The Hessian authorities require that all industries deliver their special wastes to the HIM, the so-called *Benutzungszwang* or "compulsory use". The regional authorities are primarily responsible for monitoring and regulating the transportation of wastes, as well as the actual operation of waste facilities.

6.5.2. The facilities

The HIM operates four hazardous waste facilities in Hessen. The most recent is the Biebesheim incineration facility, completed in 1981, which is considered by authorities to be the state-of-the-art in hazardous waste incineration. Two chemical–physical treatment plants are in operation in Frankfurt and Kassel, and a small landfill exists for the district of Marburg. A large landfill has been planned for Mainflingen, but has encountered extensive public and political opposition, and will probably not be constructed as planned [22]. In addition to these HIM facilities, Kali and Salz AG operates an underground salt mine deposit near Herfa-Neurode.

Transfer stations for special wastes do not exist in Hessen, yet several are planned and the HIM has been testing a waste pick-up service for small generators to encourage the safe disposal of small quantities of special wastes. Industrial customers that generate less than 500 kg/year can deliver their wastes to a special truck and pay only DM 1 per kg. Citizens bringing household toxic wastes may do so free. The system is financed by a tax on waste production, its level depending on the quantity and type of waste produced.

The Biebesheim incineration facility consists of two rotary kilns, after burners, heat recovery, and a novel scrubbing system for the exhaust gases [23]. The chemical–physical treatment plants at Frankfurt and Kassel use standard technologies for cyanide destruction, neutralization, and precipitation. Only solids or dewatered sludges, usually treatment residues, are permitted in landfills in Hessen. Lacking landfill capacity, HIM exports most of the treatment residues and other wastes for landfill to neighboring *Länder* (Baden-Württemberg, Bavaria, Lower Saxony, and North Rhine-Westfalia). Only the most toxic and persistent wastes, which cannot be easily treated, are sent to the salt mines at Herfa-Neuroda. Until recently, wastes were imported to the salt mines from all states in the FRG without restriction, but the Hessian government now restricts some wastes, particularly those containing dioxin. Foreign wastes have been accepted on the condition that the foreign authorities agree to cooperate with the FRG in other types of waste management, such as nuclear waste disposal, but this practice is now discouraged by the Hessian government.

Hazardous wastes may be treated, incinerated, or disposed of on-site by the generator only with special permission from HIM; statistics on the amounts handled on-site are not available. Existing disposal facilities may continue to operate, but no new facilities, with the exception of waste water treatment, will be permitted. Firms handling their own wastes are not permitted to accept wastes from other generators.

In Hessen, regional authorities, which have licensing authority for all industrial facilities, are encouraging waste reduction by requiring documentation of waste reduction and recycling measures by those seeking permission from the planning authorities to construct new industrial facilities and to expand existing ones. The authorities can deny this permission if the facility owner does not include up-to-date measures for pollution reduction and recycling processes. Owing to public scrutiny of new industrial facilities, some facility owners have had to produce extensive documentation concerning their efforts to reduce pollution and have been compelled to make modifications in their processes [24].

6.5.3. The financial arrangements

The original industry owners of HIM financed the first facilities, but cost difficulties forced the government of Hessen to become increasingly involved, first by supplying low-interest loans, and later by directly subsidizing investment expenditures. The Hessian government paid most of the capital costs for the Biebesheim facility (DM 100 million) and recently, contributed an additional subsidy of DM 9 million. The prices per tonne for disposing of hazardous wastes by the HIM in 1981 were as in *Table 6.5.*

Table 6.5. Disposal of hazardous wastes in 1981 by HIM [25].

	Cost	
Method	*DM per tonne*	*$ per tonne*
Landfill	120–360	(52–155)
Chemical–physical treatment	60–660	(26–284)
Incineration on land	400–3000	(172–1290)
Disposal in salt mines	180	(81)

These prices do not reflect the full costs of disposal, but are increasingly subsidized by the Hessian government. The subsidies are not passed on in full to the facility users in price increases. One reason for this is that the Hessian authorities are becoming concerned about underutilization of the facilities, especially at Biebesheim. In order to fully utilize its 60 000 tonne/year capacity, about 17 000 tonnes of waste per year are imported from other *Länder*, such as Baden-Württemberg, as well as from other countries. In addition, 20 000 tonnes per year are imported to the Herfa-Neuroda salt mines [26].

Disposal prices in Hessen are higher than those in neighboring states with less stringent regulations, as well as in other countries, e.g., the German Democratic Republic, so there is a strong economic motivation for producers to export their wastes. Except with special permission, the export of hazardous wastes from Hessen is forbidden, and export of wastes to landfills outside Hessen is performed only by HIM. This compulsory use of the facilities, which effectively creates a statutory monopoly, is crucial for their economic viability. This will be discussed in more detail in Section 6.8.

6.6 Policy Strategies for a Management Infrastructure

Bavaria and Hessen have built technologically advanced, integrated facilities for the storage, treatment, and disposal of hazardous wastes with a steadily rising investment of public funds. A high priority has been placed on relatively expensive treatment and incineration technologies with relatively little direct land disposal, and a statutory monopoly has been created by requiring that firms within the *Länder* deliver their wastes to the facilities, the costs being shared between industry and the taxpayer. This "public monopoly" strategy stands in direct contrast, for example, to states in the USA which rely on the private market to provide a network of usually specialized facilities, with an emphasis on land disposal and generally with no public subsidies. In between these two contrasting systems are many diverse economic and institutional possibilities for creating a management infrastructure. The four key variables are:

(1) The extent of public versus private ownership and control.
(2) A monopolized versus competitive market.
(3) Integrated, comprehensive facilities versus segregated, specialized facilities or on-site management.
(4) Cost allocation between industry and the public purse.

In the following, we examine the merits and drawbacks of each of these four policy variables in light of the Bavarian and Hessian experiences, as well as those of another German *Land*, North Rhine-Westfalia, and the

USA. We discuss other organizational models with different combinations of public ownership, physical and economic organization, and public subsidization. This type of analysis, or any policy analysis, requires some notion of the policy objectives, so our discussion begins with an elaboration of the environmental objectives presented in Section 6.2, so as to include implementation and allocative considerations.

6.6.1. Policy objectives

The disposal of hazardous wastes in an environmentally acceptable way, with minimum direct land disposal, a cautious use of incineration, and a first priority on the reduction or recycling of wastes, is the environmental goal for hazardous waste management that is generally agreed upon by governments, environmentalist groups, and industry. Yet, inherent in even this seemingly straightforward technical hierarchy is a fundamental strategic conflict: policies that encourage generators to reduce their wastes by making disposal difficult and expensive will simultaneously and inevitably lead generators to find loopholes in the regulations and possibly even to dispose of their wastes illegally. Especially where voluntary compliance is so important, many argue for the co-option of industry through economic incentives or lower prices. Others argue that industry should pay the full social costs of their polluting behavior, and the taxpayer should not subsidize the management of hazardous wastes.

The tension between promoting the eventual reduction of hazardous wastes and their more immediate "safe" treatment and disposal is complicated by a second contradiction, which arises from the need to create a guaranteed waste market to sustain the capital-intensive treatment and disposal (T&D) investments. The high costs of these investments, if passed on to the generator, will ultimately force the reduction of wastes (or their diversion to cheaper alternatives) and consequently undermine the financial viability of the facilities. Private investors are understandably reluctant to enter a declining market. In sum, the long-term reduction of hazardous wastes, which is the most environmentally sound waste management strategy, is in some ways incompatible with shorter term needs to create and maintain a capital-intensive, expensive T&D infrastructure to handle the large amounts of wastes presently generated.

This dialectic leads to the following four conflicting management goals, where (1) and (2) condense the environmental objectives stated above and (3) and (4) expand them by including the need for general compliance with the system and for an equitable allocation of the costs:

(1) Management of currently generated hazardous wastes in the most environmentally sound ways by promulgating a comprehensive control system and establishing a technical and economic infrastructure.

(2) The (long-term) reduction of hazardous wastes.
(3) Promotion of full compliance with the system by eliminating illegal practices.
(4) Allocation of the full costs of hazardous waste management to the generators.

Most national regulatory systems primarily address the first and third of these goals by setting up a control system to track wastes from their "cradle" to their "grave" and ensuring that the wastes are then properly handled by licensed facilities. The success of this control system rests on the existence of a network of facilities with sufficient capacity, a functioning transportation system, and an adequate number of knowledgeable personnel with sufficient resources to detect and stop illegal practices. Policies for establishing this infrastructure, the topic of this chapter, range from full government support in financing and operation of the requisite facilities to more *laissez faire* strategies, which depend on the initiative of private entrepreneurs. The second goal of reducing the quantity of hazardous wastes can be promoted through various policy measures, including high prices (even artificially high) for the treatment and disposal of wastes and other economic incentives, such as a tax on wastes (waste-end tax) or subsidies for recycling wastes, the financing and operation of a waste exchange, government-supported innovation to find alternative production processes, or an outright ban on the generation of certain very hazardous wastes.

Voluntary compliance with the system, (or the third goal) might be promoted through lowered prices to the generators made possible by public subsidies, or by other economic policies, such as an assessed tax on industry and a refund on wastes delivered to the facilities (the bottle deposit concept), or by making the generators clearly responsible or liable for any damage resulting from their wastes. Alternatively, the authorities could force compliance with more frequent inspections and other enforcement measures combined with formidable sanctions. The fourth objective of allocating the full costs of hazardous waste management to the producers of the wastes (i.e., adherence to the polluter-pays principle) would mean eliminating any public subsidies that are passed on as lower prices to the generators. It would also require measures to guarantee that industry pay the full social cost of unexpected pollution, such as implementing an industry-financed post-closure fund for hazardous waste facilities and other funds to cover pollution damages, as well as compulsory insurance or other proof of financial means.

The above goals could be elaborated still further by including institutional and procedural concerns, such as maintaining trust and confidence in the public institutions responsible for hazardous waste management and ensuring equitable procedures for implementation and oversight. These concerns are addressed elsewhere in this book; for our purposes the above list serves to illustrate the types of policy trade-off in which we are most interested.

Bavaria and Hessen have chosen to place the highest priority on the first goal, but also to promote the second and third goals by supposedly setting the price of disposal high enough to encourage waste reduction, but not so high as to create a large illegal trade. This balancing act between the need for both high and low prices has been explicitly addressed by the Bavarian and Hessian governments in trying to compromise between the demands of environmentalists to reduce wastes and the demands of facility operators to ensure a waste market. It appears that wastes have been substantially reduced, but there are huge uncertainties in any estimates of the extent of illegal dumping and exports. The fourth goal has received the lowest priority, and Bavarian and Hessian taxpayers assume a large and increasing share of the T&D costs.

In the remainder of this chapter, we focus on policy choices for establishing an economic and institutional infrastructure for the treatment and disposal of currently generated hazardous wastes in an environmentally acceptable way, or the implementation of the first goal. A detailed discussion of the range of policies for fulfilling the other three goals is beyond the scope of this chapter; yet, the inevitable compromises that will be made by pursuing any one goal must be addressed and consequently reference will frequently be made to policies that promote the other three management objectives.

6.6.2. Organizational options for a physical and economic infrastructure

There are four fundamental policy questions related to the types of physical T&D facilities and their economic organization, which are illustrated by the "policy tree" shown in *Figure 6.2*. The most basic question concerns the extent to which the government assumes responsibility for hazardous waste disposal, through such means as public ownership or subsidization. A related question concerns the economic organization of the facilities; the possibilities exist for creating and sustaining a public or private monopoly by restricting entry of other firms and assuring a waste catchment market through such means as export controls. A third branch of the "policy tree" represents the choice between integrated and specialized facilities, with the caveat that the distinction is not clearly demarcated nor is the choice for or against integrated facilities wholly separate from the choice for or against a monopoly investment venture. The primary distinction between integrated and specialized facilities is not so much the physical layout, since even integrated facilities may have specialized units not physically together, but is operational and concerns who, in the chain of hazardous waste handlers, decides on the treatment and disposal method for a particular waste. For integrated facilities, all (or almost all) wastes are handed over to the facility operators where the technology (landfill, incineration, etc.) for their disposal is *internally* decided; alternatively, for specialized, segregated facilities this choice is made by the generators and/or transporters in deciding where to deliver their wastes. A fourth choice or branch on the "policy tree", again not wholly separate from the other choices, concerns who ultimately pays the costs of hazardous waste disposal – industry or the taxpayer. The government can subsidize treatment and disposal facilities through such means as investment credits, lowered interest rates on capital investments, tax breaks, or more straightforward grants to cover capital costs and operating expenses.

As shown in *Figure 6.2* various combinations of these policy variables can lead to some 13 different forms of economic and physical organizations for a treatment and disposal infrastructure. The two extremes of government intervention are represented by the Bavarian and Hessian systems and to those operating in most US states. Bavaria and Hessen have publicly owned, monopoly facilities, which are integrated and operate with large public subsidies, in contrast with states in the USA with privately owned, competitive, and specialized facilities, which generally operate with no public money. As illustrated, the Bavarian and Hessian systems are a subgroup of four possible variations that we call the *Public–Monopoly Model,* depending on the extent of integration and public subsidy. The US model is one of four variations that we call the *Private–Competitive Model.* In between, we find different combinations of private and regulated monopolies, which we call here the *Public Utility Model.* We could also imagine hybrids, where countries or regions operate with both private and public T&D facilities, in the same way that some economies incorporate mixed nationalized and private firms. It must be emphasized again that this "policy tree" does not represent a sequence of decisions in the traditional decision-analytical sense, but serves only to dissect the three models into four interrelated policy variables. This dissection is useful in keeping separate discussions of, for example, the merits and pitfalls of public ownership from the merits and pitfalls of public subsidies, where often the two are intertwined. As shown in *Figure 6.2*, even private firms can operate with public subsidies. Also, the system of *regulation* has to be overlaid upon this economic–institutional picture, though the two dimensions interact considerably.

The success of any hazardous waste management system will depend ultimately on its compatibility with the economic, institutional, and political culture in which it is expected to work. Even seemingly independent parts of the system, such as the technologies employed, cannot be assumed transferable to other countries without sometimes substantial changes in the technology itself or in the institutions that have responsibility for its operation. In what follows, we examine the more important "policy branches" within the general context of the Hessian and Bavarian hazardous waste

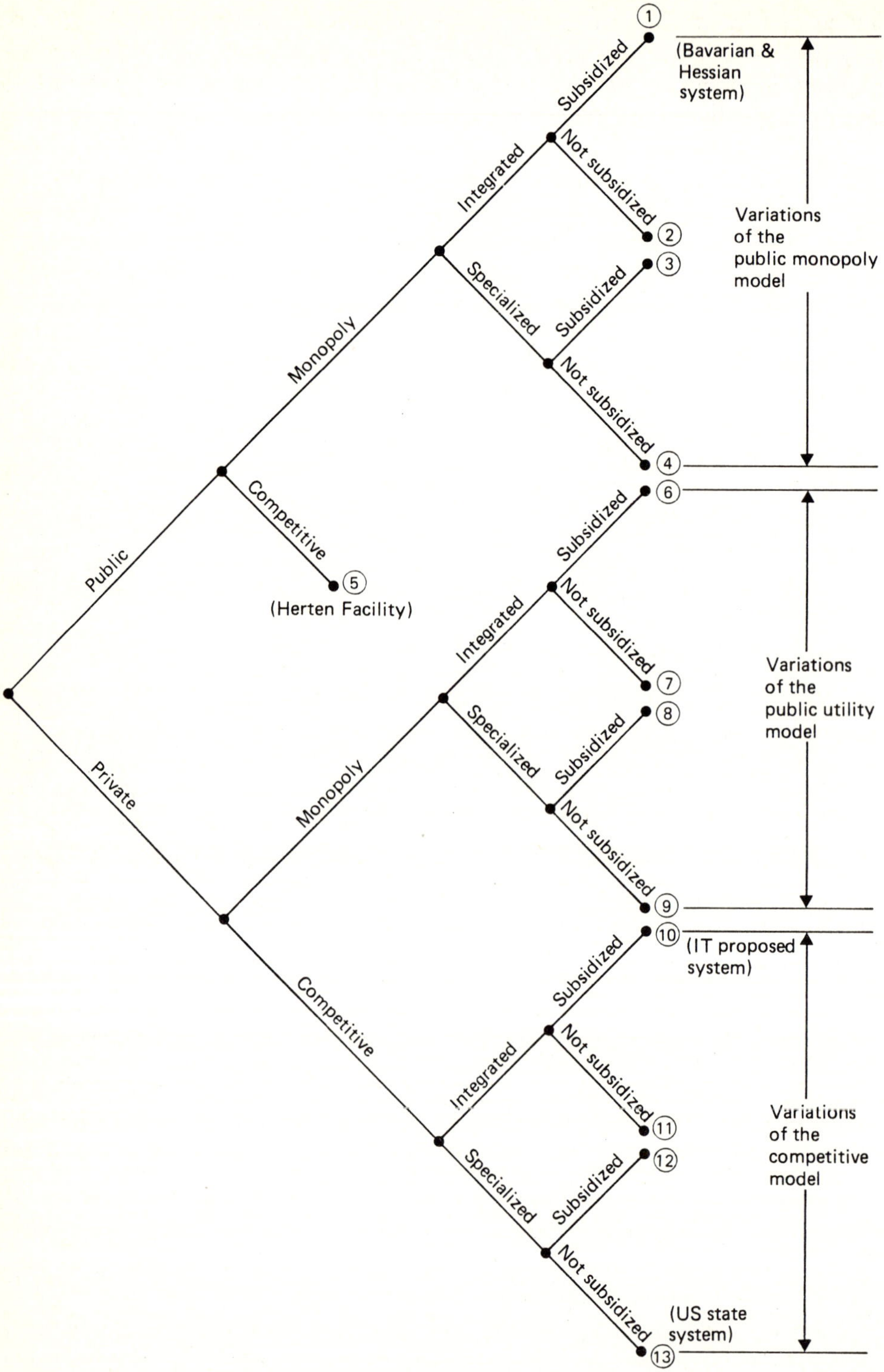

Figure 6.2. A "policy tree" for the management of hazardous wastes.

management systems and as constrained and influenced by the economic and political context of these *Länder*. Because of the different management philosophy in the state of North Rhine-Westfalia, and generally in the USA, we highlight these systems as comparative examples.

6.7. Public versus Private Ownership: The First Branch of the Policy Tree

6.7.1. The USA and the FRG

The most striking and significant feature of the Bavarian and Hessian systems, especially in comparison with other countries and other *Länder*, is the extent of public ownership of the waste disposal facilities. By financing capital investments and taking responsibility for operations, the *Land* and local governments have, in effect, assumed responsibility for the *risks* of hazardous waste management. The liability and responsibility of the generator for the long-term disposal of hazardous wastes ends once they have been ceded to the public authorities. This contrasts with the principle of generator responsibility in many other countries.

In the USA, for example, the strong tradition of private ownership has led to a predominant "stick" philosophy of government regulation. Almost without exception, hazardous wastes are collected and handled by private entrepreneurs, and a large percentage of wastes are handled on-site. The private networks of generators, transporters, and facility operators are regulated under the 1976 Resource Conservation and Recovery Act (RCRA), its accompanying Administrative Orders, and the 1983 Reauthorization. RCRA lays out a broad framework for the comprehensive control of hazardous wastes, where the details of this framework have been developed within the federal Environmental Protection Agency (EPA) and promulgated as regulations on May 19, 1980. Under Section 7003 of RCRA, the EPA can bring legal action against anyone who handles wastes in a way that presents an imminent hazard, so site operators, landowners, transporters, and even generators are all potentially liable.

The FRG, in contrast, has less severe sanctions for non-compliance with hazardous waste regulations and virtually no concept of generator liability for environmental or public health damages. The stronger emphasis in the FRG on the "carrots" versus the US emphasis on the "sticks" is firmly rooted in the political cultures of these two countries. In the FRG, as in many other European countries, there is a tradition of government–industry cooperation in formulating and implementing regulations (for a discussion of this point in the context of the preparation of the hazardous waste lists, see Chapter 5). In general, European governments have a relatively large role in supporting and sometimes subsidizing industrial development. In contrast, US government–industry relations are far more formally distant and adversarial, especially with regard to environmental issues, as shown most strongly by the lack of consultation with industry, at least in the early stages of developing environmental standards and regulations.

Irrespective of this contrast, the FRG and the USA share a "market" ideology, and the respective governments are reluctant to provide services that can be accommodated by private enterprise. This philosophy is reflected, for example, by the the FRG Council of Environmental Advisors, which has advised against the substitution of private waste disposal firms by municipal projects in government programs for the construction of new waste disposal plants [27]. It seems, therefore, that in some respects public or quasi-public hazardous waste facilities in Bavaria and Hessen are as much an anomaly in the FRG as they would be in the USA. Their present existence can be better understood by looking at the historical conditions under which they were created.

The HIM facilities in Hessen began wholly as a private venture on the part of Hessian industries to exploit a market for waste disposal; similarly, the GSB facilities in Bavaria were created by the initiative of private industry with only a small amount of government financial support. Since there did not exist at that time a well-defined or well-organized group of smaller private firms that handled wastes in these *Länder*, the individual governments did not meet with political opposition from a waste-handling industry when they increased subsidies to, and effectively took over, the HIM and GSB operations. In fact, the government was supporting the interests of private enterprise – not that of the waste disposers, but that of the waste generators. The ZVSMM has a different history, being from the start a fully public enterprise; for this reason it is an exceptional case even in the FRG. In contrast, the USA supports a powerful and growing industry of small waste handlers that could strongly oppose the creation of government facilities.

The entry of the Hessian and Bavarian authorities into what had been the clear domain of the private, commercial sector (as in the USA) switched the terms of the hazardous waste problem from the regulation and control of the private facilities to the management or "absorption" of the environmental risks presented by hazardous wastes (for a discussion of these problem definitions, see Chapter 3). The *Länder* appear to have identified themselves as "risk managers" with priority on preventing public health or environmental damage, as evidenced by their continuing efforts to support the facilities financially, to participate in their management, and, at the same time, to encourage the reduction of wastes.

6.7.2. Advantages and disadvantages of public ownership

The arguments for and against public ownership of T&D facilities are so intertwined with those for and against a monopolized market, integrated facilities, and public financing, that a complete separation is artificial. In theory, smaller, specialized facilities could also be publicly owned [(3) and (4) in *Figure 6.2*], but the advantages of this organization are not so clear. From the Bavarian and Hessian experiences, we find that public involvement was necessary both to finance the large capital investments needed for the rotary-kiln incineration facility and the secure landfills (ZVSMM), and to cover the losses these facilities incurred once in operation (GSB and HIM). Since financial subsidies were deemed essential, the public authorities tied these subsidies to ever-increasing public management and control.

The gradual slide toward public ownership of the GSB and HIM facilities in a country with a strong market ideology was viewed positively by the *Länder* for promoting their environmental objectives. Public management (and public subsidies) would allow greater attention to be given to meeting environmental needs and still provide T&D services to industry at "affordable" prices. This argument for public ownership is tied to public subsidies, and from *Figure 6.2* it is apparent that many institutional arrangements can be coupled with public support to keep the prices (in this case) competitive with out-of-state management alternatives.

The rationale for greater public control of the T&D facilities in Bavaria and Hessen, however, went beyond that of transferring some of the disposal costs to the taxpayer through public subsidies. Among many *Land* officials there is a deeply entrenched mistrust of private enterprise in the hazardous waste business, rooted in a history of bad experiences [28]. Misgivings about the ability of the private market to handle wastes in the most environmentally correct way and to assure the long-term security of hazardous waste land disposal motivated the authorities to opt for public ownership. Furthermore, as seen in the ZVSMM case in Bavaria, public ownership actually pushed the development of more environmentally acceptable management technologies.

Another argument for public ownership and control, which was part and parcel of the underlying philosophy that hazardous waste management exists for environmental control and not as a profit-oriented enterprise, was the obligation on the part of the public authorities to accept *all* wastes delivered to the facilities, not simply select those with the highest economic value. Authorities feared that the private market would not have the flexibility and breadth to serve the whole range of waste generators, including those producing "dirty" wastes with little or no fuel value for incineration. A public enterprise with the philisophy "environment before profit" could take these wastes and spread the losses incurred. Besides serving only part of the waste handling needs, the Bavarian and Hessian authorities also had reservations about the ability or will of private enterprise to ensure the long-term security of hazardous waste land disposal. A public authority, it was thought, would have a longer term interest in the continuing measures necessary to protect future generations.

These reservations expressed by the *Länder* about private enterprise serving environmental needs can, in theory, be countered by appropriate regulatory measures, such as requiring post-closure funds to cover the costs of long-term maintenance for land disposal facilities. The "dirty" wastes could also, in theory, be accommodated by private enterprise, although at substantially higher prices, which might create greater temptations for illegal disposal.

While a regulatory program may, in theory, serve the same environmental needs, experience in many countries has shown the difficulties in planning, permitting, and monitoring private disposal facilities, which are complicated by a lack of funds for hiring the requisite personnel and the general nonavailability of trained personnel. A central, public, or quasi-public facility concentrates expertise on the surveillance and control of a comprehensive management facility, where the management decisions can be made with environmental goals as a priority. On the other side, the absence of competition and the profit motive in waste management may also lead to the well-known tendency for public operations to increase their bureaucracy and operate inefficiently.

The choice between private and public facilities thus appears to be viewed as a trade-off between the almost inevitable inefficiencies of public enterprise and the sometimes substantial difficulties in monitoring private facilities to assure that environmental objectives are met. The assumption underlying this trade-off, however, is that public authorities and public managers are themselves environmentally conscious or that they have the will and ability to promote the environmental objectives discussed above. This appears to have been the case in Bavaria and Hessen, but cannot be assumed to be universal. In the UK, for example, local authority disposal sites are allegedly worse managed and less supervised than their privately run counterparts. Public enterprises generally cannot claim to be forerunners in environmental innovation, and a mismanaged public enterprise can be environmentally more damaging since public industries are notoriously subject to less inspection and control from other governmental authorities. As an official in North Rhine-Westfalia has pointed out, a private facility can be threatened with closure [29]. The lack of regulatory scrutiny may be seriously compounded by the fact that citizens generally have more limited recourse to challenge or receive compensation against actions of government than against actions of private firms. This may be most serious in the USA, where liability on the part of waste handlers has become a strong motivation for environmental protection.

In sum, the environmental advantages of a public waste-management enterprise hinge critically on the experience and consciousness of the civil servants in charge, especially since they are subject to few external checks. The lack of direct public liability may lead to considering combined public and industrial financing and control. Industrial cooperation in the management of hazardous waste facilities might serve as a valuable policing influence within the community of generators, especially if the waste generators retain some joint liability for the quality of disposal. Firms that practice careful management will then have a strong stake in ensuring that other users do not expose them to liability charges or a bad press. In addition, many problems encountered in gaining the cooperation of the large industrial firms can be avoided if they become partners in the disposal venture. A waste expert from the Bavarian Bureau of Environmental Protection sees this cooperation as the essential ingredient in creating regional facilities, and even more difficult to establish than the financial base. The good cooperation between industry and government in Bavaria allowed a consensus regarding a hazardous waste management system to be reached quickly, and the GSB was founded after only a little more than a year of negotiations [30].

Alternatively, this close partnership between government and industry, which is more typical of European than US regulatory practices, can effectively diffuse control through co-option of the regulators. Advocates of more cooperative forms of environmental practices argue that this disadvantage is overcome by the implementation advantages of collaboration and compromise between industry and government [31, 32]. A more adversarial system must also reach such compromises, but this is usually accomplished through lengthy and expensive court proceedings, often in a climate of bad faith.

A closely related alternative to both the public enterprise and the hybrid public–private enterprise is the *public utility model* shown in *Figure 6.2.* Here the T&D facilities are privately owned, but closely regulated, and their market is assured. We discuss this model below within the context of monopoly strategies.

6.8. Monopoly versus Competition: The Second Branch of the Policy Tree

Bavaria and Hessen have created systems for disposing of hazardous wastes which, by emphasizing treatment and incineration over direct land disposal, have reduced the environmental risks of hazardous wastes below those of neighboring *Länder* and countries. This system has its price and, although the costs are spread between the citizens of the state and industry, hazardous waste disposal is still more expensive in these *Länder*. Even considering transportation costs, it would often pay generators to ship their wastes across the borders for land disposal rather than send them to the regional facilities within Hessen and Bavaria for incineration. To counter this problem the *Land* authorities have passed legislation forbidding the export of hazardous wastes without special permission and requiring all generators who ship their wastes off-site to deliver these wastes to the *Land* facilities. This *Benutzungszwang,* or compulsory-use clause, has been one of the most crucial, and most controversial, measures taken by the *Land* legislatures in their support of the waste management system. The importance of creating a monopoly for the public facilities can be appreciated by examining the near failure of a similar facility in the neighboring *Land* of North Rhine-Westfalia.

6.8.1. Experience in North Rhine-Westfalia

North Rhine-Westfalia offers a useful contrast to its neighboring *Länder*, Hessen and Bavaria, in that it has attempted a mixture of private and public facilities for hazardous waste management. North Rhine-Westfalia is the FRG's major hazardous waste producing *Land*, generating nearly twice as much hazardous wastes as the other ten *Länder* combined [33]. The highly industrialized and densely populated Ruhr region of North Rhine-Westfalia, which is organized administratively into autonomous cities and counties, has instituted a regional structure, the *Kommunalverband Ruhrgebiet* (KVR), to deal with common problems such as hazardous waste management. Recognizing a need for facilities with higher environmental standards than those offered by private entrepreneurs, the KVR has planned and partially carried out the construction of a central incineration facility (Herten), central multicomponent and special-waste land disposal facilities, and four collection centers. This system is fully financed by the KVR.

This plan for an independent public system to supplement the private waste-handling sector has not been fully successful. The incineration facility at Herten has experienced serious financial problems, due partly to unanticipated technical difficulties, but more importantly to, until recently, a lack of financial or regulatory assistance from the *Land* government. The highly priced services offered at Herten could not compete with less advanced alternatives offered by the private market and by other *Länder*. For example, a generator of varnish residue could legally choose whether to pay DM450/tonne for incineration at Herten, DM200/tonne for its storage at the Herfa-Neuroda underground deposit, DM80/tonne for its deposit in a special waste land disposal facility, or, after mixing it with sand, DM5/tonne for its deposit in a construction rubble landfill [34]. Even strict standardization and regulation at the *Land* level would not prevent generators from taking advantage of laxer regulations in other *Länder*, unless an export ban was also successfully imposed.

The crucial difference between the Herten operation and those in Hessen and Bavaria has been the lack of compulsory use (*Benutzungszwang*) of the public or semi-public facilities. The problem has been partly administrative in that the KVR has no legal authority to impose restrictions, and the *Land* authorities have not passed supporting legislation to create an effective public monopoly at the expense of the many smaller private handlers. Subsidies from the *Land* government, which reduce the Herten prices to competitive levels, were the only remaining option to keep the facility in operation, and this option has been recently taken. The current price subsidies do not, however, result in competitive prices with other legal management possibilities. For this reason, the government has appealed to hazardous waste generators and collectors to send a proportion of their wastes to Herten or risk the possibility of stronger legislation requiring full use of the facility.

The experience in North Rhine-Westfalia underscores the importance of the historical conditions in which the relatively successful waste management systems in Hessen and Bavaria were put into place. A desire on the part of public authorities to finance an environmentally comprehensive system of management facilities, combined with the explicit cooperation of waste generators and little political opposition from the initially nonexistent private waste-disposal industry, were important, though not entirely sufficient. In addition, the *Land* authorities needed the legal authority to assure a market for their facilities, which is also proving critical in North Rhine-Westfalia. In the absence of standardized regulations across the *Länder*, as was the explicit goal of the US Congress in passing the RCRA legislation (and is now planned for the *Länder*), it is essential for a state system to exclude waste exports as competition for its facilities. The legal power of Hessen and Bavaria in establishing compulsory use, though controversial in the FRG and in Europe generally, reflects the greater powers of the *Länder* compared with their US counterparts. In the USA, a compulsory-use clause would violate the interstate commerce clause of the constitution, which was recently upheld in a famous case involving New Jersey and Philadelphia.

GSB, HIM, and ZVSMM have, therefore, operated as statutory monopolies. With the free-market ideology in the FRG, it is not surprising that the *Benutzungszwang*, and especially the export restrictions, has come under attack both within the FRG and from the outside. The European Community (EC), in keeping with its mandate to promote free trade among member countries, is presently issuing a legal challenge to the Bavarian and Hessian export controls. This could be one of the first cases in which a country or region will not be allowed to keep its pollution within its own borders, and shows a direct conflict between the aims of the free market and environmental concerns.

The FRG is responding to pressures to allow interstate transfers of hazardous wastes by changing the system and, in so doing, promoting uniformity between the *Länder*. As planned, the hazardous waste lists will be revised such that a management method is mandated for each listed waste. Insofar as the hazardous waste lists conform, this will eliminate the current cost advantages found in exporting wastes to other *Länder* with less stringent requirements. Under this revised system, relaxation of the *Benutzungszwang* is not likely to create out-of-state competition for the Bavarian and Hessian facilities, given their subsidized prices. However, the prospect of sending wastes out of the country, especially to the GDR where disposal prices are considerably lower, would remain problematic.

6.8.2. A natural monopoly?

An important question arises whether the cost advantage of one large integrated system, such as HIM, GSB, or ZVSMM, would effectively result in a monopoly if lower cost disposal methods were disallowed, i.e., if the new listing system in the FRG went into effect, and if prices were not subsidized. In other words, are the economies of scale sufficiently great to create a natural monopoly? Some recent evidence suggests that there are only small economies of scale with respect to incineration, indicating that several smaller facilities might operate with the same average costs as one large facility [35]. If this is indeed the case, there are no economic grounds for operating on a large scale, although there may be compelling institutional reasons, which are discussed below. This also means that if the government did not intervene with capital subsidies, etc., yet did require expensive incineration, treatment, and secure land disposal of hazardous wastes, then a configuration of smaller scale, competing facilities would likely develop to meet this demand.

If there are no economies of scale in the treatment and disposal of hazardous wastes in an environmentally acceptable manner, then there are no *economic* grounds for the government to subsidize the starting capital necessary for large-scale operations or to regulate the monopoly prices. If, alternatively, significant economies of scale can be shown to exist or if there are other reasons for operating as a monopoly, then the government could, as shown in *Figure 6.2*, opt to finance and operate the integrated facilities as a public venture [(1) and (2)] or alternatively as a privately owned and operated monopoly regulated by government [(6) and (7)].

This latter "public utility model" has a long history and precedent in providing such goods as electricity, water, gas, and other public utilities. Generally, these privately owned firms are profit maximizers, but their prices (and other practices) are closely regulated by the public authorities.

In many respects, it appears that the management of hazardous industrial wastes would fit naturally into this model, and the wealth of experience and tradition with regulated public monopolies in many countries would greatly facilitate acceptance of this type of organization for hazardous waste management. However, experience with public utilities regarding environmental protection is mixed, and more research would be needed to ascertain the conditions under which this operating model would best serve the public interest. Alternatively, the system could be organized into several, segregated facilities specializing in the treatment of particular wastes, where these facilities are each assured a collection area in a similar way as oil is collected in the FRG [(3), (4), (8), and (9)]. These "monopolies on a smaller scale" would operate in the absence of competition with the ensuing inefficiencies, but might create some advantages of an assured market and of government oversight.

Ultimately, the question of a monopolized hazardous waste system versus a competitive one hinges on both the economies-of-scale question (yet to be definitively resolved) and on the question of organization – whether there are inherent disadvantages in the competitive multiple-actor model for hazardous wastes life-cycles. Here we cannot easily divorce the arguments from a related concept of "economies of scope" across the disposal and treatment of several related hazardous materials. There may be compelling reasons to centralize the *location* and *management* of hazardous wastes unrelated to the technical economies of scale. The key to these economies of scope lies in the economic and political costs of *siting* hazardous waste facilities and in the *governance* of the ongoing waste disposal activity. These important considerations are discussed below.

6.9. Integrated versus Segregated Facilities: The Third Branch of the Policy Tree

6.9.1. The general problem

An important choice facing many regional or *Land* authorities is whether to encourage the construction of regional, integrated facilities to serve a diverse clientele of hazardous waste generators or whether, instead, to encourage a greater number of more specialized facilities. The distinction, however, is not entirely clear-cut, since specialized facilities may serve a large number of generators, and regional systems may not be integrated but consist of several more specialized facilities. Nor is the choice unrelated to the question of public financing and ownership since, as we have pointed out above, the large capital investment in regional facilities will probably, though not inevitably, require public financial support as well as government intervention to assure the operation of the monopoly. Specialized facilities may either operate competitively or as monopolies with an assured area of operation or waste catchment region.

Arguably, the most important differentiating characteristic of an integrated facility concerns the point in the waste life-cycle at which the decision is made as to how the waste will be handled: whether incinerated at high temperatures, treated and disposed of on land, mixed with other wastes, etc. Where the facilities include all or most of the possible handling techniques integrated under one management, this decision is logically made internally; if the facilities are specialized, the choice is logically made by the generator and/or transporter in deciding where to ship the waste (although clearly it can be shipped only to facilities licensed to take it). This distinction becomes important when considering the extreme complexity of a waste-handling system made up of thousands of autonomous generators, transfer station operators, and transporters with sometimes limited knowledge of the properties of their wastes or the full treatment network available. It may be that the most important element of control is then *simplification*, and an integrated facility, where the only rule is that the generators deliver their wastes, significantly reducing the system's complexity. An integrated waste management facility, by condensing the hazardous waste life cycle into a relatively comprehensive regulatory–management organization, significantly reduces waste life-cycle indeterminacies (see chapter 3), thus enhancing effective control. The authorities can better oversee the operation of a small number of facilities, which operate with a staff large enough to include environmental experts, than a multitude of smaller operations. The technical staff may also be better equipped to cope with any novel problems that may arise.

Another very cogent argument for concentrating facilities at one location is the increasing difficulties encountered in siting hazardous waste facilities [36]. There may be significant costs in obtaining permission for and in legitimizing the decision for a site. According to many observers, the siting of new disposal facilities is the most urgent problem in the area of hazardous waste management. Even in the FRG, with present overcapacity in some states, projections of hazardous waste generation and the lead time necessary for capital-intensive facilities continue to make the siting problem salient. Obtaining public approval to site waste facilities may be the most difficult obstacle in implementing the preferred waste management hierarchy.

While central facilities may be better equipped and easier to monitor, they have some drawbacks. One important problem concerns the distances that transporters must haul wastes, with the consequent increase in transportation costs and risks (the relative additional costs are considered to be fairly insignificant, but the risks may be substantial) [37]. The surveillance of waste transfers is perhaps the most difficult element to manage with regard to the requisite manpower and individual checks [38]. Increasing the hazardous waste traffic will undoubtedly raise significant administrative difficulties.

A second disadvantage of larger facilities is the potential dangers of dealing with wastes in large quantities. The bureaucracy of the facility becomes more enmeshed in administrative detail as waste traffic increases, especially at that point where there is no longer an identifiable and stable relationship between the handler and his clientele. In other words, small specialized firms have the advantage of a relatively constant stream of wastes, which they are experienced in handling and of which they are generally clear about the composition.

6.9.2. Experience in the USA

In 1974, the EPA prepared a report for the US Congress which advocated regional, centralized processing facilities for hazardous wastes [39]. No suggestion was made that the government might actually build or run these facilities, but it might give assistance by creating a franchise system with territorial limits [variations (6) and (7) of the public utility model]. Although hazardous waste legislation was subsequently passed and is now being implemented, there exist to date no integrated facilities in the USA of a kind similar to those existing in Hessen and Bavaria. The hazardous waste industry remains in private hands [variation (13) of the private competitive model], and neither private entrepreneurs nor public officials have sensed a secure market for large, regional facilities sufficient to justify the substantial capital investments necessary. This does not mean that there are no large operations, however. By 1980, four major firms took nearly half the sales revenue of the entire industry. [40] Chemical Waste Management, Inc., for instance, has an estimated 40% share of the US market and provides a variety of treatment, disposal, and storage facilities; yet, it largely offers land disposal and consequently has not invested the large sums of capital necessary for more incineration and treatment technologies. Large-scale, rotary-kiln incinerators do exist, but they usually serve a fixed and identifiable number of industries requiring a specific service, especially the incineration of PCB wastes.

There have been continuing attempts to establish privately financed, regional processing facilities. Rollins Environmental Services (RES), which began operations in 1969 in New Jersey, planned in the early 1970s a national network of 25 integrated facilities, as well as extensive on-site efforts to reduce waste-stream volumes at the source [variation (11) of the private competitive model]. From the start RES favored incineration, though it offered (and offers) other methods. Its ambitious plans for comprehensive and environmentally superior facilities were not realized, since it became apparent from the lack of government interference that there was not a market for expensive treatment and incineration of hazardous wastes.

More recently, the California-based IT Corporation received the necessary permits to build a major integrated processing facility that offered land disposal, chemical and physical treatment, and rotary-kiln incineration in Louisiana, with a large initial capital investment of approximately $100 million [variation (11) of the private competitive model]. IT recognized that this was a high-risk venture, but believed the relatively strict hazardous waste regulations in Louisiana would create a market for its comprehensive and expensive services, including advisory services to industries to help them reduce their wastes. After vocal and organized public opposition, however, IT has indefinitely postponed the project.

Experience in Bavaria, Hessen, and North Rhine-Westfalia gives reason to be pessimistic about the commercial chances of ventures such as IT's, unless the state governments offer strong regulatory (and financial) support. It may prove necessary to give franchise rights to establish a private monopoly by restricting the entry of smaller firms that could capture specialized niches of the market. A ban on landfill for priority wastes would assure a certain market for an incineration facility. Yet, in the absence of standardized practices in neighboring states, these measures would be of little help if the state cannot restrict exports of wastes. Finally, the long-term prospects for a large, integrated facility with many wastes being incinerated at high costs appear limited by the capacity shown in Hessen and Bavaria for generators to reduce their wastes.

The future of comprehensive, regional facilities in the USA, thus, depends, on the developing role of government regulation. Since it is unlikely that state or local governments will provide large subsidies, there must be a predictable market for private investors to exploit. The most recent amendments to RCRA will eventually ban land disposal of a significant number of hazardous wastes, and will thus promote more expensive technologies with reduced environmental risks. The new regulations, if enforced uniformly across states, coupled with an industrial structure characterized by many smaller heterogeneous generators, may provide such a market. For other industrial structures, we can expect increased investment in on-site facilities as well as smaller entrepreneurs capturing the market offered by the many developing niches of hazardous waste streams.

6.10. Allocation of Costs: The Fourth Branch of the Policy Tree

At the core of any hazardous-waste management system is the question of who pays the costs. Despite agreements in the OECD [41] that industrial polluters should pay the full cost of their pollution, in Europe there is a strong tradition of the government picking up the tab for environmental protection, in contrast, for example, to the USA, where the government hesitates to aid industry directly. In the Netherlands, for instance, government officials generally agree that it would be politically untenable to institute a "superfund" tax on industry as a way of paying the clean-up costs for polluting dump sites [see Chapter 4].

In the FRG, there was an early reluctance on the part of the Hessian and Bavarian authorities to provide public subsidies to the hazardous waste facilities. The early intentions of HIM, ZVSMM, and GSB were to operate as nonprofit enterprises, but to charge industry the full cost of disposal. The incineration of wastes, however, proved so costly in comparison with land disposal that it was politically difficult to require the firms in these two states to pay significantly higher disposal costs than in the rest of the FRG. Subsidies from the public purse were not, however, in the spirit of allowing industry a "free ride" since treatment prices remain relatively high in these two *Länder* as compared with disposal costs elsewhere.

As emphasized above, passing large disposal costs on to the generators, which may have the very positive environmental effect of reducing the generation of wastes, will create a dilemma to the facility investors by reducing their market of wastes and a dilemma to enforcement agents by encouraging generators or handlers to dispose of their wastes illegally. There is a very delicate balance between financial viability, incentives for waste reduction, and enforcement. We now discuss each of these problems in turn.

6.10.1. The declining waste market

The GSB and HIM facilities, and to a lesser extent the ZVSMM, are troubled with serious overcapacity. Predictions of waste generation that formed the basis for construction of the plants greatly overestimated the quantities of wastes to be generated, which have remained constant for the past several years. Although stricter air pollution and water pollution controls have resulted in the generation of more solid wastes, waste reductions due to the higher management costs and slow economic growth have combined to counteract any increase in the waste generation rate. The shortfall in waste delivered to the facilities is further aggravated by the fact that the wastes delivered are becoming dirtier. The high cost of treatment and incineration is prompting firms to find their own uses for the cleaner wastes and to utilize more efficient recycling technologies.

From the waste-reduction perspective, a measure of success of a high-priced facility is how quickly it manages to put itself out of business. But this strategy presents an obvious dilemma to the public authorities and underscores the conflicting institutional objectives. From the point of view of a government agency trying to raise funds for a large capital investment, the prospect of medium- or long-term bankruptcy of the operation is less than helpful. It is not surprising that in the Netherlands, where the authorities are trying to promote privately financed, integrated facilities, there is a great deal of ministerial interest in creating *and sustaining* a "waste industry", in direct conflict with the stated aims of environmentalist groups for reducing wastes. In contrast, the Hessian government has encouraged regional planning authorities to take direct measures to reduce pollution at the source by making it a condition that new industrial facilities or modifications incorporate the best available technologies to reduce pollution and wastes. This ministerial policy has aggravated the serious problem of overcapacity of the existing facilities and (illegal) export to other *Länder*, which has resulted primarily from cost differences in different *Länder*. According to the prestigious Federal Council of Environmental Advisors, there is a significant difference in price structure between the older and newer facilities, which is one reason why the advanced installations are underused, while a "special waste traffic" to inadequate waste disposal facilities is in full swing [42].

There are some partially effective stop-gap measures to the dilemma of waste reduction facility maintenance, as illustrated by the following measures taken by Hessen and Bavaria to offset decreasing supplies of hazardous wastes:

(1) Expansion of the universe of hazardous wastes by adding wastes to the list.
(2) Restriction of the amount of wastes that are exported by requiring all wastes, unless officially exempted, to be brought to the public facilities (*Benutzungszwang* or "compulsory use").
(3) Restriction of on-site disposal.
(4) Importing wastes from other states and countries.

These measures represent a compromise between the conflicting objectives of reducing wastes and sustaining the capital-intensive facilities. Yet, they will not suffice in the long term if the costs to the waste producers remain high [43].

The opposing institutional objectives prescribed by sustaining highly qualified facilities and, at the same time, reducing hazardous waste generation are even further complicated, since any price incentive for reducing wastes is also an incentive to dispose of wastes illegally. Inevitably, high prices for waste disposal encourage both activities. The success of an environmentally adequate, but costly, management system depends ultimately on whether it encourages more legitimate recycling and reduction measures while controlling illegitimate dumping, burning, or exporting.

6.10.2. The enforcement gap

With thousands of diffuse firms producing and transporting hazardous wastes and the easily accessible opportunities for evasion, full control is impossible, especially considering the limited funds available in state budgets for this purpose. As we have spelled out, a serious disparity arises between the legislative efforts to control hazardous wastes and the implementation of this legislation; this has been referred to as the "enforcement gap" [44] or "executive deficit" [45].

The most important regulatory "stick" for enforcing compliance with the hazardous waste laws are fines which, in the FRG, can be as high as DM 100 000 (US$44 000); however, this may not be high enough to dissuade some generators from profitably abusing the law. In the FRG, there are no criminal penalties. The Waste Disposal Act does not explicitly address questions of liability for environmental damages resulting from improper waste disposal practices. Violation of the waste regulations is, however, a violation of civil law, which may result in liability for damages. Special regulations apply to accidents during the transport of waste which assign strict liability, i.e., it is not necessary to prove the fault of the driver. In addition, the law requires that transporters and disposers be insured against possible accidents.

In the USA, the "sticks" available to the EPA and the states to enforce the regulators are of two types: punishment and liability. The RCRA provides rather stringent civil and criminal penalties; for example, a company or individual found to be criminally violating the regulations is subject to penalties of up to $50 000 per day and imprisonment for up to two years. More effective even than the threat of civil or criminal punishment is the strict liability that RCRA imposes on generators, transporters, and facility operators. Recent actions taken by the government in cleaning up past dumping sites have alerted the business community to the huge sums of money required and the possibility of their liability. In addition, large established firms are concerned about the damage that a civil or criminal law suit would do to their public image. Many observers of the implementation of RCRA feel that these sanctions, coupled with strict liability, are motivating firms, especially the larger ones, to comply with the regulations [46].

In both the USA and the FRG, however, the difficulties are so great in policing the complex system of waste management that regulators repeatedly stress the importance of information dissemination and moral persuasion in encouraging industry to comply [47]. Indeed, figures show that the cost of *control* may be a significant portion (20–40%) of overall disposal costs [48]. With limited financial resources for enforcement, the regulators can at most inspect and monitor the large generators. In the USA, the largest 5% of hazardous waste generators produce approximately 98% of the wastes [49]. But this may be where efforts are least needed since, as a regulator at the FRG's Federal Environment Ministry, UBA, has pointed out, the large firms are nowadays highly sensitive to hazardous waste problems, having made substantial investments in their management, and can generally be trusted to comply with regulations [50]. The director of the ZVSMM makes a similar point. The economic incentive created by high prices tends to encourage large companies to invest in low-waste technologies and recycling; however, this economic lever is not so effective for small business, such as paint shops or gas stations, which have the (illegal) option of mixing their toxic waste with household waste.

As might be expected, there are opposing opinions on the extent of illegal disposal practices in Bavaria and Hessen. Officials at the *Land* and local levels have expressed the opinion that illegal disposal practices, as opposed to semi-legal practices, are no longer a significant concern; the problem of "midnight dumping", they claim, is a problem of the past [51]. Alternatively, civil servants at UBA are not so sanguine about *Land* practices, in general, stating that the extent of illegal practices could be anywhere from 10 to 90%, or highly uncertain [52]. Yet the general perception on the part of most officials is that the high disposal prices in Hessen and Bavaria have not increased the amount of illegal midnight dumping. The extent of these practices cannot be readily estimated, but many experts feel that illegal disposal is generally less of a problem in the FRG than in other EC countries [53].

The more serious problem, it seems, is the transport of hazardous wastes over the *Land* or national borders to less-qualified facilities. In theory, a waste management firm in another *Land* should not accept wastes that are transported without permission from Hessen or Bavaria. In practice, officials state that it is not difficult for generators to find out-of-state firms willing to take their wastes. There is also a lucrative waste traffic to the GDR.

Waste experts at the *Land* and federal level are also in agreement that there is a large "gray area", which is not strictly illegal, yet is not entirely in the spirit of the law. Schenkel, from the UBA, compares hazardous waste regulations in the FRG with tax regulation where, in both cases, elaborate formal rules and bureaucracy exist, but because these rules are systematically scrutinized for loopholes the regulations are only partially effective [54]. One such problem is the declaration of a waste as an economic good, relieving the entrepreneur of his obligation to register the waste, which may eventually be sold or burned in facilities that lack the necessary environmental equipment. This practice does not come under the direct control of the *Länder*. Generators may also mix their wastes with used oil or discharge their wastes as wastewater. Additionally, it is important to recall that a large amount of hazardous wastes are disposed of on-site and, for reasons that may be other than intentional abuses, these practices may be insufficiently controlled and result in environmental pollution. For this reason, Hessian and Bavarian authorities are attempting to curtail on-site disposal of hazardous wastes.

Hucke [55] has shown, in connection with enforcing air pollution controls, that under the surface of relatively clear regulatory standards the actual practice of implementation involves a significant amount of bargaining between regulators and firms, rather than the command and compliance usually assumed. One important reason for this is the separation of function between state officials, who have the better technical knowledge, and local officials, who in the case of air pollution are charged with issuing the licenses and sanctions for non-compliance. Waste management officials have the same, if not more, discretionary powers as regulators of air emissions. There exist precise lists of hazardous wastes; yet, officials have to use their judgment in determining whether a firm's waste is equivalent to the waste listed with respect to chemical composition, concentration levels, and so forth. Similarly, judgment comes to bear in the process of licensing public facilities and transporters, in approving on-site facilities, and even in allocating the time of the limited number of inspectors. As in other areas of environmental regulation, the process of bargaining between regulators and firms in the FRG is ever present, and significantly extends the "gray area" of hazardous waste practices.

6.10.3. Reducing hazardous wastes

A theme through this chapter has been the inevitable dilemma presented by a pricing policy aimed at providing incentives for the safe disposal of waste (thus reducing illegal practices), and which, at the same time, encourages producers to reduce wastes. Lowered costs to the producers of wastes can *only* be supported if coupled with other measures for waste reduction. Cost or price incentives, though proved as being effective, are not the only option for promoting the reduction or recycling of hazardous wastes. Governments might, for instance, tie tax advantages or low-cost financing to firms that take initiatives in reducing wastes. Or, the government might reduce costs to firms that invest in new capital equipment by offering low-cost financing, higher depreciation allowances, and direct government assistance and advice. The UBA in the FRG finances research for this purpose and directly advises firms about the technical possibilities for recycling waste reduction. As another example, mentioned earlier, the Hessian government is tying construction permits for new facilities to a demonstration that the firm is using the best available technology for waste reduction.

The USA, alternatively, almost exclusively relies on price incentives to promote waste reduction and recycling. A private initiative in this respect was taken by Dow Chemicals in the mid-1960s, when it began charging its plant managers $100 for each drum of hazardous waste deposited on land, which led to more intensive recycling efforts. An important initiative, however, now being considered by the federal government and already in place in 11 states, is the waste-end tax, or a tax placed directly on waste generators. This tax may be used to replace or supplement the feedstock tax that presently finances state and federal "superfunds", since the tax serves the second purpose of reducing wastes [56]. It is too early to judge fully the experience of the states with this tax, but there appears to be a major drawback. The tax has generally failed to raise the funds anticipated, and an unexpectedly large drop in reported waste volumes (a 28% decline in California within one year) resulted. This decline, which was too rapid to be explained by waste reduction and recycling, appeared to result largely from under-reporting and possibly even illegal practices [57].

6.11. Concluding Remarks

Any country or region intent upon shifting hazardous waste practices from land disposal to more environmentally acceptable treatment and incineration methods, as well as waste reduction generally, must give serious consideration to choosing and promoting a physical and economic infrastructure that promotes these alternatives. The choices are not easy since they involve difficult trade-offs between what have been identified in this chapter as the goals of a waste management system:

(1) The promotion of environmentally sound treatment and disposal practices.
(2) The support of the longer term reduction of wastes.
(3) The assurance of full compliance with the system.
(4) The allocation of the full costs of hazardous waste management to the generators.

These system objectives are conflicting since, for example, promoting the first goal using more costly and environmentally superior methods calls for a sacrifice in the fourth goal (by allowing government subsidies) or in the third goal (since generators will attempt to evade the significant rise in disposal costs). None of the German or US systems examined here have been successful in fulfilling all four of these objectives.

Hessen and Bavaria have placed a high priority on the first objective by constructing expensive treatment and incineration technologies with relatively little direct land disposal, but at the expense of the fourth objective, since the capital-intensive facilities have been built with a steadily rising investment of public funds. The authorities have assured a market for this high-cost system by requiring that all wastes treated off-site are delivered to these facilities, thus creating a public monopoly to discourage the illicit disposal and export of hazardous wastes with a significant portion of the costs being borne by the taxpayers. This public monopoly model stands in sharp contrast to the private competitive model found in other FRG *Länder* and in the USA, where scattered and usually specialized facilities are owned and operated by private entrepreneurs with little or no direct government subsidization. The USA, with its emphasis on generator responsibility and liability, has placed high priority on the fourth management goal, but with a serious sacrifice in the first objective, as shown by the large percentage of wastes that are disposed of on land.

An important conclusion of this chapter is that the private competitive model cannot be expected to fulfill the first goal by providing the infrastructure for the higher cost treatment and incineration technologies for a large number of regulated wastes, even if government regulation requires this type of management. The reason for this is the anticipated decline in the market for the higher cost methods, which ensues from the export of wastes to those countries or regions with less stringent regulations, as well as from the remarkable capacity for generators to reduce their wastes (as witnessed in Bavaria and Hessen). For the market to be attractive to private investors, direct government intervention and support will be called for by restricting the export of regulated wastes as well as covering the financial losses due to the reduction and recycling of wastes. Price subsidies to the generators may also be indispensable as a means of discouraging the illicit disposal of wastes, which aggravates further the declining waste market.

Hessen and Bavaria have responded to the failure of the private sector by sliding toward the public ownership and financial support of large, integrated waste-handling facilities; the authorities have thus "absorbed" the risks and the public has to some extent "absorbed" the costs. The success of this model, which may or may not be feasible or desirable in other regions or countries, must be viewed in the historical and political context in which it emerged and operates. The evolution of a publicly financed monopoly to accommodate a market that is generally in private hands is as much an anomaly in the FRG, which stands committed to competitive enterprise, as in other capitalist, market economies. Historically, the Hessian and Bavarian systems (with the exception of Mittelfranken) were set up by consortia of private industries with little competition from the almost nonexistent private waste-handling sector. The financial failure of these early initiatives, combined with a pervasive mistrust on the part of *Land* officials of private entrepreneurs for handling hazardous wastes, led to government intervention with the full support of the waste-generating industry.

Direct government subsidies were consistent with the FRG tradition of public support for environmental protection. In addition, the importance of restricting both the export of waste and on-site disposal, and of legislating the compulsory use of central facilities, cannot be over-emphasized, as the near failure of the Herten incineration facility in North Rhine-Westfalia illustrates. The choice of dedicated and environmentally concerned public managers is also an important prerequisite. Finally, the construction of integrated, comprehensive facilities proved appropriate for the Hessian and Bavarian states, with their heterogeneous industrial structures (including many middle-sized or small generators), but may be less appropriate for regions with larger firms that are better prepared to treat their own wastes.

The apparent success of the Hessian and Bavarian systems, therefore, hinged on several factors:

(1) The absence of a private waste-management industry to resist increasing public support of the central facilities.
(2) The ability of the government to restrict exports and on-site management, and to require compulsory use.
(3) The traditional public acceptance of taxpayer support for environmental measures.
(4) The presence of environmentally committed civil servants to manage the facilities.
(5) The existence of a heterogeneous industrial structure of waste generators.

Where one or more of these conditions is lacking, this public monopoly model may be dismally inappropriate. In the USA, for example, any attempts to create a public monopoly would probably meet strong resistance from the increasingly powerful private waste-management sector, and the restriction of exports to other states might be deemed unconstitutional. US industry has traditionally been held to be responsible for environmental costs, and the financing of facilities with public funds may meet strong political obstacles. Also, the importance of appointing good and motivated public managers is critical; a badly managed public enterprise can be more damaging from an environmental perspective than its private counterpart, since it is less accountable to other government agencies and to the public through requirements of strict liability and the possibilities of legal damage suits. For these reasons, direct responsibility or "absorption" of the risks by the government may not be feasible or desirable in the US political culture.

With the possible failure of the private competitive model in providing a long-term and comprehensive alternative to land disposal and the questionable feasibility or desirability of the public monopoly model, what options remain for promoting environmentally sound treatment and disposal practices? One possibility for consideration is the hybrid public utility model, which combines private ownership with government control. In the face of disparate national regulations, a market for the expensive and comprehensive treatment and disposal offered by an integrated, public utility might be assured by providing low-priced services. The losses incurred might then be covered through a general industrial waste tax. This model has the advantage of providing comprehensive treatment and disposal, which condenses the waste management decisions to one geographic location and, in so doing, reduces the possibility of intentional or unintentional errors, and retains the responsibility and liability in private hands. The low-priced services would discourage illegal practices, and the general waste tax would assure full absorption of the costs by industry. Many questions concerning this model, of course, remain unanswered; the challenge ahead is to explore this and other options for providing an economic and physical infrastructure that meets all the goals of an environmentally sound and equitable waste management system [58].

Notes

[1] US Environmental Protection Agency (1980), *Subtitle C: Resource Conservation and Recovery Act of 1976*, Draft Final Environmental Impact Statement, pp. 5–37 (EPA, Washington, DC).

[2] California Air Resources Board (1982), *An Assessment of the Volatile and Toxic Emissions from Hazardous Waste Disposal in California*; and *Suggested Control Measures to Reduce Organic Emissions Associated with Volatile Organic Waste Disposal* (California Air Resources Board, Sacramento, CA).

[3] US Environmental Protection Agency (1980), *Everybody's Problem: Hazardous Waste*, p. 15 (EPA, Washington, DC); and Prurot, M. (Rapporteur) (1984), *Report on the Treatment of Toxic and Dangerous Substances by the European Community and Its Member States*, p. 8–9 (CEC, Brussels).

[4] Brasser, L.S. (1982), Acting Deputy Director of the Netherlands Organization for Applied Scientific Research, personal communication.

[5] UK Department of the Environment (1978), *Cooperative Programme of Research on the Behaviour of Hazardous Wastes in Landfill Sites, Final Report of the Policy Review Committee* (HMSO, London).

[6] Szelinski, A. and Nels, C. (1983), FRG Environmental Agency, West Berlin, interview.

[7] Brickman, R., Jasanoff, S., and Ilgen, T. (1985), *Controlling Chemicals: A Cross-National Study of Policy and Politics*, Cornell University Press, Ithaca, NY.

[8] For instance, as discussed in Chapter 5, lists of hazardous wastes range from the 86 mandatory wastes in the federal regulations to 299 in Hessen; and only a few *Länder*, notably Hessen, Bavaria, Hamburg, and Baden-Württemberg, are attempting to computerize the trip-ticket system and, in effect, make it operational. This system puts a large burden on the regional authorities, who, in contrast to many states in the USA, are responsible for checking that each shipment of hazardous waste reaches its intended destination. In practice, the paperwork is too excessive for effective control. For example, the municipality of Kassel in Hessen is relatively lightly industrialized with around 200 firms generating hazardous waste, 50 to 70 hospitals, and numerous gas stations. According to an official in the Kassel *Regierungs-*

bezirkspräsidium, there are over 12 000 trip-tickets per year, which is more than enough for three full-time employees to control effectively (Herold interview (1983), hazardous waste expert in the Regional Government of Kassel, Hessen).

[9] Stolpe and Weingren (1982), *Wohin mit dem Giftmüll?* Bundesverlag Bürgerinitiativen Umweltschutz (BBU), Bonn.

[10] Furmaier, B. (1983), *Organization und Stand der Sondermüllbeseitigung in Bayern*, unpublished mss, Bavarian Bureau of Environmental Protection, Münich.

[11] *Ibid.*

[12] *Ibid.*

[13] Rückel, H.-G. (1985), Director, ZVSMM, interview.

[14] Rückel, H.-G. (1983), *Zweckerverband Sondermüllplätze Mittelfranken*, unpublished mss. (Schwabach, FRG).

[15] Rückel, *op. cit.* [13]

[16] ZVSMM (1985), *Price List for Special Waste Management* (ZVSMM, FRG).

[17] Defregger, F. (1983), The Bavarian hazardous waste system, illustrated by the Ebenhausen treatment plant and Gallenbach landfill site, in Lehman. J.P. (Ed), *Hazardous Waste Disposal* (Plenum, New York, NY).

[18] GSB (1985), *Price List for Management of Special Waste* (GSB, FRG).

[19] Der Hessische Minister für Landesentwicklung, Umwelt, Landwirtschaft, und Forsten, (1983), *Sonderabfallbeseitigung,* Argumente in der Umweltdiskussion, June.

[20] The state of California has also classified selected wastes as priority wastes. With some exceptions, the California priority list is remarkably similar to the Hessian category III list.

[21] *Op. cit.* [19].

[22] Vahrenholt, F., Hessian Environment Ministry (1983), interview.

[23] Although the facility employs good environmental control equipment, it would not, e.g., meet all of California's air quality requirements. California Foundation on the Environment and the Economy (1983), *Alternatives to the Land Disposal of Hazardous Waste: The European Experience* (CFFE, San Francisco, CA).

[24] Vahrenholt, *op. cit.* [22].

[25] Vahrenholt, F. (1984), The status of hazardous waste management in Hessen, in *Concepts and Methods of Hazardous Waste Management* (Wiesbaden Special Waste Seminars, FRG).

[26] C.O. Zubiller, Hessian Environment Ministry, interview.

[27] FRG Council of Economic Advisors (1978), *Summary of the Environmental Report*, February.

[28] Furmaier, B. (1983), Bavarian Bureau of Environmental Protection, Munich, interview.

[29] Mertens, B. (1983), North Rhine-Westfalia Ministry for Nutrition, Agriculture, and Forestry, interview.

[30] *Op. cit.* [23], p. 4–23.

[31] Majone, G. (1982), *Reforming Standard Setting*, Working Paper WP-82-90 (International Institute for Applied Systems Analysis, Laxenburg, Austria).

[32] *Brickman et al., op. cit.* [7]

[33] Schenkel, W. (1984), Sonderabfallbeseitigung in der BRD, in Kemerling, W. (Ed), *Sonderabfall und Gewässerschutz*, 19th Seminar Österreichischer Wasserwirtschaftsverband (Gmunden, Austria).

[34] *Ibid.*

[35] Kragg, B. (1983), *The Hazardous Waste Management Industry*, unpublished mss. (Harvard University, Cambridge, MA).

[36] Construction of the early facilities in Hessen and Bavaria went virtually unopposed by the public. Yet, a great deal of local protest occurred during the siting of the Biebesheim facility in Hessen, although this facility is considered by experts to be technically advanced and, from an environmental point of view, far preferable to past operations. Similarly, citizen action groups heatedly opposed the construction of a landfill for treatment residues by the ZVSMM in Bavaria. Other countries are facing equally difficult obstacles in siting facilities. In the USA, for instance, Massachusetts has passed novel siting legislation that involves all the interested parties, including the public, in the siting process by requiring mediation. In spite of this new siting law, hazardous waste firms continue to have problems in gaining permission to site facilities. See O'Hare, M., Bacow, L., and Sanderson, D. (1983), *Facility Siting and Public Opposition* (van Nostrand Reinhold, New York, NY), and chapter 11 of this book.

[37] Linnerooth, J. (1985), *The Transportation of Dangerous Goods: An International Comparison of Legislation and Implementation,* draft report (International Institute for Applied Systems Analysis, Laxenburg, Austria).

[38] Andrews, R. and Pierson, T. (1984), Hazardous waste facility siting processes: Experience from seven states, *Hazardous Waste*, **1**, 377–386.

[39] US Environmental Protection Agency (1974), *Report to Congress on Hazardous Waste Disposal* (EPA, Washington, DC).

[40] Chemical Waste Management has recently purchased the treatment facilities of SCA, Inc., a large waste management firm, but this purchase did not add capacity.

[41] Organization for Economic Cooperation and Development, (1975), *The Polluter Pays Principle: Definition, Analysis, and Interpretation* (OECD, Paris).

[42] Wolbeck, B. (1983), Political dimensions and implications of hazardous waste disposal, in Lehman, J.P. (Ed), *Hazardous Waste Disposal* (Plenum, New York, NY).

[43] In the very long term, this dilemma might partly resolve itself as the facilities become obsolete. Indeed, Bavaria plans to phase out slowly these investments as waste reduction strategies are progressively fulfilled.

[44] Wolbeck, *op. cit.* [42].

[45] Spilker, R. (1982), Germany: A black hole in the North Sea for toxic wastes, *Ambio,* **11** (1), p 57.

[46] Hirschhorn, J.S. (1984), US Office of Technology Assessment, Washington, DC, interview.

[47] Diver, C. (1980), A theory of regulatory enforcement, *Public Policy,* **28**, 257–301.

[48] Hirschhorn, *op. cit.* [46]; Schenkel, *op. cit.* [33], p. 5.

[49] Senkan, S. and Staufer, N. (1980), What to do with hazardous waste, *Public Policy,* **28**, 257–301.

[50] Sutter (1983), FRG Environmental Agency, Berlin interview.

[51] Furmaier, *op. cit.* [28].

[52] Szelinski and Nels, *op. cit.* [6].

[53] Wolbeck, B. (1983), FRG Ministry of the Interior, Bonn, interview.

[54] Schenkel, *op. cit.* [33].

[55] Hucke, J. (1982), Implementing environmental regulations in the Federal Republic of Germany, *Policy Studies*, Vol 11, No 1, 130–140.

[56] Hirschhorn, J.S. (1983), Difficult to base expanded superfund on existing feedstock taxes: Point and counterpoint, *The Environmental Forum,* **2**(8), 18–20.

[57] Greer, L. (1983), The waste end tax: An idea before its time: Point and counterpoint, *The Environmental Forum,* **2**(8), 18–22.

[58] We are especially indebted to Jobst Conrad for collecting valuable interview material for this chapter. We also thank Richard Andrews, Paul Kleindorfer, and Howard Kunreuther for their very useful comments.

CHAPTER 7

Decentralized Regulation and Technical Discretion: The UK

Eryl Mädel and Brian Wynne

7.1 Introduction

The previous chapter examined two cases in which public regulatory authorities not only finance facilities, but also manage the treatment and disposal of wastes. These authorities were at an intermediate level between national and local government. This contrasted with the Netherlands, where enforcement at the local and provincial government level takes place against standardized, precise, central norms, with a mainly private treatment and disposal (T&D) industry. According to experience in the FRG, the recent Dutch moves to establish a joint public–private T&D infrastructure will be threatened by the high levels of export of wastes from the Netherlands to cheaper alternatives.

In this chapter we examine yet a different combination of overall institutional elements in the UK. Here the responsibility for regulatory policy is highly decentralized and, indeed, focuses mainly upon a different part of the hazardous waste life-cycle. Central government enacts framework legislation with deliberately imprecise norms, leaving local authorities, at a far greater degree of decentralization than the German *Länder*, to convert these into practical regulatory standards and practices. In addition, the actual management of hazardous wastes is nearly totally (*ca.* 98%) in the hands of private industry, regulated by the local authorities through a combination of site licensing, inspection, waste registration, and legal enforcement mechanisms. The idealized model of the UK system is that technical imprecision allows flexibility and local discretion. This is reflected in the idea of harmonious collaboration between government and industry, with local government being allowed, as is usual in the UK, to "optimize" regulation to fit local circumstances:

> The framework for hazardous waste disposal is thus a joint venture between the private and public sectors, with considerable cooperation between the two, in which the private sector provides the service and the public sector provides the control [1].
>
> It may therefore reasonably be said that the national policy for the disposal of all controlled waste – including hazardous waste – is a summation of the policies of the individual waste disposal authorities developed within the legal framework and on the basis of technical and administrative advice provided by the Department of Environment [2].

The British system has seen a flurry of major inquiries and official reports in recent years [3], with sharp expressions of public protest ironically affecting mainly the high-technology and "cleanest" end of the T&D spectrum. Against the official ideal, one eminent inquiry Chairman referred to the UK system as "Ramshackle and antediluvian", and the criticisms from the new Hazardous Waste Inspectorate have indicated a degree of impatience and conflict within central government, over alleged complacency. Prominent in this public debate have been questions about the local authority network's heavy responsibilities for risk assessment and regulation, and differences of perspective between the local authorities and most other parties. Also of recurrent and growing concern is, for certain hazardous wastes, the heavy reliance on landfill *codisposal* with domestic refuse as part of official UK strategies.

Underlying this uncertain climate are questions about the realism of leaving a highly flexible, discretionary framework to be implemented by an extremely decentralized and uneven network of mainly small, local government agencies that have to regulate a fully privatized, very ill-defined private commercial T&D sector. What is in principle a sound and defensible policy (flexibility and technical imprecision) may be a bad one if the conditions of its proper implementation are not being fulfilled (and may, indeed, be incapable of fulfillment in present circumstances). This general issue is focused in the UK on landfill codisposal, where what may be a sound policy under ideal conditions of site selection, licensing, and management, may be unsound if the practical local social realities of implementation are nowhere near such idealized assumptions.

In this chapter we provide another example of the relationships between risk assessment, technical regulatory definitions, uncertainties, and the institutional and cultural setting of regulation. To achieve this, we first describe the main features of the formal regulatory framework. Following this we point out several features of the formal framework that are significant in implementation, especially the interaction between the way technical norms and methods are formulated, and the institutional roles and relationships in regulation. The main focus of attention in this respect are the implications of the UK approach to listing and classifying wastes, as already outlined in Chapter 5, and of the UK regulatory focus upon site licensing of T&D facilities as the main mechanism of overall control in the hazardous waste life-cycle. These considerations lead us finally to examine the problems of practical enforcement and the resilience or otherwise of the whole system toward possible future developments when seen in a comparative perspective.

An important factor in the UK, which links technical T&D practices with institutional frameworks, is the strong dependence on landfill, especially codisposal of "special" wastes with domestic wastes. The practice of codisposal creates a natural connection between local authorities, with their traditional responsibility for domestic wastes, and the regulation of hazardous wastes. The espousal of controlled landfill in UK policy therefore automatically implies regulatory dependence on the local authority, as does the strong regulatory role of local planning control (in site licensing). Furthermore, whilst nearly all systems elsewhere are attempting to reduce landfill dependency, in the UK this sector seems to be *developing* at the expense of the (more expensive) high-technology end of the T&D industry, which is not in a healthy condition. We also ask whether this might be related to the institutional mechanisms of UK regulation.

7.1.1. Current practices in the UK

Unlike other countries, the UK operates with two different schemes for defining hazardous wastes. Control of a broader population, including those considered dangerous to the environment, is exercised at the very back end of the life-cycle, by the selection and licensing of T&D sites, which is the responsibility of local authorities. For this purpose there is an *advisory* list of "difficult" wastes containing about 200 items, produced by central government. This is said to be "for record-keeping purposes." A smaller population is defined as "special wastes", namely those thought harmful to human health. This definition follows closely the EC's list of "toxic and dangerous" wastes and is intended to control *transportation* of wastes in a cradle-to-grave notification system, so as to control their consignment only to licensed and "safe" sites. This control is retrospective.

Figures and definitions for waste in the UK are, as usual, highly approximate. In addition to the two classification systems noted above, several different but overlapping terms are in common usage; these are of different legal, practical, or historical standing. The Public Health Act (1936) [4] uses the descriptions *house* and *trade* refuse. The Refuse Disposal Act (1978) [5] uses two classifications, of *business* and *residential* waste.

The term *notifiable* waste of the now-superseded 1972 Deposit of Poisonous Wastes Act (DPW) [6] is still widely used in practice. DPW's successor, the Control of Pollution Act (COPA) (1974) [7], distinguishes *household*, *commercial*, and *industrial* wastes, collectively known as *controlled* wastes. Its Section 17 Regulations (implemented in 1981) also define *special* wastes. The government advisory paper on site licensing (introduced in 1976) gives a list of wastes called *difficult* [8]. In practice, also, many operators use the trade association's classification scheme of *black*, *gray*, and *white* wastes [9]. The term *hazardous* wastes is also in common practical usage; e.g., in the name of the government Hazardous Waste Inspectorate (HWI) set up in 1983. In these circumstances, and with the usual problems surrounding even single definitions, the following data are rough guides only.

Excluding sewage and in-house disposal, total UK wastes are thought to amount to nearly 450 million tonnes per year (mte/year), of which nearly 60% is agricultural. Mining and related wastes provide a further 20% or more (nearly 100 mte/year). Industrial wastes, excluding mining, etc., are about 50 mte/year (12%) and domestic and commercial about half of this [10]. Notice that much of the agricultural and mining wastes, as well as household wastes, can be hazardous, and in the nature of their life-cycles may have more potential for giving exposures than do industrial wastes. As emphasized in Chapter 3, the boundary of the regulated problem is highly uncertain.

Of the 50 mte/year or so of industrial wastes, about 3.5 mte/year are estimated by Hawkins and Bailey [11] to be hazardous waste according to the terms of notifiability as understood under DPW, but about 4.5 mte/year according to HWI estimates [12]. According to the HWI, also, "special" wastes under COPA Section 17 register about 40% of those wastes originally notified (until 1981) under the DPW [13]. In other words, UK *special* wastes amount to about 1.7 mte/year. To emphasize the uncertainties, the government *Review of the Section 17 Regulations* reached an equivalent estimate for 1981 of 0.7 mte/year [14]. The 1985 Royal Commission Report [15], although quoting similar figures, reminds us baldly that "the quantity of industrial wastes generated in the UK in any year is not known."

There are 68 waste disposal authorities (WDAs) in England, 37 in Wales, and 58 in Scotland. Yet an indication of the degree of unevenness in the system is that about 75% of the total hazardous ("notifiable") wastes in 1983 was generated or disposed of within only 12 WDA areas. There are some 5000 waste T&D facilities overall (the vast majority being landfills), of which anywhere between 500 and 1700 are thought to be licensed to receive special wastes; but more than 90% of the total hazardous waste arisings were taken by only about 100 of them [16]. Of these 5000 sites many are municipal waste landfills, where the *codisposal* of certain hazardous wastes with municipal refuse is allowed. This is based on the understanding that properly managed domestic waste can help to transform and attenuate the hazardous wastes more than in a concentrated "containment" hazardous waste landfill. We come to the landfill question later. *Table 7.1* gives a summary of the rough proportions of hazardous waste that go to the different T&D options.

Table 7.1. Approximate UK T&D routes for hazardous wastes by percentage (figures rounded to nearest half-unit).[a]

T&D route	*Percentage*
Landfill	85
Mineshafts	2.5
Solidification	3
Sea disposal (incl. incineration at sea)	5
Chemical treatment	3.5
Incineration	1.5

[a] Total, 4.5 mte/year: for comparison, "Reclamation", which is excluded from the hazardous waste total, is estimated to treat about 1 mte/year of material [17].

In addition to the very significant geographical unevenness, which is related to the inconsistency of regulatory practices and prices (as well as to geological factors), a further uncertainty in the figures for the high-technology options is whether they signify available capacity or actual usage. Bailey and Hawkins, for example, give figures for incineration and chemical treatment of 0.4 mte/year out of their total of 3.5 mte/year, i.e., about 11% of the total [18]. The first HWI Report of 1984, from which *Table 7.1* is derived, gives only 5%. The larger estimate appears to represent *available* capacity, not actual use.

In fact, the health of the higher technology range of the T&D industry in the UK is a matter of some concern. There is barely enough high-temperature incineration capacity (61 000 te/year) to meet the existing "market demand" (56 000 te/year). This demand is recognized to be artificially depressed by lax landfill site licensing, and is expected to grow "as improved landfill practices are achieved [19]." The last ten years or so have seen the closure of five UK high-temperature incinerators for commercial reasons. Another closed because of planning consent renewal problems. The chemical treatment industry has also suffered sharp reductions in demand (way beyond the industrial recession), which are also attributed to the development of new and cheaper landfill competition. The incinerator closure at Bonnybridge, Scotland, in 1984 left no such commercial facility available north of Liverpool.

7.1.2. The landfill problem

A major consequence of the closure of the Bonnybridge plant is that industry in the north of Britain faces great problems in disposing of waste that should not be landfilled. Two of the remaining incinerators have already increased their prices, although the supply of wastes to both facilities is still high, with the result that there is a queue several months long. Apart from actual disposal costs, transport costs are becoming so high as to be a significant part of investment plans for companies, especially for some firms in the north of Britain, who send to the south loads in the region of 4000 gallons every two days, at a unit cost of 1500 pounds sterling.

One of the outcomes of this is that firms are naturally tending to invest in their own incinerators, e.g., Fine Organics is building an incinerator near Durham. It is estimated that this will cost 150 000 pounds sterling and will initially burn 2000 gallons of mixed organic waste daily. With the high transport costs, as indicated above, the initial capital outlay would be covered in a very short time span, so long as operating and maintenance costs do not spiral, which is a significant uncertainty.

The other major, and possibly more worrying, outcome of incinerator closure is that some firms have now been landfilling wastes that were previously incinerated, diluting them first to acceptable levels. Aware of these problems, the HWI noted that "following the closure of the facility at Bonnybridge, there are already certain hazardous waste arisings for which there is no environmentally acceptable disposal route available: these are PCB-contaminated capacitors and transformer carcases." The HWI report went on to say that as a consequence of the reduction in available incinerator capacity, "perhaps inevitably the first cases of landfilling of these wastes are coming to light [20]".

Progressive T&D options in the UK are in difficulty, and underdeveloped. The blame is laid at the door of the easy availability of cheap landfill and the inadequate site selection and licensing control exercised by a significant number of WDAs. 1984 discount prices for landfill ranged from 2.80 pounds sterling per tonne for solid hazardous waste, to 8.00 pounds sterling per tonne for liquid hazardous wastes. In *Table 7.2* we indicate the estimated amounts of waste disposed of by various methods and the cost per tonne.

From *Table 7.2* it can be seen at a glance why industry prefers landfill to other disposal methods. Furthermore, lax WDA control and the codisposal commitment allows landfill to develop its competitive advantage legally:

> The wide variation in standards and practices adopted by disposal authorities has had a profound effect on the economics of hazardous waste management. Intense competition among disposal companies has meant that the economic viability of the individual disposal

facility can be significantly influenced by the attitude of the disposal authority. Our impression is that some major operators, who generally try to adopt current best practice, are steadily losing business to other operators who are able to exploit variations in standards and enforcement. Major operators are known to be substantially discounting disposal charges to retain even a reduced level of business. While fair competition is desirable in the waste disposal industry, the current situation does not always allow for this. Neither are the current level of charges sufficient, in our opinion, to sustain good practice.

We have identified two principal reasons for this situation. Firstly the last few years have seen the proliferation of new landfill sites licensed for hazardous wastes and located in areas of the country where the disposal authority has failed to specify and enforce adequate standards. Secondly, the influence of the relatively recent public sector move towards co-disposal of wastes at their own sites, has been to force prices down. The most notable example of this was the pricing decision of one English disposal authority to halve the prices charged for wastes at one of their sites which grossly distorted the market: in May 1982 this disposal authority reduced its price for acids from 11 to 5 per tonne pounds sterling [21].

Table 7.2. Estimate of amount/cost of waste disposed of by the major disposal methods.

Method of disposal	*Amount (mte)*	*Price charged for disposal (pounds sterling/tonne)* (excludes transportation costs)
Landfill	2.7	2.50–35
Incineration	0.4	40–900
Treatment		10–390
Sea disposal	0.4	2–95

The UK system has reached a point where its Inspectorate can worry that some crucial, high-technology T&D options are inadequately available.

In addition to a system that encourages cheap landfill, the virtually 100% private commercial structure of the UK T&D industry leaves no financial and institutional support for more capital-intensive T&D options.

7.2. The Regulatory Framework

7.2.1. The main legislative framework

The general UK approach to regulation was summed up by a DOE official as follows:

> Because the effects of pollution are usually experienced first within the confines of particular localities, one of the principles followed by successive Governments has been that the primary responsibility for dealing with pollution problems should rest, as far as is practicable, with authorities operating at a local or regional level, principally local authorities and the water authorities. Thus, central Government lays down the statutory framework for pollution control, but implementation is delegated to a large extent to local level. Authorities may in many areas exercise a considerable degree of discretion as to the limitation they impose on the release of local pollutants, so that account may be taken of local resources and social priorities, the uses to which surrounding areas are put, and the capacity of the environment to absorb pollutants, although in practice they often work to fairly uniform standards or widely accepted limits [22].

This philosophy is also practiced for hazardous wastes, even to the extent that different parts of the hazardous waste life-cycle are controlled by different "sectoral" regulatory traditions. The main legislative framework is COPA, passed in 1974 and enacted in several stages, some of which are still awaited. The Act invests legal responsibility for waste disposal in WDAs. In England these are the county and borough councils, in Wales and Scotland the district councils, and elsewhere the island councils. Thus, hazardous waste disposal, legally called special waste, but also variably entitled "notifiable", "controlled", or "difficult" waste (as explained before), is *institutionally* an extension and adaptation of municipal waste disposal responsibilities defined in the Public Health Act of 1936. The major difference is that, although WDAs are responsible for the collection of domestic waste, they are not obliged to collect industrial waste. In addition to COPA, the 1970 Town and Country Planning Act (updating long-standing traditions in land-use planning law) controls the location of disposal sites through the requirement for local planning permission for all new uses of land.

The relevant sections of COPA relevant are briefly summarized here:

(1) *Section 1* (not yet implemented) requires WDAs to ensure that adequate arrangements exist in their areas for the disposal of controlled wastes.

(2) *Section 2* (implemented in 1978) requires WDAs to investigate what arrangements are needed for the purpose of disposing of controlled waste in their areas. In addition, they have to prepare and periodically revise a waste disposal plan.

(3) *Sections 3–11* (implemented in 1976) define the site licensing system. Site licensing is the most fundamental regulatory instrument for controlled waste disposal in the UK – all sites that receive controlled wastes must be licensed by the WDAs. Site licenses, which should specify allowed waste types and operating conditions for the disposal

sites, are issued by the WDAs after consultation with the Health and Safety Executive (HSE), the Regional Water Authority (which has the power of veto over the license if it is thought that there is any risk of polluting the water supply), and (in England), the local district council (for planning permission).

(4) *Section 17* (implemented in 1976) provides for the introduction of regulations that apply controls to the more difficult wastes and their disposal, i.e., "special" wastes as defined by an *inclusive* list combined with certain excluding characteristics and definitions (a switch from the exclusive list under DPW), as laid down in Part I, Regulation 2 (see *Table 7.3*). The Control of Pollution (Special Waste) Regulations (1980) were drawn up in 1980, but came into force on 16 March 1981. At the same time the 1972 Deposit of Poisonous Waste Act (discussed below) was repealed.

These regulations were intended to fulfill the UK obligations under the 1978 European Economic Community (EEC) *Council Directive on Toxic and Dangerous Wastes* [23] by providing for tighter controls over the *transportation* of dangerous wastes. The purpose of the regulations was to:

(1) Maintain a registration system for the disposal of special wastes.
(2) Institute a consignment note system for the disposal of these wastes, giving a "cradle-to-grave" record for each waste transfer, treatment, and disposal life-cycle.
(3) Require the keeping of records at landfill disposal sites showing the location of special wastes deposited there.
(4) Give an emergency reserve power to the Secretary of State enabling him to direct a consignment of special waste to a specific site.

"Special" waste is defined in Regulation 2, part II [24] (see *Table 7.4*). It is important to note that in the UK the definition of special wastes relates to the potential risks of diversion during transport, because the list only applies to the coordination of consignment notes between dispatch and disposal. Thus, definition and listing of hazardous wastes, viewed in many countries as a first step toward designating the best T&D options from the point of waste arisings, has been regarded very differently – in particular, as much less of a *directive* measure – in the UK. This may change with the gradual introduction of the concept of the best practicable environmental option (BPEO) as a more defined means of designating T&D sites for given wastes.

The remaining sections of COPA are not directly relevant to the present discussion.

In summary, the two main provisions of COPA are the licensing of T&D facilities and tighter controls on transportation via the definition of

Table 7.3. UK inclusive list under COPA (1974).

Acids and alkalis
Antimony and antimony compounds
Arsenic compounds
Asbestos (all chemical forms)
Barium compounmds
Beryllium and beryllium compounds
Biocides and phytopharmaceutical substances
Boron compounds
Cadmium and cadmium compounds
Copper compounds
Heterocyclic organic compounds containing oxygen, nitrogen or sulfur
Hexavalent chromium compounds
Hydrocarbons and their oxygen, nitrogen and sulfur compounds
Inorganic cyanides
Inorganic halogen-containing compounds
Inorganic sulfur-containing compounds
Laboratory chemicals
Lead compounds
Mercury compounds
Nickel and nickel compounds
Oganic halogen compounds, excluding inert polymeric materials
Peroxides, chlorates, perchlorates and azides
Pharmaceutical and veterinary compounds
Phosphorus and its compounds
Selenium and selenium compounds
Silver compounds
Tarry materials from refining and tar residues from distilling
Tellurium and tellurium compounds
Thallium and thallium compounds
Vanadium compounds
Zinc compounds

special wastes. These aspects, some of the many subtleties of their operation, and enforcement problems that arise for WDAs are discussed in the following sections.

Other relevant pieces of UK legislation are The Dumping at Sea Act (1974) [25], The Health and Safety at Work Act (1974), [26], and Alkali, etc. Works Regulation Act (1906) [27]. Dumping at sea is governed by the London and Oslo Conventions (1971) and is implemented in the UK through the Dumping at Sea Act of 1974. Under this Act it is an offense to dump waste in UK waters or for British ships to dump waste in other seawaters without a license. The licensing authorities are the Ministry of Agriculture, Fisheries, and Food (MAFF), the Welsh Office, and the Department of Environment (DOE) in Scotland and Northern Ireland.

Table 7.4. Definition of "special" waste under Section 17 Regulations.

MEANING OF "DANGEROUS TO LIFE"

1. Waste is to be regarded as dangerous to life for the purposes of these regulations if –

 (*a*) a single dose of not more than five cubic centimeters would be likely to cause death or serious damage to tissue if ingested by a child of 20 kilograms' body weight or

 (*b*) exposure to it for fifteen minutes or less would be likely to cause serious damage to human tissue by inhalation, skin contact or eye contact.

Assessing effect of ingestion

2. (1) The likely effect of ingestion is to be assessed by the use of reliable toxicity data in the following order of preference:

 Class 1: information about the effect of oral ingestion by children;
 Class 2: data derived by extrapolation from information about the effects of oral ingestion by adults;
 Class 3: other information about human toxicity;
 Class 4: information about animal toxicity;
 Class 5: information about the toxicity of analogous chemicals.

 (2) Where conclusive information falling within one of the classes set out in sub-paragraph (1) is available no regard shall be paid to information falling within a class bearing a higher number, and the reference to using data in an order of preference is to be understood accordingly.

3. Where the waste is in such a form that –

 (*a*) the ingestion of less than five cubic centimeters is not possible, or

 (*b*) there is no risk that a toxic constituent could be assimilated if the waste were to be ingested.

 then it is not to be regarded as dangerous to life by reason of sub-paragraph 1(*a*) of this schedule.

Mixed waste: samples

4. Waste is to be regarded as dangerous to life if a sample of five cubic centimeters taken from any part of a consignment falls within either of the descriptions in paragraph 1 of this schedule.

The major consideration before the granting of a sea dumping license is the protection of the marine environment, especially shellfish and other inshore fisheries and breeding grounds. Currently, there are 13 licensed sites for the sea disposal of industrial waste. The other criterion that is supposed to operate is whether an alternative land T&D method has been considered. The amount of waste officially dumped at sea is approximately 5% of the total hazardous waste disposed of, though some WDAs rely on it very heavily. Therefore, this method does play a significant part in overall UK waste disposal.

The Health and Safety at Work Act is, as its name suggests, related to any risks to the health and safety of employees and comes under the control of the Health and Safety Executive. The Alkali, etc., Works Regulation Act [amended by the Alkali, etc., Works Order (1966) and (1971) and Health and Safety Regulations (1983)] requires all works to be registered by the Industrial Air Pollution Inspectorate.

The UK planning control system is important in the regulation of hazardous wastes. This broader framework of land-use planning and development control overlaps with the particular issue of hazardous wastes and licensing for T&D sites, under Sections 3–11 of COPA. The Town and County Planning Act (1947) [28], amended in 1970, requires local authority planning consent for the development of facilities. Consent can be refused, but appeal can be made by the developer to the Secretary of State for the Environment, who may hold a local public inquiry into the case. As part of the planning consent, the authority can insert conditions (also appealable), which may include specific restrictions and exclusions on certain waste types, or required management practices to control their risks. In the case of Sections 3–11 of COPA, implemented in 1976, WDAs are required to license all T&D sites. The advisory Waste Management Paper (WMP4), *The Licensing of Waste Disposal Sites*, was issued by the government in 1976, and contains an illustrative list of nearly 200 "difficult" wastes [29]. As usual for the UK, however, neither this list nor the other principles expressed in WMP4 have legal standing. In fact, this list is officially "for record-keeping purposes only", not for the control of disposal [30]. As we see later, the dependence on T&D site licensing without attention to points further upstream in the hazardous waste life-cycle have had important consequences for UK hazardous waste management.

7.2.2. Other policy actors

At central government level, responsibility for hazardous waste lies with the Land Wastes Division of the DOE in England, the Scottish Department in Scotland, and the Welsh Office in Wales. Legal responsibility for proper regulation and waste control falls on the waste producer and the WDAs. The role of central government is indirect, falling into three categories:

(1) To provide a general oversight and framework of waste disposal *legislation*.

(2) To *adjudicate* on appeals against refusal of site license applications and on planning appeals involving waste disposal sites.
(3) To provide administrative and technical *advice* to disposal authorities – this is largely carried out via a series of advisory Waste Management Papers (WMPs) [31], by informal consultation, and by commissioning research.

A general sense of rapport with industry is based on a long-standing relationship within the general philosophy of collaboration that has evolved over many decades. It has, in fact, been argued that it began with the Alkali Acts of 1863 [32]. In addition, Harwell operate a Waste Management Information Bureau, which is mostly funded by the DOE.

In 1984, following the recommendations of the Gregson Committee, the HWI was set up within the DOE. Its first annual report in June 1985, was unusually frank and critical of various aspects of UK regulation, especially the uneven and inadequate nature of much local authority licensing and inspection.

The regional water authorities and the Scottish River Purification Boards are responsible for controlling water pollution, and have a right of veto over T&D site license applications if they can show a direct risk to water supplies. This is also subject to appeal to the Secretary of State. Historically, the protection of water supply has been one of the strongest areas of environmental protection, and the water authorities have built up a high level of technical expertise.

The private sector accounts for 98% of all industrial waste disposal, half of this being conducted "in-house" at the site of production. Such sites still require a license from the local WDA. The waste T&D industry is in the hands of several large and many small companies, whose trade association is the National Association of Waste Disposal Contractors (NAWDC), which includes about 75% of the T&D industrial sector. Members subscribe to a code of practice aimed at raising the standards of the industry. NAWDC also has its own classification of waste: *white wastes* (least hazardous with characteristics similar to those of domestic refuse leachate); *black wastes* (extremely hazardous and generally not acceptable for landfill); *gray wastes* (which do not fall into the more clear-cut categories) [33].

7.2.3. The development of UK regulation

By the early 1970s the laws governing hazardous wastes were no longer specific nor rigorous enough to deal adequately with the growing problems. Attention was devoted to the general problem of waste disposal in the early 1960s, with the setting up of two committees – the Ministry of Housing and Local Government Technical Committee, whose report *Pollution of Water by Tipped Refuse* was published in 1961 [34], and the Technical Committee on the Disposal of Solid Toxic Wastes (the Key Committee) [35]. For nearly a decade the efforts of these committees were greeted with apathy from the government and public alike, until events overtook the authorities and enforced a hurried, official acknowledgment in legislation. The following account of events that led up to the first specific legislation on hazardous waste disposal was given by Lord Ashby [36]:

> The Key Committee was set up in 1964 as a result of an incident in 1963 when some animals died because of a fluoroacetamide leak from rusty drums which had been dumped by a local pesticide factory. The committee worked in a leisurely not to say glacial manner and it did not report for six years, not until 1970, and only had 20 meetings in the whole of the six years suggesting that little governmental pressure was being exerted. However, the findings of the committee were disquieting and it listed 17 serious incidents where toxic wastes had been dumped and caused damage. The Committee made 38 recommendations but received little attention from the Press and none at all from the Government; despite the fact that this was the beginning of the period when public opinion was being aroused about the environment.
>
> At the same time the Standing Royal Commission on Environmental Pollution was set up and produced its report one year later in 1971 [37] expressing concern that nothing had been done to implement the recommendations of the Key Report. The Government's response was that it was preparing to reorganize local government.
>
> The Royal Commission repeatedly approached the Secretary of State on the need for legislation to curb the indiscriminate dumping of toxic wastes and to have some type of control similar to that for air and water. The response was that more information was needed by the Department of Environment (DOE). The Royal Commission provided the information giving seven cases of dangerous handling of toxic waste – and this by a nationally known firm of waste contractors. There was still no response from the Government.
>
> In 1971 the Royal Commission drafted another report strongly criticizing the Government for lack of action – the Government's reply this time was that there was no parliamentary time for more legislation. But in January 1972, the Birmingham *Sunday Mercury* revealed that employees of the same nationally known firm of waste contractors were dumping wet waste and drums containing cyanide, phenols, caustic soda and other materials – some of the drums were accompanied by a delivery ticket describing them as innocuous. This was disclosed by the Conservation Society. On February 22, 1972 a member of the Conservation Society visited parliament to try and get some action on the part of the Government. However, two days later on 24 February, there were headlines about drums of cyanide with labels being scratched off being found on waste land in Nuneaton where children played. One week after this was disclosed, a Bill was hurriedly pushed through Parliament – i.e., the 1972 Deposit of Poisonous Waste Act [38].

In an explanation of the relevance of the above story, Lord Ashby went on to relate a comment that appeared in an editorial in *The Times* under the headline "How to move a Government":

> It is instructive to note what did and did not prompt the Government to squeeze a Bill ... into an already crowded legislative program. The urgent representations of an official commission ... moved by the "disturbing cases which have come to our knowledge", did not. Headlines about drums of cyanide waste on derelict land in the Midlands did [39].

According to informal Whitehall lore, the DPW act was written over a weekend, so urgent was action now perceived to be. Not surprisingly, perhaps, it has often been called a holding operation only. In reality it became something more than this, despite being superseded within a few years. The main provisions of the act were:

(1) A prohibition on the tipping of poisonous, noxious, or polluting waste where it was liable to give rise to an environmental hazard. An environmental hazard was defined as subjecting persons or animals to material risk of death, injury, or impairment of health, or threatening the pollution or contamination of any water supply.

(2) A requirement that local and river authorities be notified before the removal or deposition of wastes. All wastes had to be prenotified at least three clear working days before removal or deposit, unless they were specifically exempted from the regulations. In other words, there was an *exclusive* list system, which failed safe in the case of ignorance. The regulations provided a schedule of non-hazardous wastes, and if a particular waste was included in the schedule *and* did not contain any hazardous quantity or hazardous concentration of a poisonous, noxious, or polluting substance, then it was exempted from the regulations. Notice that these criteria were qualitative, involving no quantitative definitions or tests.

This act was not meant to be more than an emergency measure; in 1974 COPA was passed, partially repealing certain DPW sections. The DPW was finally repealed by the COPA Section 17 Regulations in 1981. The main contrasts between DPW and COPA are given in *Table 7.5*. Although it was a shotgun affair, and did not live long, DPW has had continuing effects, not least upon the approaches of the central actors in regulation – the local authorities.

The classification method for "notifiable" wastes under DPW defined a larger population of wastes than the Section 17 Regulations of COPA. In fact, COPA distinguishes *household*, *commercial*, and *industrial* wastes, calling them collectively "controlled" wastes, but only "special" wastes are subject to the regulations.

Table 7.5. Major differences between DPW 1972 and COPA 1974.

DPW 1972	*COPA 1974*
No site licensing	Site licensing (implemented 1976)
Prenotification of disposal	Consignment note system (implemented 1981)
Exclusive list	Inclusive list
Qualitative definitions	Quantitative definitions (implemented 1981)
Local authority and regional water authority joint responsibility	Local authority has sole responsibility
Environmental and human hazard define notifiable waste	Human hazard only defines "special" waste

The DPW *exclusive* list was an understandable way of dealing with what had been allowed to slide into an urgent need for a classifying mechanism of notifiable wastes. If the job had to be done so hastily it was safer to list what could be excluded than to try to be accurate about what should be included. However, industry was dissatisfied with the ensuing need, borne of uncertainty, to notify larger amounts of waste than were hazardous by more "accurate" criteria. When COPA was formulated, therefore, a much shorter *inclusive* list was established (in 1981). A major reason for this was to cut down on unnecessary paperwork dealing with the large number of marginal wastes. However, a further reason was that the list and registration was no longer seen by national officials as a means of *control*, except via *post hoc* information. The DOE thought the new approach would cut the notifiable waste population – and associated documentation – by 70%. In reality it has not done anything like this. According to one government official, industry continues to err on the side of caution – "notifiable and special are synonymous in the waste producers' eyes" [40] – even though the burden of proof lies with a WDA to demonstrate that a waste is special.

WDA perspectives on hazard have also been shaped by DPW. They resisted the reduction of the population of regulated waste (from "notifiable" to "special") because, they argued, the exemption of a large population of uncertain wastes would only reduce the overall standards of management. Waste T&D contractors, who admittedly have an interest in retaining a larger population of controlled waste, have also asserted that in their experience hazardous waste has been lost to the system by narrowing the regulated list. This was despite the government's argument that the reduction would allow them to cut paperwork and focus (scarce) attention on the more limited amount of hazardous wastes, their producers, and handlers.

Thus DPW controlled a larger population of wastes than COPA on at least three grounds, before considering finer points in the definitions:

(1) The list was *exclusive*, that is, fail-safe.

(2) Even something on the list of excluded wastes had to be controlled if it met criteria of public health risk or water pollution. COPA dropped the water pollution restriction, but *also* allowed for the *exclusion* of any waste on the list that did not fulfill defined hazard criteria.

(3) DPW expressly incorporated the synergy principle, that a waste be deemed hazardous if it could "by itself or cumulatively with other deposits of the same or different substances," pose a risk to water or health.

This reduction of the hazardous wastes list was accompanied by a change in the registration system and its role. In 1976 site licensing was introduced as *the* main organ of overall control, and this had its own (advisory) list of "difficult" wastes. Thus, the new COPA Section 17 list was not only officially reduced in size, but also in its role, against the perception of its implementing agents. The burden of regulation was placed instead on T&D site licensing, which was said to end the need for prenotification. The new hazardous waste notification under Section 17 of COPA was only ever intended to apply to the *post hoc* tallying of outgoing *transport* consignments with registered receipts at (domestic) T&D facilities.

To summarize, specific regulation of hazardous wastes in the UK was shaped by a political crisis over uncontrolled toxic-waste dumping. This was soon followed by a radical change in the regulatory framework (COPA), which sharply reduced the population of controlled wastes, destroyed the regulatory effect of DPW's prenotification of waste movement to T&D, and placed the control burden on the very back-end of the waste life-cycle, namely T&D site licensing. Since this site licensing framework was said to control hazardous waste T&D, the only requirement of the 1978 EEC Directive that needed to be fulfilled was the control of the transport phase, with a "cradle-to-grave" notification system. It was for this purpose only that the main UK hazard classification (Section 17 of COPA) was brought in, with a waste list similar to the EEC list.

As we see below, both site licensing and controlled waste registration in the UK have been severely criticized, calling into question the combination of free market T&D and highly dispersed regulatory responsibility. Before we move to these issues, however, we first examine the way in which the technical criteria for defining special waste in the Section 17 Regulations increase other uncertainties and influence the overall climate of enforcement.

7.3. Hazardous Waste Definitions and Criteria – Their Structure and Role

The UK method of legally defining hazardous wastes is compared in Chapter 5 with that of the USA, the FRG, the Netherlands, and Austria. The main points are therefore only summarized here, the purpose being to analyze their role and implications in the UK context.

(1) An inclusive list of 31 waste types defines a population of wastes that *prima facie* may be special wastes.
(2) Special waste is confined to that waste which is listed *and* "as a result of its presence is *dangerous to life*." The crucial "*and*", indicates that the UK, although using a lists-plus-characteristics approach, does not use characteristics to supplement the list, but to *shed* wastes from control. Low concentration could be one such characteristic, and concentration limits do enter the framework, though in a very particular, indirect, and flexible way.
(3) The interpretation of the key definitions of special waste is the responsibility of the waste producer, T&D contractor, and local WDA in mutual consultation. However, where differences of view exist, producers are free to define their waste as *not* special according to their interpretation of the criteria, and it is then up to the WDA to prove otherwise in court. Thus, the technical clarity or otherwise of the legal definitions is important. This burden of proof problem is important in implementation, as discussed below. In the voluntaristic UK regulatory climate, the autonomy and acknowledged variability of WDA practices and the free market movement of wastes between WDA areas and T&D contractors are very significant. Apart from the economics of transport, there is freedom for a producer or handler to "shop around" for a weak WDA and a "flexible" T&D contractor and site.
(4) The lack of front-end control of waste arisings and the freedoms within the system are justified by reference to the compensating innovation of site licensing of T&D facilities by WDAs. Thus, site licensing is crucial.

The "dangerous to life" criterion of Section 17 is defined by four criteria. These are given below with an indication of some practical ambiguities:

(1) *Does the waste have a flash point of 21°C or less?* The use of a flash point criterion means all wastes that are petroleum spirits or low flash point solvents are special wastes (unless, of course, they are defined as for recycling). Difficulties arise when these materials are mixed with

varying amounts of other compounds, which could raise the flash point above 21 °C. For example, mixtures of acetone and water may or may not have a flash point of less than 21 °C depending upon the relative concentration in the mixture. It would be quite possible for a producer to decide that *normally* the flash point is above 21 °C and, therefore, the waste is not special. It is then up to the WDA to prove that the material is a special waste if it wishes to take enforcement action.

(2) *Does the waste contain known or probable human carcinogen(s) at a concentration of 1% or more?* There is enormous dispute about the carcinogenicity of many chemicals. Although there is central advice, WDAs may ignore it; e.g., two WDAs have excluded asbestos as non-carcinogenic. It is also difficult to measure definitively whether the concentration at which the material is present is in excess of 1%.

(3) *Is the waste likely to cause serious tissue damage on exposure for a period of up to 15 minutes?* The only guidance available refers to injury of sufficient severity to threaten life or cause permanent physical impairment or disfigurement. This requirement is difficult to quantify in terms of attack on the eyes, which are by far the most sensitive organs quoted. In some cases even water can cause more damage than a special waste. The 15-minute exposure rule, individuals' differing reactions to chemicals, and "serious", are also very difficult to quantify; i.e., there is again plenty of scope for discretion and disagreement. The key term "likely" is also subject to uncontrolled assumptions not only about physical processes but human behavior, e.g., is a drummed waste "likely" to be broken open? [41].

(4) *Is ingestion of 5 cm^3 of waste likely to cause death or serious tissue damage to a 20 kg child?* Use of the explicit reference to a child in this criterion has been criticized; yet it is defensible. The regulations were drafted with the critical scenario in mind that children playing on open land might ingest waste that had been wrongly dumped there; 5 cm^3 was regarded as a typical mouthful.

There are several points to note about these toxicity definitions:

(1) They are based on a postulated exposure scenario that is different from that of any other country; in fact, each country is different.

(2) Unlike other countries, the UK scenario involves a *single*, *acute* exposure. There is no explicit provision in the special waste regulations for long-term harm from multiple exposures or for chronic low-dose exposure (as, for example, from contaminated water). Indeed, this has been expressly rejected by the Section 17 Review Committee [42].

(3) As suggested in (2), the scenario, like any other, inevitably generates problems as to what it excludes. A topical example in the UK concerns drummed wastes. The regulations stipulate that to be defined as special a waste must be in a condition that could create exposure. But if the scenario is acute exposure from ingestion, *drummed* toxic waste may be exempt, even if the drums may in future burst or corrode. Examples of this practice being allowed have caused demand to tighten the regulatory definitions.

A problem for producer and WDA is to obtain toxicity data that are reliable enough to be credible *and* of clear relevance to the waste in a given situation. Indeed, much of the "intrinsic" data necessary to determine acute toxicity of chemicals to humans, let alone small children, is simply not available. It is always necessary to extrapolate from animal data, which itself is usually scanty and conflicting (see Chapter 10). The DOE advice ranks different kinds of evidence and gives worked examples, but, as usual, these are expressly only illustrative, and have no legal standing. It should be emphasized how much these technical definitions are open to variable interpretation in real cases, and that the burden of precision and proof rests with the WDA, because these definitions are for *excluding* wastes *prima facie* on the hazardous waste list.

In addition to these limitations and discretionary freedoms, there is a major problem of implementation for WDAs – that the definitions are only legally enforceable at the moment of deposition of a waste. In other words, if an inspector finds a waste deposited in an unlicensed site and can prove the depositor, he or she still has to prove the waste was special at the moment of deposition. In practice, this is tantamount to a requirement to catch depositors red-handed (and still prove that their waste is technically a special waste).

7.3.1. The burden of proof and technical clarity

Given that scientific risk analysis and its application in specific cases are uncertain, when there is disagreement who should have the burden of proof in defining whether a waste is to be included or excluded under regulatory controls? The Association of County Councils, representing the WDAs, who shoulder the burden of proof, understandably find the lack of clarity objectionable:

> ... the theoretical foundation of the calculations remains highly unsatisfactory ... The reliability of human toxicity data is extremely poor ... Extrapolation from figures for a rat or rabbit to a 20 kg child is impossible [43].

> ... waste disposal authorities will have the greatest difficulty in proving in court that certain wastes are special [44].

This combines objection to both the "intrinsic" risk uncertainty and the uncertainty of projection into real situations. The regulatory options toward these uncertainties are either to build in discretion so as to introduce them into the routine decision-making process, or to "short-circuit" or "suppress" them by the expedient of fixing simpler, more clear-cut decision rules, such as precise concentration limits for listed chemical constituents – the Dutch method.

The Gregson Report noted that the UK Section 17 Regulations do not call for a "pass" or "fail" toxicity test, but a toxic hazard assessment, and any prosecution relating to a disputed waste will have to be judged on the basis of professional opinions about the likely effects of ingestion or exposure [45]." (It might also have added the phrase "about the *likelihood* of ingestion or exposure.")

Others, in evidence to the Gregson Committee, described the discretionary license of the UK regulatory criteria as offering "a field-day for lawyers" [46] in the endless legal argument that imprecise definitions entail.

Thus, the "dangerous to life" criterion is regarded as especially open to endless conflict of interpretation in practical cases. The flexibility which is supposed to be an asset of the UK system in tailoring risk management to real situations, thus entails a corresponding cost in terms of enforcement problems. Legally speaking, the responsibility for deciding whether a waste is special lies with its producer, and it is no defence to claim ignorance, unless "due diligence" has been shown in the determination of a waste's status. However, the crucial point about the UK role of characteristic definitions is that they are delisting tests; therefore, the burden of proof in listing a waste lies with the regulator, the local WDA, and not on the producer to delist it, as is the case in the USA. The WDAs evidently find the *technical* uncertainty of regulatory criteria unacceptable, because they do not have the *institutional* resources to negotiate acceptable practical solutions with industry.

At present, if a case is brought to court the prosecution must establish a *prima facie* case, then prove it "beyond reasonable doubt." Thus, the clarity of the definition of special wastes may become crucial (though in cases involving site licenses, the license conditions may also stipulate the legal responsibility). Although the waste producer may have to take on the burden of proof of showing that he or she did operate with due diligence, this is on the lesser test, of balance of probabilities. The evidence of local authorities to the Gregson Committee and their stance in the Review of the Section 17 Regulations [47], was that prosecution has been repeatedly undermined by the vagueness of the toxicity criterion in the regulations.

The burden of proof issue is entangled in broader considerations of consistency with natural justice, that parties be innocent until proved guilty. Proposals by the local authorities to shift the onus of proof on to the disposer or producer of wastes have therefore been rejected in the Section 17 Review. Yet there is an important principle underlying the debate, involving the interactions of technical uncertainties and institutional realities.

7.3.2. The burden of proof and ideal-world assumptions

Precedents for placing the burden of proof on the defendant (here a waste producer) concern cases where the defendant has *exclusive knowledge*, that is, those where the plaintiff could not be expected "to prove an objectively ascertainable fact". The local authorities assert that the "exclusive knowledge" criterion is fulfilled in the case of producers and wastes. Producers argue that this is not true, because local authority officers can legally enter premises to analyze wastes whenever they see fit: they therefore share in the knowledge and the attendant responsibility. In the producers' view, the status of wastes is therefore "an objectively ascertainable fact".

This conflict of legal principles reflects two opposed views as to the status of ideal and real, or formal and informal, institutional relationships. In *practice* the local authorities do not have access to the requisite industrial knowledge to make the status of wastes an objectively ascertainable fact (quite apart from any technical disagreement over interpreting toxicity and exposure information into actual cases of wastes). But producers can legally benefit from the disparity between the ideal situation and the real one, by claiming that formal access equals real availability of knowledge. Thus, unrealistic assumptions about ideal worlds are built into the regulations and enforcement arrangements.

The same factors that cause WDAs to wish for more precise technical criteria for the enforcement process, namely their lack of expertise and other resources, and the lack of institutional control of the prevailing uncertainties, also causes them to seek *less* discretion and flexibility in the risk management, decision-making process. What is useful knowledge to them is very different from that of other actors in the system. Thus, curiously, in the UK a flexible, discretionary technical system is *least* favored at the very points in the system where it should be exercised if its theoretical benefits are to be realized.

As noted before, however, the restrictions of the formal hazardous waste classification framework are defended on the grounds that T&D site licensing is the main control mechanism.

7.4. Site Licensing and the Landfill Dimension

Given that the vast majority of T&D sites are landfills, UK site licensing really means the selection and control, through explicit licence conditions, of the use of landfills – e.g., what wastes are to be excluded altogether? What treatment or volume restrictions and required management practices should be stipulated for others, etc.? In theory, a sophisticated situational "degree-of-hazard" risk management framework is possible here, especially if proper landfill control also sustains "high-technology" T&D options for certain wastes. However, it depends utterly upon the rigor of the initial site selection process, and upon the quality of specific controls and practices at a multitude of sites that cannot possibly all be directly, independently, and professionally supervised. Since landfill and the site licensing system interpenetrate so heavily in the UK, we examine the landfill situation, then look at site licensing, before returning to general WDA enforcement questions.

7.4.1. Landfill

About 85% of all notifiable ("hazardous") wastes are disposed of by landfill, of which 50% are landfilled in-house. According to the DOE, apart from being the cheapest method of disposal, "sensible landfill is realistic, and an ultracautious approach to landfill of hazardous and other types of wastes is unjustified [48]." This conclusion is based on the results of a study carried out from 1973 to 1977 on 19 landfill sites, which were supposed to represent the main geological types found in the UK. The UK is said to be lucky with its geology in this respect, as there are many clay areas and much of the country, especially in the North and West, uses water from runoff rather than from groundwater systems.

In general, there are two broad methods of land disposal:

(1) Concentrate and contain.
(2) Dilute and disperse or codispose with "ordinary" municipal waste.

"Concentrate and contain" may refer to fully engineered "landfill", of the kind specified in the USA, and envisaged in the Netherlands, where the containment physically isolates the wastes from surrounding land. This is indistinguishable from permanent storage. It may also refer to the use of natural liners, such as clay strata, which is a cheaper and much more common practice. Plastic liners are also used, sometimes to enhance natural properties. "Concentrate and contain" does nothing to actually reduce the inherent toxicity of deposited materials, so if containment is accidentally ruptured at some future point, the hazards are still there to be released in concentrated form. Artificial liners have had limited use, and thus no experience of their long-term behavior exists. Clay and other naturally "lined" landfills merge into the "dilute and disperse" category, since some permeability may occur, allowing percolation away of diluted and, thus, less toxic materials. Codisposal with domestic refuse is thought to actually encourage chemical and biochemical degradation and transformation of some toxic materials into less toxic products, and (in properly chosen and operated sites) to allow only acceptably transformed or diluted materials to leach away, diluting even more as they go.

The DOE report concluded that "extensive experience in the UK over a long period of time has shown that very few documented cases of significant groundwater contamination due to landfills have occurred, thus indicating that the controlled disposal of wastes by landfill is acceptable [49]."

Even with the official unpopularity of landfill in other countries, there is pretty much a consensus in the UK on the acceptability of co-disposal landfill, so long as "sensibly" operated. However, the conditions of "sensible" practice are pregnant with question marks, and beneath the consensual style and language of the UK system, suitably encoded criticisms of the lack of control over "licensed" practice, and of the lack of central pressure to improve standards, can be traced back over several years.

The Environmental Safety Group at Harwell, for example, in their evidence to the Gregson Committee [50] qualified their remarks on the merits of landfill. They argued that:

> ... for *certain* hazardous wastes their limited codisposal in a controlled fashion would result in no serious pollution hazard over and above that presented by domestic and light industrial waste on the site [but] it is imperative for the DOE to produce guidelines on the extent to which the codisposal of given toxic wastes was considered to be an environmentally safe practice ... For some materials codisposal is not considered advisable and alternative technological approaches are advocated. For others the data is still somewhat imprecise and research is still in progress to provide improved guideline data ... Since the Landfill Research Programme was initiated in 1973 there has been a considerable change in the structure of the waste disposal industry and in waste disposal technology. The investigations carried out then showed how certain types of wastes such as heavy metal sludges and cyanides behaved in both landfill sites and lysimeter type experiments. Since then the guidelines on the best disposal practice for most, but not all, of the major groups of compounds have been produced.

Despite such differences as to emphasis or pace, Harwell joins with the DOE in supporting sensible codisposal, on the recommendation that more work be carried out on landfill. In addition, the Gregson Committee, found landfill codisposal acceptable, but not without qualms:

> The safety of landfill, including codisposal depends vitally on good management. The scope for abuse is considerable and the waste disposal industry has sometimes been skating on thin ice ... Accordingly landfill must not be used in marginal cases just because it is cheap – the "cheapest tolerable means" approach – and all hazardous waste disposal must be subject to rigorous control [51].

These explicit questions as to whether "sensible" operation of landfill codisposal is practically achievable were underlined by the HWI's first highly critical oversight of the situation in 1985. This was joined by findings of groundwater contamination in the UK, presumed due to landfills, and by open criticisms of the scientific quality of the 1970s research program by which the UK landfill codisposal commitment is justified [52].

Especially with the growing international policy interaction on hazardous waste management, the UK is in a somewhat embattled position, as reflected in the observations of a DOE official:

> Landfill means all things to all nations. It may be to a carefully selected, well-engineered and properly managed facility, either for codisposal with other wastes or solely for hazardous waste, or it may be to an uncontrolled dump. As improved control measures are implemented the uncontrolled dump is gradually disappearing in most European countries; *in England it has disappeared* [emphasis added]. There are also national differences in the concept of controlled landfill: the UK, based on its research findings, promotes a landfill philosophy which acknowledges and utilizes the beneficial effects of codisposal and the natural mechanisms of degradation, attenuation and dispersion. An entirely opposite view prevails in much of continental Europe, where controlled landfill disposal for hazardous waste means the concentration of such wastes in sites wherein they are contained, either naturally or by artificial liners, and where all leachate produced is collected and removed for treatment.
>
> The international perception of the UK's attitude to hazardous waste management is interesting. Whilst acknowledging the degree of control achieved, the expertise applied and the relative lack of problems experienced by the UK, our European partners are nevertheless critical of what they see as the indiscriminate consignment of wastes to landfill, simply on the grounds of its relative cheapness. The facts are very different: the UK may fairly be said to have achieved high standards in respect of environmental protection and public health related to waste management long before most of the other European nations. Hazardous waste disposal has been specifically regulated since 1972, all disposal sites for controlled wastes have been subject to licensing since 1976 and codes of practice covering a wide range of hazardous waste have been published by the DOE, commencing in 1974. It is ironic in the criticism received to reflect that in the UK there is little evidence of the inheritance of problem sites reported by some of our European neighbours [53].

As we saw in Chapter 4, the Dutch could not site a national landfill and expressly banned it as an option for chemical wastes. Other countries in less special positions than the Netherlands are trying to move away from landfill except, perhaps, for other waste treatment and destruction residues. The FRG environmental agency has explained, for example, why it is moving away from codisposal:

> The FRG tends towards separate treatment even though this method is clearly more expensive than the former (codisposal). The motive ... is undoubtedly the principle of prevention [54].

John Lehman, a former Director of the Solid Waste Program of the US Environmental Protection Agency (EPA), said that:

> Our philosophy, as the land protection group within EPA, is to minimize hazardous waste disposal to land. Consequently we strongly support hazardous waste recycling or detoxification treatment prior to land disposal wherever possible [55].

Lehman and others indicate that US policy is directed ultimately at no landfill anywhere at any cost, i.e., landfill is made difficult and costly. Where it is allowed, it will be in containment landfills engineered to precise, legislated specifications.

The US attitude to landfill codisposal appears to reflect a significantly different response to natural uncertainties from that of the UK (see Chapter 12), where the uncertainties of codisposal are recognized but not treated as automatically negative. UK officials argue, against international coolness toward their policy, that many countries are just as dependent on landfill as the UK, even if their official figures and policies say otherwise, because, for example, treatment residues that go to landfill may be accounted against the treatment method, not against landfill.

However, the arguments for or against landfill, or codisposal in particular, tend to be carried out with little attention to the crucial questions of *implementation* and the worlds of real practice which are hidden away in cryptic qualifications such as "*sensible* landfill is acceptable". The UK approach to landfill as a control strategy reflects a principle (even if it does not *fulfill* it) of fine-tuning regulation to situational variations and differential risks. In doing so it may assume too much of a local fine-tuning capability in implementation. In this case a *theoretically* sound policy may be *in practice* questionable, because it asks too much of implementation. Straightforward bans on landfill codisposal, whilst in some cases overcautious, may be more realistic because institutionally simpler to implement in the field.

In the UK (as elsewhere) the crux of the landfill issue is not so much whether or not landfill *in principle* is good or bad, but what conditions control its use, including especially what is allowed to be landfilled at all, and in what forms. This leads directly to the question of T&D site licensing, since in the decentralized and private market system of the UK this is the designated key mechanism for controlling the routing of wastes to particular T&D options and for managing their risks satisfactorily, once there.

7.4.2. Site licensing

The sections dealing with site licensing under COPA were brought into force in 1976. Waste Management Paper No.4, *The Licensing of Waste Disposal Sites* [56], lays down the general policy:

(1) To ensure that waste treatment and disposal are carried out with no unacceptable risk to the environment and to public health, safety, and amenity.
(2) To place at a suitable local level the responsibility for deciding what conditions should be imposed at a given site, so that local circumstances can be fully accounted for.
(3) To ensure that changing patterns of waste disposal do not prejudice objective (1) and, equally, that those responsible for waste treatment and disposal take proper advantage of technical progress.
(4) To give waste disposers a clear idea of what operating standards are required of them.
(5) As a result of (4), to secure the provision of sufficient facilities for the treatment and disposal of waste.
(6) To ensure that sufficient information is available to the responsible authorities to enable them to fulfill their statutory duties.

Until the creation of the HWI in 1984 there was no central scrutiny of local licensing. Occasional *ad hoc* scrutiny took place when WDA decisions were appealed to the Secretary of State.

Once a site has planning permission a WDA can only refuse a landfill license on the grounds of (a) water pollution or (b) danger to public health. It therefore has the burden of proving why, if it refuses a license. The conditions in licenses are restricted to certain aspects and are also, to some extent, governed by the T&D operator's license application, i.e., type of waste, amount, choice of disposal method, etc.

There are two aspects to site licensing as a regulatory mechanism: the quality of the license conditions themselves and the quality of their *enforcement*. The Waste Disposal Engineers Association conducted its own survey of T&D licensing by English WDAs in 1982 and concluded that "site license conditions are fairly standard throughout England [but] their enforcement and interpretation varies considerably [57]." This is a fairly common picture, implying that the licensing mechanism is sound, but for the lack of proper policing. Yet the HWI's scrutiny produced a disquieting picture of the *quality* of licenses too, not merely their enforcement. This was graphically illustrated in the HWI's attempt to create a register of all UK T&D facilities, public and private, in each region, with a classification of sites according to the waste they received. The results revealed an unsuspected mess:

> Realization of this objective proved impossible. As data from site licences was accumulated it rapidly became apparent that any meaningful classification of sites by waste types would be extremely difficult. The primary reason for this was the vast array of descriptive nomenclatures used by WDAs in their site licences. The list of group names found (not specific waste types) encompassed 385 entries. Names assigned to waste types were often such as to defy anyone, other than the WDA that wrote the licence, from identifying what types of waste were included, either because of the general nature of the description given, e.g. sites licensed solely for "controlled waste", or because of the use of meaningless descriptive terms, a prime example being "industrial semi-inert". The only way to resolve this would have been to go back to the WDAs with a series of questions on hundreds of sites and this would have taken very much longer than the six months it took to collect, analyze and collate the information from site licences [58].

The overall conclusion was that:

> Regrettably, what the preparation of this register has clearly and unequivocally demonstrated is the extensive disparity between site licenses issued by different WDAs in England and Wales [59].

In fact, the inital impression of HWI from those sites that the inspectors themselves visited was that the site licenses are:

> ... in general unhelpful documents which do not provide sufficient guidance to the operators. Many of the licences seen place too much reliance on the operator, giving him an enviably free hand as far as operations are concerned, but fail to place sufficient emphasis on the standards to be achieved or the practices necessary to achieve them [60].

Adequate enforcement is thus preempted by inadequate site-licensing, brought about by a combination of:

(1) Historical neglect of the importance of the problem.

(2) Lack of resource and expertise at local level.
(3) Institutional restrictions on WDAs which undermine their ability to control and enact thorough local management reinforcing (1) and (2).

Thus, the overall control and reassurances supposedly offered by T&D site licensing are seen to be largely vacuous. This control cannot, as it stands, justify the official claim that it allows avoidance of "front-end" control via closer hazard definitions, stricter prenotification requirements, more waste data, and more positive designation of waste T&D routes. Furthermore, these inadequacies and inconsistencies "are being exploited by some waste disposal contractors [61]," who may legally deposit wastes at sites where the conditions are so vague as to mean anything.

Multiplying the system effects of this T&D licensing failure are:

(1) The freedom mentioned earlier for waste producers or handler–brokers to find legal cheap landfill, which in a free market undermines the demand for and investment in higher technology T&D facilities.
(2) The very large scale of this free-enterprise zone in UK waste life-cycles makes definition of these life-cycles all the more difficult.
(3) The general uncertainty created by these inconsistencies undermines all-round confidence and trust, which are especially vital ingredients for the viability of a highly voluntaristic system like that of the UK.

The design and enforcement of T&D licenses is the central element of implementation, so we look next at enforcement.

7.5. Enforcement

Enforcement involves consultation, informal advice and persuasion, monitoring, record-keeping, development of site licenses, issuing of formal compliance notes, and, ultimately, prosecution. This must be performed with many kinds of industry and company, with wastes being freely transferred in and out of the area of responsibility, and with a need to keep up not only with all the immediate practical tasks, but also with scientific, technical, and other developments in policy advice. As public opinion is more actively expressed, the conflicts that WDAs must resolve become more difficult. There are further statutory duties also, such as the gathering and preparation of data for a waste disposal plan, site licensing, appeals, etc.

All of these functions are normally delegated to a "waste disposal department" in the local authority, which is usually also responsible for municipal waste collection, treatment, and disposal. However, the strength of these departments, and thus the quality of enforcement, varies enormously throughout the UK. In some WDAs it is part of the duties of a single person. The HWI noted that "inadequacies and inequalities are commonplace [62]" with site license breaches going uncorrected by WDAs. Furthermore,

> Very different attitudes towards the policing of hazardous waste disposal sites prevail amongst English and Welsh WDAs. The two extremes are exemplified by one English WDA, where until recently no staff provision was made for inspection duties and other WDAs, particularly some of the metropolitan counties, which deploy teams of inspectors, with provision for out-of-hours working.
>
> Between these extremes, there remain considerable differences. Comparing for example, two major hazardous waste landfills of broadly similar inputs, both in the south of England, at one, WDA inspectors made two or three unannounced visits each week and took spot check samples for analysis from waste arriving at the site. This WDA also imposes rigorous documentary prenotification requirements on the hazardous waste disposal sites in its area. In the second WDA, which has only one major site licensed to accept a wide range of hazardous waste, it is normal practice to make an inspection visit once every three weeks, usually by prearrangement. In this case no attempt is made to sample wastes: normal practice is for the WDA inspector to provide the site operator with empty sample jars for the operator to fill over the next three weeks, and to collect the filled containers left for him. This takes the basis of trust too far [63].

Lack of manpower and lack of expertise tend to run together. To illustrate, during interviews in our research, one environmental health officer in Wales, when asked what he would do if asked where an old car battery should be disposed of, answered that he did not know; yet another environmental health officer in a nearby WDA, has been conducting surveys on his own initiative into the amount of hazardous waste arising from local school laboratories, pharmacies, etc., and what they do with this waste. The former could be thought of as incompetent; however, handling of waste (all waste) is only a small part of his overall responsibility. In another district – a net receiver of hazardous waste – the total staff in the *whole* pollution control section now consists of two people. Formerly there were four, but with a shift of political emphasis to housing, the pollution control department was reduced to two and the housing department boosted instead. This arbitrary designation of personnel indicates the institutional blocks to professionalism in the field.

The replacement of DPW by Section 17 of COPA in 1981 was supposed to reduce the population of registered wastes by about 70% and thus cut the administrative costs proportionately for WDAs and industry. The new regulations implied a shift from bureaucratic administration of the waste notification system to field inspections, sampling, etc. In actuality however, the expected reduction of consignment notes has not occurred, and WDAs are still heavily burdened with an *administrative* mode of regulation. In one Welsh district council, with two staff dealing with all pollution, 2000–2500 notifications, sometimes with five-page attachments, are received per year, each needing to be read, understood, collated, and some followed up, in addition to all the other regular responsibilities.

Contrary to the DOE's official position that local authority regulatory resources would be spread less thinly by the new approach, the WDAs were hostile to it because of what they saw as a substantially *increased* workload caused by the extra field-monitoring required, despite the lack of resources to do it. They argue that the new special waste definitions and consignment arrangements leave a large undefined middle-ground of environmentally hazardous wastes that are now excluded from registration, without which the WDAs feel unable to exercise responsibility for ensuring that no special wastes are included in that uncontrolled population. This conflict was one of the central issues of the Section 17 Review (1985).

The key issues in the review were the scope, administration, and enforcement of the regulations, with the focus of attention on breaches of control. The report warned that "the public's perception of and sensitivity to such breaches represents an urgent stimulus to improve effectiveness of control [64]." Although it is obviously impossible to give any definite figure of breaches, the DOE estimates that, on the evidence currently available, there are 150 significant incidents of illicit disposal of toxic and dangerous wastes each year [65]. To this must be added the broader effects of lax legal disposal noted earlier.

In the UK the enforcement failures caused by a widespread lack of local resources, expertise, and attention are compounded by the complexity of interpretation of the regulatory definitions, for example, of what constitutes special waste. Although difficulties arise from the imprecision and complexity of the technical definitions of hazardous wastes (see earlier), enforcement is undermined by legal statements in the regulations too. The worst of these in local authority experience is the definition of unlawful disposal as "causing and knowingly permitting" controlled waste to be deposited on an unlicensed site. The practical consequence of this legal phrasing is that illegal disposal virtually has to be discovered *in the act*, for prosecution to be possibly enforceable. There may then *also* be complex and expensive technical uncertainties as to whether the waste falls into a controlled, hazardous category, and whether the site license (if any) legally permits that waste consignment or not.

In addition to these difficulties and the sheer lack of resources, enforcement is undermined by organizational dislocations even within the local WDA:

> The waste disposal officer typically has responsibility for all aspects of waste management within his Authority's area, i.e. the waste survey and disposal plan, the acquisition and development of local authority disposal sites, their operation, for site licensing and inspection and for the administration of the Special Waste Regulations. Decisions on the issue or modification of site licenses are occasionally delegated to the waste disposal officer but more usually retained by the appropriate Committee. Enforcement actions are normally the province of the council's legal department.
>
> In a few disposal authorities in England and Wales site licensing, inspection and enforcement are located separately from the waste management function, e.g., in a department with a traditional enforcement role, such as weights and measures. Thus the waste disposal officer is responsible only for the authority's operational role in waste disposal [66].

Council legal departments are responsible for *all* legal functions, and so do not specialize in waste regulation, nor even necessarily in pollution control generally. They are naturally more cautious in calculating the costs, risks, and benefits of prosecuting a case than enforcement officers might be, and this caution is etched even deeper by the vagueness of the legal definitions involved. Waste disposal officers have complained about the difficulty of persuading their legal departments to act; yet the legal officers are also in a difficult position over prosecution, its uncertainty, and expense. They need to have clear-cut cases, which are hard to find. Typical enforcement scenarios involve prosecution to secure the fulfillment of site license conditions, e.g., to avoid landfilling an allegedly special waste, or prosecution of a disposer of allegedly special wastes in an uncontrolled site, both rarely clear-cut.

WDAs, as far as possible, rely on advice and/or persuasion to deal with license problems, using prosecution of offenders as a last resort. Success has usually been with operators who wish to act responsibly anyway, but who have made mistakes, and there is now a feeling that tougher measures have to be established for persistent offenders. The problem is that even if a legal department decides to prosecute a case, it takes three months to two years to move a case to court, the outcome is uncertain, and then the sentences are often too light. Most offenses fall under magistrates courts, where the maximum fine allowable is only 1000 pounds sterling. Should a case even reach the court, the defendants usually have good scientific experts and can employ a specialist hazardous waste or pollution lawyer, while WDAs have to rely on the nonspecialist local authority lawyer. Even the largest authority, the Greater London Council (GLC, now disbanded), found this a continual frustration.

Even if a significant shift away from desk-bound "enforcement" to field inspection were to occur, these enforcement problems would not diminish (indeed, they would probably multiply, at least until inspection began to take effect in better practices), because the costs of following up direct inspection with tests, sampling, and chemical analyses are very considerable and, as discussed in Chapter 10, are not definitive anyway. In fact, the weakness of many WDAs with respect to expertise and technical resources leaves them effectively dependent upon the firms they are supposed to be regulating for advice and analysis. In the collaborative British system this is not of itself a matter for official concern.

The typical British attitude is reflected in the comment by a senior official at the DOE, "If someone from the US EPA has discussions with an industrialist, there usually has to be at least one lawyer present. The fact that almost anyone, and especially industrialists, can telephone the head of the Land Wastes Division of the DOE for advice on a certain waste problem is something incomprehensible in the American situation [67]." This informal collaborative network is supposed to operate at the local level too. For example, in one district in Wales, a major multinational oil firm has its own in-house landfill on its refinery site and until recently an incinerator (purpose-built for acid tars). The local WDA licenses and "monitors" the site. The local environmental officer makes an *occasional* visit but, as one such officer emphasized, the whole network is based on trust. In fact, if there is likely to be a problem, usually associated with smells, the company telephones the WDA beforehand so that if a member of the public does complain the WDA will already know the cause of the problem. This naturally helps their public credibility. If the WDA needs any scientific advice or laboratory tests it automatically calls the head of the company laboratory, who readily helps out. The company naturally has far more technical expertise than the officials at the WDA, and is far better equipped to control its own site. Previously, the company did take some waste for processing from an adjoining WDA, but after an incorrect declaration was made, it has not accepted any more. In fact, it is a significant comment on transaction costs that, despite under capacity, they claim there would now be far too much extra work to accept wastes from outside. The company also has its own consignment note system within the refinery. The person in charge of waste disposal at the company site could not foresee any problems that he had not already dealt with in the past and could not see any action necessary should anything occur, other than to inform the local WDA as a matter of courtesy. A strong climate of institutional trust and a relaxed attitude toward uncertainty prevails.

In another district a similar situation exists, but local expert help is provided by a major international waste disposal company, instead of the waste producer. Again, the same level of trust prevails, and the WDA official was confident that he could trust the firm, despite the fact that its public credibility had suffered through allegations of malpractice. The foregoing examples are taken from district council WDAs, but this kind of trust and dependency is not confined to them. Although in such a complex, decentralized system relationships are varied, the overall culture is summed up in the official advice given by the DOE in Waste Management Paper No. 23, that conflicts or ambiguities in the definition of special wastes for regulated T&D can be resolved by mutual informal consultation of the relevant parties, without recourse to elaborate legal or technical rules and procedures.

When the regulated company is responsible this cultural climate is acceptable, but even with responsible companies differences of interest and judgment on risks can occur, and mechanisms for handling these must be available. Most of the officers in the smaller district WDAs thought that, apart from the problem of resources, they were more in command of the situation than a county-level WDA in that they know better "who is who" and "what is what" in their area. As our interviews showed, however, and as other sources have confirmed, this self-confidence is not entirely soundly based.

In the larger councils, the GLC, and other metropolitan county councils (now abolished) the situation on the face of it was quite different, and yet in some basic elements, such as legal enforcement, it was not. Some of these WDAs had large, full-time teams consisting of chemists, engineers, inspectors, and enforcement officers; they often helped out smaller, less well-equipped WDAs. With the political abolition of the GLC and metropolitan county councils in 1986, the HWI advocated that hazardous waste management should remain the responsibility of a common unit in each area, thus maintaining these pools of expertise. It warned that the disbandment of such teams could lead to an increase in already widespread "cowboy activities and fly-by-night tipping":

> In Merseyside [one of the metropolitan WDAs] it has sometimes proved necessary to deploy inspectors in twos, and even in threes, equipped with personal radios, when investigating the activities of some local asbestos strippers and their "transfer" stations [68].

There is a widespread feeling that the abolition of the larger WDAs may only hasten a general perception that regional units of regulation on something like their scale are necessary.

We have seen that practical inspection and regulatory relationships within and around the WDA vary widely. It is necessary to recall that this is a fragmented system of regulation, at the very back-end of the waste cycle, mainly at the disposal site. WDAs, in particular, have advocated more systematic attention at the front-end of the waste life-cycle to reduce the uncertainties besetting them at the back-end. Registration of special waste arisings is a central instrument of any front-end regulatory attention. However, a waste arisings registration scheme has been ruled out, trip-ticket notification being regarded as adequate. The large distinction between the two arises from the free market of waste handling, which means that the small and fragmented WDAs have no control over waste movements in and out of their area, and so experience great uncertainty. A national register of waste generation would, if it were practicable, reduce these uncertainties considerably. It is therefore worth looking at some of the practical realities of notification.

The trip-ticket notification system was established under the earlier DPW regime, for a wider population of wastes as already explained. Attitudes formed under DPW have proved difficult to adapt to the new philosophy and framework of COPA, leading to conflict and confusion over the precise purpose and practical implementation of the waste notification requirements.

The producer of a waste defined as special must, before the waste is removed, prepare six copies of a standard form, fill in the waste description and point of production and destination, and send a copy to the receiver ("away") WDA, at least three days prior to despatch. When the carrier collects the waste, the name and date are recorded by the carrier in the next section, and the waste producer completes a further section confirming that appropriate advice was transferred on collection. A copy of this goes to the "home" WDA. Finally, the receiving T&D operator fills in a section confirming the validity of licensed disposal at his or her facility. Copies are returned to the home WDA, who is responsible for "closing the loop" on proper disposal by tallying this copy against the first copy completed by the waste producer.

The main confusion over this system is whether it corresponds with the national policy of placing control emphasis upon T&D site licensing. The local authorities have argued that it would be more consistent if the *receiving* or away WDA closes the loop on the trip-ticket, because the receiving WDA in the end needs to control what is coming into its area's T&D facilities, if the regulatory strategy is back-end site licensing control. But this argument assumes that the trip-ticket system is meant to be a real-time *control* mechanism (as opposed to a *post hoc* information system) and this has been vehemently resisted by industry and, more gently, by the government. A CBI spokesman expressed industry's frustration with the WDAs' attempts to retain a more rigorous framework of control:

> The need for records has been enforced through the EEC Toxic Wastes Directive but extensive "cradle to grave" documentation was not envisaged in the drafting of Section 17. Unfortunately, these sensible new requirements were also lost sight of in an increasingly emotive debate of whether the documentation provided WDAs with a means of control over the point of disposal. Such polemic diverted energy which could have been used with greater benefit to educate those who would be involved with the new Regulations and, above all, completing the guidance literature, Waste Management Paper No. 23 etc., on time. Thus the Regulations were launched in a climate of hostility to be operated at grass roots level by individuals who were anxious, ill-

prepared and often confused. Little surprise that their reception has been so mixed [69].

Thus, again, a fundamental conflict about the meaning and role of key information has weakened the coherence of social relationships on which the UK lays so much emphasis in justifying its system. To show the ramifications of even apparently minor factors, however, let us look at the prenotification requirement. The theory behind prenotification is that the receiving WDA and the T&D operator should know when which kinds of wastes are due, evaluate whether they should be received, and make proper arrangements for really difficult consignments. In practice, "prenotification" is often mailed, second class, on the day of waste despatch. Strictly speaking this is illegal, as COPA Section 17 requires that it be posted so as to *arrive* at the T&D contractor at least three days in advance of delivery. But with all the other day-to-day pressures on plant operators, late prenotification is not unusual. When it does happen, the receiving officer at the T&D site may not know about a waste until it arrives at the gate. He then has to decide on the spot :

(1) Whether the waste fits the copy of the documentation, which the driver should carry.
(2) Whether the documented waste description fits the site license conditions.
(3) What to do if these are in doubt, e.g., phone someone responsible.

When both site license conditions *and* documentation of the waste are vague, as is often the case, the site operator is under immense pressure to let it through. It has been pointed out that in some WDAs a site clerk would need to have a degree, not only in chemistry, but in law too.

Several other normal factors, some "structural", others ordinary organizational realities, mean that the information or "control" offered by waste consignment notes falls far short of any theory:

(1) Loads defined for recycling are exempt, yet often end up as wastes for which the cheapest disposal is being sought.
(2) The consignment document sent from the producer to the home authority only states where the waste was produced and where it was taken. But in the commercial structure this is often an intermediate transfer station. The producer will not know what happens beyond this point, further responsibility for transformation and/or disposal being in the hands of the broker. Thus, the "cradle-to-grave" document may only record an intermediate "grave", which may not be the ultimate disposal point. In theory every stage and new life-cycle is fully recorded, but in reality this becomes less likely with intermediate commercial agents and handling points.

(3) Widespread "branching" of waste life-cycles mean that some consignments quite legally disappear. If a broker receives a "waste" that contains valuable materials and sells it on as a good to a recovery merchant or transforms the material on-site, part or whole of it is legally removed from notification. Under the present special wastes definitions, even simple dilution could remove a waste from the control and notification system. If it later notices a lack of correspondence in the records, the WDA will not know, unless it follows up with time-consuming specific inspections and inquiries, whether the waste was illegally disposed of or legally processed.

(4) The size, weight, whether tankers or skips are half-full or complete, mixed, etc., are often not defined. No units are specified on the form, nor would they be enforceable.

(5) There are supposed to be six copies of the documents, the last one returning to the home WDA for "closing the loop". Yet this bottom copy, in particular, may be simply illegible.

(6) Information necessary for an on-the-spot check-up of waste descriptions or sources may be simply left off or illegible (names, phone numbers), making proper control decisions, e.g., for a T&D site manager, very laborious and unattractive.

(7) Collection and transport contractors are sometimes left to fill in the documents on behalf of the waste producer. They are allowed to amend the document if necessary.

(8) The "season-ticket" system whereby regular consignments are given a single document covering several loads, whilst obviously sensible in many respects, depends upon the regular consistency of these loads, and thus multiplies the above uncertainties proportionately.

(9) Finally, the sheer volume of documents to receive, decipher, read, collate with partners, extract data from, prepare follow-up questions, file, and retrieve when needed, is overwhelming. Informal reports circulate of WDA offices littered with mountains of unprocessed consignment notes (the estimated number of such documents in circulation per year in the UK approaches half a million).

One of the official justifications of the restricted definition of special waste under COPA is that DPW's earlier documentation of a larger waste population had already provided the necessary regulatory information base. In fact, the WDAs did not process the data that was available to them on the DPW consignment notes, because of the lack of resources to do so and the difficulties of translating the notes into meaningful data. Many of these difficulties remain as outlined above, and still obstruct efficient use of the consignment note system, even for its designated purpose of *post hoc* information and identification of mismatches and problems, let alone for any system of actual control. The transaction costs of the system are already high and an attempt at more rigor would escalate them wildly, with questionable returns in terms of regulatory quality.

One possible way of reducing uncertainty would be to register all special waste producers, so that only this identified population would be generating consignment notes. Yet in the DOE's estimation:

> ... if special waste producers alone had to be registered (about 100 000) the initial direct costs for industry would be of the order of 10 m pounds sterling and those of authorities 4–5 m pounds sterling a year ... If *all* commercial and industrial waste producers (between 1.5 and 2 m) were included, the initial costs to industry would be over 150 m pounds sterling and those of authorities over 60 m pounds sterling [70].

Although these figures were disputed by the WDAs, when combined with the inevitable uncertainties outlined above in description, communication, coordination, and checking, the orders of magnitude involved demonstrate the limits of enforcement via "definitive" information bases.

It is a symptom of the deeper conflicts of perception as to the basic relationships involved in regulation that the consignment note system should have ever been regarded as potentially a direct method of control. The whole idea behind the home WDA closure of the loop was that the WDA's should thereby gather data on hazardous waste arisings in their area; whereas the away WDA closure implies more emphasis on the back-end control of wastes via T&D *site* management. But the central designers of the UK regulatory framework have regarded it all along as a means of generating necessary information, a *supplement*, not an alternative, to their model of informal collaborative relationships at relevant points, particularly for T&D site licensing.

It is remarkable how even an elaborate and apparently precise system of notification can be subject to so much informal confusion and uncertainty in practical enactment. If mutual trust prevails these uncertainties remain in check, but once it begins to crumble they become a repertoire of further conflict, requiring escalating legal or technical rigidity.

The belief that the consignment note system has not even generated adequate *information*, let alone control, is reflected in the WDAs' insistence that waste producers should be required to provide more detailed information on wastes. Central and local authorities have basically different views of what technical information is *necessary* for regulation; these derive from different views and experiences of institutional relationships in regulation. Institutionally fragmented, socially marginal to the commercial T&D networks, and feeling marginal to central policy negotiation, the WDAs naturally seek recourse in more comprehensive and precise technical information and criteria.

7.6. Conclusions

The UK distribution of regulatory responsibility is one of the most decentralized in the industrialized world. Combined with private commercial control of waste life-cycles and imprecise regulations, this adds up to a fragmented and extremely flexible system, with attendant benefits and costs.

Hazardous, or "special", waste is defined by an inclusive list of generic substances, but with *exclusions* based on waste characteristics. These are scientifically defined, but imprecisely so. Discretion is built into the technical framework, corresponding with a collaborative, informal institutional climate that encourages negotiation from case to case according to situational variations,

Nearly 200 local waste disposal authorities have separate responsibility for interpreting the imprecise regulatory terms, applying and enforcing them, and administering the special waste trip-ticket system. Under various constraints, such as limited criteria for refusal, they are also responsible for T&D site selection and licensing, and imposing conditions that, supposedly, specify allowed and disallowed wastes, acceptable operating practices, etc. They are also responsible for inspection and enforcement; in effect, their range of responsibilities bridges regulatory policymaking and implementation, but within key formal and informal constraints, and at such a level of decentralization that inadequate concentration of resources is a serious problem.

T&D site (i.e., landfill) licensing is the mainstay of overall control in the UK system. However, the special waste definitions apply only to trip-ticket transport requirements, while a different, longer list of wastes "for record-keeping purposes" relates to disposal. The UK appears to be unique in having two different lists for different parts of the waste cycle, which underlines the extreme disaggregation of its regulatory system. However, the technical content of the classifications gives no indication of their informal and flexible *institutional* role, and thus of the variable consequences.

Much of the UK system of hazardous waste management can be understood from its institutional history. It arose from concern over uncontrolled toxic chemical dumping, and local authorities were the automatic organizational units to take over responsibility. This integrated hazardous waste disposal with domestic refuse disposal: *institutionally*, via regulatory dependence upon disposal site licensing and established land-use planning; and technically, via codisposal of "hazardous" with "nonhazardous" domestic wastes in landfills. However, whereas the public authorities have a responsibility to collect and manage domestic wastes right from their point of production, they are "outsiders" to most if not all of the hazardous industrial waste life-cycle, and do not control what leaves or enters their area. This is a cause of uncertainty and insecurity, especially when they view it in close proximity with domestic waste management.

This is also probably a significant obstacle to the development of stronger professionalism, since it is a major institutional constraint (one of several) upon the exercise of a coherent professional managerial role at the WDA level.

The most important general observation about the UK system is its remarkably strong dependence upon informal trust and collaboration amongst the various institutional actors, at all levels. Apart from anything else, this makes it more difficult than usual to pick out specific parameters, such as hazard definitions or other regulatory mechanisms, for comparison with other systems. More extremely so than in other systems, these are only the tip of an iceberg of implicit, culturally established understandings of responsibilities and relationships. Thus, connoisseurs of the UK system will point out with some justification that criticisms, for example, of the restricted scope and clarity of UK special waste, or of its apparently unbridled enthusiasm for landfill codisposal, omit the full story; this would include a very considerable, and more flexible, *supplementary* context of confidence between parties, self-regulation and mutual regulation, collaboration, and trust. Informal norms and pressures may be at least as substantial and effective as more formal ones, especially when the issue is intrinsically ill-defined and calls for flexibility.

The UK's technical imprecision and uncertainty in regulation is not merely a result of its strong institutional traditions of mutual confidence and trust; it is a *necessary* dimension of wider UK processes, especially intragovernment and government–industry relationships. This cultural style does not merely allow, but *needs* technical uncertainty as a currency of informal institutional interaction, demonstration of good faith, and willingness to compromise – the hallmarks of UK regulation. The technical uncertainty is unthreatening (unlike, for example, in the Netherlands or the USA), because of a lack of institutional uncertainty; there is confidence that negotiations between interest groups will resolve uncertainties into acceptable compromises. There are few other countries in which the decision process for defining the full, legal hazard status of a waste could be described in an official government document as follows:

> Where a waste producer does not have access to such [suitably qualified] staff, guidance may be available from certain of the larger specialist waste disposal contractors: failing this it is suggested that the waste producers and waste disposal authority (and other parties as appropriate, e.g., a water authority) should hold joint discussions to establish the status of a particular waste [71].

Under these conditions introducing precise and inflexible definitions, criteria, and controls would be a statement of lack of trust in the other parties. This would corrode the whole edifice and its forms of social control. It would be a tacit breach of a whole labyrinth of interdependent commitments that make up the viability of the overall system.

Even the definition of a "regulator" in the national system is nowhere near so clear-cut in the UK as elsewhere. The Land Wastes Division of the DOE is proud of the 200 or so man-years of industrial experience amongst its staff of about 12. They see themselves more as consultants than policemen, even acting, on occasion, as honest brokers for firms seeking T&D services, putting them in touch with T&D operators who could help. They are keen to emphasize their extensive personal network throughout the industrial world that they regulate, and believe very strongly that regulation only works because industry wants it to work, and that mutual regulation through this elaborate informal network is the crucial component of the system. Furthermore, expert review committees hardly demarcate roles between industry, science, and government agencies, because such formal distinctions do not matter. Thus, scientific inputs to regulation are naturally imprecise and informal because *institutionally* "science" is embodied in industrialists, government staff, and others, in an organic mix with economic, organizational, and other pragmatic interests and perspectives. Interestingly, compared with other systems this not only creates an operationally and technically different system, but even different definitions of what *science* is. This is discussed in relation to processes of public legitimation in Chapters 11 and 12. Flexibility in definition and *interpretation* of technical criteria are essential in this form of decision process. They design uncertainty *into* UK regulation, but uncertainties which the insider actors are confident can be negotiated. This culturally rooted perspective creates relaxed responses to *natural* uncertainties, such as the effects of landfill codisposal (see Chapter 12).

Therefore, imprecise regulatory criteria and restricted formal powers are not only a pragmatic response to the ill-defined structure of hazardous waste management as an issue; they are a necessary function of the wider UK institutional setting, which shows a healthy disrespect for formal controls. However, there are deeper conflicts that emerge within the appearance of a wholly informal collaborative UK system. The very actors who ought to be *implementing* and benefiting from its discretionary opportunities for local optimization are the ones who are least happy with its lack of precision and certainty. It is ironic that the justification of flexible criteria and *situational* risk management emanates from central government, not the local WDAs who implement those criteria into actual regulation. Yet the justification is supposed to be about their local autonomy to shape regulation to their own circumstances. In other countries, such as the USA and the Netherlands, central government seeks control via standardization and precise formal controls. It does not trust that it can manage the behavioral uncertainties that imprecise regulations would involve. In the UK, central government holds the opposite view, but local authorities are left bearing the brunt of institutional uncertainties over wastes and their production, exchange, movement and disposal.

In fact, the local authorities consistently complain about the lack of precision and certainty in the technical standards they are supposed to enforce, and the lack of formal controls further up the waste life-cycle (such as licensing of waste producers and handlers, and more precise information on industrial waste generation). The reason is obvious in the light of this chapter. They are in no position to *manage* the behavioral uncertainties into consequences they find acceptable. Contrary to the central model, they are beset by institutional uncertainty and need technical–legal certainty to compensate. In other words, they feel outside or, at best, on the margins of the informal social networks of central policy bargaining, and of the private commercial arena of industrial waste handling. There are, of course, extensive informal relationships between parties at the local level, and some between local and national levels; but the overall sense at the WDA level is of fragmentation and marginalization, largely disconnected from the idealized collaborative and trusting social world portrayed at national level. The very fact that the HWI revealed such a surprising mess at local level, at the very point of regulation officially defined as the crux, namely site licensing, indicates both a previous lack of effectiveness of this informal, trust-based framework at this level, and a lack of central awareness that its model is not necessarily a realistic portrayal of life at the sharp end.

The logical conclusion is that the WDAs are too small, have too many combined regulatory responsibilities and constraints weighing upon them, and have too little expertise or other social resources to be confident about converting the technical uncertainties into satisfactory results. Their lack of power to focus regulatory attention earlier in waste life-cycles is a major impediment because of the great uncertainties they face before wastes actually arrive at a disposal site, where WDAs effectively begin to regulate. Strictly speaking, it is up to a local authority, e.g., the district council of Torfaen in South Wales, where the Rechem incinerator receives wastes, to interpret the scientific literature in toxicology and carcinogenicity into practical regulations [72]. In the light of the complexities only *outlined* in Chapter 10, this is an awesome and unrealistic responsibility. In practice, of course, the WDA is dependent upon others.

The options would seem to be:

(1) To consolidate the extreme regulatory decentralization by matching the degree of technical rigidity and precision to the enforcement agency's institutional situation – this would mean a parallel to the Dutch approach, of specific, listed constituents in a waste at fixed numerical concentration thresholds. In defining *by fiat* more standardized risk situations as control tests, this would reduce the *theoretical* degree of refinement in the risk discriminations, but it would also

reduce the degree of uncertainty faced by the WDAs. The improvement in implementation might more than compensate for the reduction of *theoretical* situational optimality.

(2) To leave the regulatory definitions, etc., alone, but regionalize the WDAs into larger institutional units, concentrating more expertise, legal, and other resources there. This would potentially give the ensuing regulatory bodies the formal and informal capacity to *manage* the prevailing uncertainties flexibly (the theoretical justification of imprecise and informal approaches) into adequate results more in accordance with their perceptions and interests.

Either of these general options could be complemented by stronger powers to regulate more upstream, e.g. by registering waste producers and/or waste arisings.

At present, the vigor of the local authorities' more adversarial, precision-based regulatory stance is muted, but the situation is changing fairly rapidly. The HWI's first (1985) report contained unusually sharp and stinging prose for an official government agency in a culture that is used to heavily muted public criticism, accompanied by private bargaining and in-fighting. Furthermore, public concern has been aroused by international incidents, repeated allegations of health risks around high-temperature incinerators, and a succession of inquiries and reports. Local public opposition to landfill sites has thus far not coalesced into a national movement, but with bodies such as Greenpeace now taking an interest, and with waste imports rising sharply in the mid-80s, this could change. The House of Lords Gregson Report warned against the complacency that is sometimes a companion of informal collaborative networks:

> There is a belief that because in this country there have been comparatively few major incidents, we can, without too much effort, maintain this position in the future ... This complacency is, at times, tinged with the arrogance that "We know best", and has probably given rise to a very serious loss of public confidence in the whole activity of waste disposal [73].

In these circumstances the balance of strengths and weaknesses of the UK institutional approach may tilt away from the flexible, informal, and trust-based framework. Whatever the actual quality of environmental end-results of UK hazardous waste management compared with other countries, it is an important problem for the *credibility* of regulation that there are little or no overall points of reference by which legitimately interested, nonspecialist "third parties" such as environmentalists or other public organizations, committees of inquiry, labor unions, international regulatory bodies, such as the EC (and we would have to include as outsiders, partly, even local WDAs), can obtain a clear view of the whole system of regulation and its rationale. Such nonspecialist bodies cannot have the degree of involvement that understands variations of risk management from one situation to another. They need to see consistent and clear decision rules, but in the UK these do not exist.

Another consequence of the collaborative, informal social networks of UK regulation is that it cannot institutionalize a more investigative, critical role as part of the regulatory system itself. For example, what in the USA would be regarded as normal "critical research" on the scientific analysis of PCBs in illegally dumped electrical capacitors, was performed in the UK by an investigative national TV program (BBC Newsnight), rather than by regulators or scientific groups in the normal system. A more resilient approach to public credibility of regulation would have this critical expert "watchdog" role built into regulatory institutions and processes themselves, rather than leave it to the inevitably somewhat haphazard polarized and *ad hoc* style of the media.

Therefore, a question against UK flexibility in technical criteria and institutional rules is whether it can reconcile itself to the growing needs of external accountability and corresponding demands for synoptic, precise standards and formal controls. One can argue that these are more symbols for public reassurance than effective regulatory tools, but the fundamental conflict is there, and its terms are changing.

It may well be also that even if they could be extended to include the local authorities (who are, after all, the main regulators), the benefits of a relaxed and flexible institutional system incur the associated cost of being unable to move the focus of waste management further upstream in the hazardous waste life-cycle, to preempt many later uncertainties. UK dependence upon private commercial T&D, with no specific restrictions other than the special waste trip-ticket, and back-end site licensing, leaves a marked lack of overall coordination between waste arisings and the provision of appropriate T&D facilities. Although the low-cost "flexibility" of landfill control has allowed this system to work, it has created difficulties for high-technology T&D. Whether the free market can provide adequate alternative T&D capacity if landfill becomes more expensive, controversial, and restricted is open to question.

The UK already focuses more strongly than most upon the very back-end, namely T&D site licensing, rather than on the point of waste arisings. Whereas some countries have created incentives to reduce waste arisings by increasing T&D costs all-round, UK landfill prices have been actually falling, undermining even the high-technology back-end T&D options, let alone providing any upstream pressure to recycle or reduce waste arisings. Thus, one report has concluded that:

> ... there have been three EEC Directives on waste [74] placing emphasis on recycling, but judging by their impact on practice in the UK they seem to have been totally ineffective ... In several other countries, e.g., Denmark, FRG, France a strong commitment has

> developed and this is enshrined in legislation and encouraged by appropriate financial measures to increase the amount of waste recovered ... Of all major EEC states, the UK has stood out in its diffidence in this area.
>
> ... there is no substantial or continuing commitment to these goals of the kind that has emerged elsewhere in the Community, nor is there any real prospect of such a commitment developing in the near future [75].

Of course, differences of view exist in this point as on all others. The UK's reclamation industry claims to be an important wealth creator in its own right [76].

The UK approach offers no chance of a coherent management of *wastes* from the production point (as would occur in the rejected National Waste Register and in more elaborate regulatory waste lists and treatment designations). The site based approach instead creates *de facto* pressure to stretch site license "restrictions"; rejection simply leaves a waste "out in the open" again, out of effective regulatory control until it appears at another site. (The trip-ticket notification system, in addition to being restricted to special wastes, is not a real-time regulatory mechanism.)

A further ramification of site licensing dependence is the express reliance upon the "sensible" landfill of hazardous wastes. The *scientific* justification of policies seriously underestimates the importance of analyzing the social realities behind such easily glossed-over conditions like "sensible". If institutional realities make the conditions nonachievable, then the "scientific justification" is reversed, and the "optimal" policy turns out to be a bad one. The dependence on landfill is extremely vulnerable to potential local public opposition, inspired by evidence that "sensible" landfill may indeed be socially unenforceable. Thus, what appears to be an optimal *technical* regulatory design in view of the structural properties of the hazardous waste issue, because it tailors risk assessment to local situations, may be violating the narrowed margins of error because the embedded social assumptions of ideal situational *management* are not realistic. The evidence points in this direction.

How a regulatory system handles its basic institutional uncertainties and conflicts is important even if its material environmental and economic effects are difficult to compare with other systems. To a significant extent, if institutional uncertainties are left to fester they breed a lack of confidence, which leads to corrosion of the climate of self-regulation and "independent" compliance that have to be nurtured in any system. The British institutional economy of flexible, mutual regulation via informal relationships, tacit norms, and imprecisely formulated rules is envied by some observers from opposite conditions, notably from the institutionalized mistrust and adversarial culture of the USA. Informal networks of institutional confidence, however, tend naturally to create boundaries of exclusion, even if also not formalized. These are becoming significant in the UK, and new balances between institutional distribution of powers and technical strategies may need to be evolved. As an example, a more determined and coherent management of waste *arisings* would take some pressure off local site licensing, with its attendant enforcement problems (and rising public-acceptance difficulties), whilst still allowing flexible technical risk definitions to be retained, but perhaps implemented by larger, regional WDAs.

Since the WDAs find it technically and institutionally impossible to examine and classify wastes adequately at their point of production, they are forced onto the very back end, namely the disposal site. Thus the "barrier" philosophy of risk management prevails, where the emphasis is on selecting a good disposal site; once this is achieved it is thought that what goes in is not very important, because (it is thought) a good site will deal adequately with virtually anything. Thus upstream waste discrimination and direction (more like probabilistic risk analysis and risk-ranking) comes to be seen as unnecessary.

Whilst there is some technical justification to the "barrier" approach, especially since domestic wastes are known to produce leachates as hazardous as many special wastes, it is difficult to justify over-reliance upon it. Analysis of the UK case creates the distinct impression that this technical approach is inspired as much by the *institutional* realities of WDA impotence to intervene further upstream in waste life-cycles, as it is by carefully and critically developed technical knowledge of site assessments and long-term risk potentials. It is also likely that although at present the ambiguities (described in section 7.3.1) in special waste definitions do not give rise in practice to extensive technical disagreements, this may be due to the fact that waste producers privately recognize that the present approach suits their interests, and, in this context, the costs of being generous in giving the benefit of the doubt over special waste definitions to regulation, are less than those that would be involved in a change of regulatory focus to further up-stream. No doubt more stringent upstream regulatory attention to waste "arisings" would also generate more conflict as to their precise legal definitions and technical risks. But more concentrated institutional resources would probably be able to negotiate these with industry without undue weakness, and without resort to excessive technical standardization and rigidity.

Notes

[1] House of Lords Select Committee on Science and Technology, Chairman: Lord Gregson (1981), *Hazardous Waste Disposal*, 3 vols (HMSO London). This quote is taken from Vol. I, p 28.

[2] *Ibid*, Vol. III, p 27.

[3] For example, Welsh Office (1985), *The Incidence of Congenital Malformations in Wales, with particular Reference to the District of Torfaen, Gwent*, review of notifications made to the Office of Population, Censuses & Surveys (Mid Wales Litho Ltd., Griffithstown, Pontypool). See also Denny (1984), *Bonnybridge Report* (Scottish Office, Edinburgh). House of Lords, *op.cit.* [1]; UK Hazardous Waste Inspectorate, 1st Report (June 1985), *Hazardous Waste Management: an Overview*; and 2nd Report (July 1986) *Hazardous Waste Management: Ramshackle and Antediluvian*, both London, UK Department of the Environment; UK Royal Commission on Environmental Pollution (Chairman R. Southwood), 11th Report (1985), *Managing Waste: the Duty of Care*, London, HMSO. See also Haigh, N. (1986) *Comparative Report: Water and Waste in Four Countries, A Study of Implementation of the EEC Directives in France, Germany, The Netherlands and United Kingdom*, Graham & Trotman, London; and Lawrence, D. (1987), The Department of the Environment: Ramshackle or Antediluvian?, *Waste Management*, **77**, 28–34. The description of UK waste management as ramshackle and antediluvian was from Sir Richard Southwood, at a press conference launching the Royal Commission's (1985) Report.

[4] The Public Health Act, 1936.

[5] The Refuse Disposal Act, 1978.

[6] Deposit of Poisonous Waste Act, 1972.

[7] Control of Pollution Act, 1974.

[8] Department of the Environment (1976), *The licensing of Waste Disposal Sites*, Waste Management Paper No.4, pp 52–58 (HMSO, London).

[9] National Association of Waste Disposal Contractors (1982), *Trade Directory*, p 6 (Sandwell, London).

[10] Bailey, G. and Hawkins, R. (1983), *The Future for Rubbish – Waste Management Options for the Environment Reviewed*, p 10 (CPC, London).

[11] *Ibid*, p 9.

[12] Hazardous Waste Inspectorate (1985), *op.cit.* [3], p 7, 3.2.1.

[13] *Ibid*, p 21.

[14] Department of Environment (1985), *Report of a Review of the Control of Pollution (Special Waste) Regulations 1980*, Annex 2.2 (DOE, London).

[15] UK Royal Commission on Environmental Pollution 1985, *op.cit.* [3].

[16] HWI (1985) *op. cit.* [3], p 14; Bentley, J. (1982), Controlled disposal of special wastes, *Chemistry & Industry*, 17 April.

[17] *Op. cit.* [3], p 15.

[18] *Op. cit.* [3], p 23.

[19] *Op. cit.* [3], p 35, 6.1.13.

[20] *Op. cit.* [3], p 35, 6.1.14.

[21] *Op. cit.* [3], p 27.

[22] *Op. cit.* [1], Vol II, p 92.

[23] EEC (1978), Council Directive on Toxic and Dangerous Wastes, 78/319/EEC, *Official Journal*, L 84, March.

[24] *Op. cit.* [7].

[25] Dumping at Sea Act, 1974.

[26] Health and Safety at Work, etc., Act, 1974.

[27] Alkali, etc., Works Regulation Act, 1906.

[28] Town & Country Planning Act, 1947.

[29] *Op. cit.* [8]

[30] Shiels, A.K. (1982), *Legislation on Hazardous Wastes* (DOE, London).

[31] Department of Environment (1976–1981), Waste Management Papers Nos 1–23, (HMSO, London).

[32] Bentley, J. (1981), *The Philosophy behind the UK DOE Waste Management*, paper presented at the US National Solid Wastes Management Association Conference on Waste Technology, Boston, Mass., October 19–21.

[33] *Op. cit.* [9].

[34] Ministry of Housing (1961), *Pollution of Water by Tipped Refuse*, Joint Report of the Ministry of Housing and Local Government Technical Committee (HMSO, London).

[35] Key, A. (1980), *Disposal of Solid Toxic Wastes*, Technical Committee on the Disposal of Solid Toxic Wastes, 1970, (HMSO, London).

[36] Lord Ashby is a leading environmental policy actor in the UK. He was Chairman of the 1971 Royal Commission on Environmental Pollution and an active member of The House of Lords Select Committee on Hazardous Waste Disposal, 1980–1981.

[37] UK Royal Commission on Environmental Pollution (1971), *First Report*, (HMSO, London).

[38] Ashby, Lord (1981), House of Lords Debate on hazardous waste disposal: select Committee Report, *Hansard*, pp 440–441 (HMSO, London).

[39] *Ibid.*

[40] *Op. cit.* [30].

[41] The HWI discusses this particular example. *Op.cit.* [3], 1986.

[42] *Op. cit.* [14].

[43] *Op. cit.* [1],Vol. 1, p 32.

[44] *Ibid.*, p 33.

[45] *Ibid.*, p 27.

[46] *Ibid.*, p.29.

[47] *Op. cit.* [1] and [14].

[48] Department of Environment (1978), *Cooperative Programme of Research on the Behaviour of Hazardous Wastes in Landfill Sites*, p XV. Final Report of the Policy Review Committee (HMSO, London). Also, evidence to Gregson Committee, *op.cit.* [1].

[49] *Ibid.*

[50] *Op. cit.* [1], Vol. II, p 249.

[51] *Op. cit.* [1], Vol. I, p 62.

[52] Harris, R. and Lowe, D. (1984), Changes in the organic fraction of leachate from two domestic refuse sites, *Quarterly Journal of Engineering Geology*, **17**, 57–69; and see Environmental Data Services Ltd. (1985) *ENDS Report No 129*, 12–15, October, Southwell Press, Surrey, UK.

[53] Mills, D. (1983), *National Practices in Hazardous Waste Management in Western Europe*, paper presented at the Society of Chemical Industry, Water, Environment Group Meeting, January.

[54] Schmitt-Tegge (1979), *Treatment Disposal of Industrial Wastes in the FRG*, ACS/CSJ, Chemical Congress, Honolulu, April.

[55] Lehman, J.P. (1974), Federal program for hazardous waste management, *Waste Age*, (Washington, DC).

[56] *Op. cit.* [8].

[57] Waste Disposal Engineers Association (1983), Licensing and enforcement policy, *Waste Management*, vol 17, pp 183–194.

[58] HWI (1985) *op. cit.* [3], p 57.4.

[59] *Ibid.*, p 58,12.

[60] *Ibid.*, p 28, 5.3.11.

[61] *Ibid.*, p 28, 5.3.13.

[62] *Ibid.*, p 28, 5.4.2.

[63] *Ibid.*, p 29, 5.4.6–7.

[64] *Op. cit.* [14], note 11.

[65] *Ibid.*

[66] HWI (1985) *op. cit.* [3], p 23, 5.2.2–3.

[67] Personal communication, April 1983.

[68] HWI (1985) *op. cit.* [3], p 25, 5.2.13.

[69] Marsh, R. (1981), *Do the Regulations Require Amendment: The Industry View*, paper presented at a Seminar on The Management of Toxic Waste from the Cradle to the Grave in London. (Gemprint, Leighton Buzzard, Bedfordshire).

[70] *Op. cit.* [14], note 35.

[71] Department of the Environment (1981), Waste Management Paper No. 23, London, HMSO, p.17.

[72] *Op. cit.* [3].

[73] *Op. cit.* [1].

[74] EEC (1975), Council Directive on the Disposal of Waste Oils, 75/439/EEC, *Official Journal* L 194, July; EEC (1976), Council Directive on the Disposal of Polychlorinated Biphenyls and Polychlorinated Terphenyls, 76/403/EEC, *Official Journal* L 108, April; EEC (1978), Council Directive on the Disposal of Toxic and Dangerous Waste, 78/319/EEC, *Official Journal*, L 84, March.

[75] Environmental Data Services Ltd. (1983), Progress in resource recovery: an EEC perspective, in *ENDS*, Report 100, p 14–16, May, Southwell Press, Surrey, UK.

[76] *Op. cit.* [10].

CHAPTER 8

Hazardous Waste Management in Hungary

Ernö Kiss

8.1. Introduction

The Hungarian system of waste management is underdeveloped, largely due to starvation of capital and other resources required to actually build the facilities. There is, in fact, an active public awareness of the need for regulation and a sophisticated appreciation amongst regulators of the technical needs and possibilities. Given that communication is normally not well developed between East and West European countries, a main aim of this chapter – more prominent than for the others – is simply to provide data, and to build a descriptive account of what is happening, in technical and institutional terms, in the Hungarian context. Given its relatively open social and economic interactions, especially with Austria, Hungary is an important conduit of information between the countries of Eastern and Western Europe. We first give an overview of the present situation, then trace the emergence of policy perceptions and responses to the issue, before going into more detail about the procedures and regulations being developed. We then give an account of some of the problems encountered in Hungary, before finishing with a more comparative discussion.

8.1.1. An Overview of the Hungarian Waste System

To provide some perspective we give here some rough figures for the various types of waste produced in Hungary. The usual uncertainties apply, as outlined in Chapters 3 and 9.

Institutionally, the Hungarian system distinguishes fairly sharply between municipal and/or domestic wastes, production wastes, and agricultural wastes. Although in practice these waste life-cycles interpenetrate a great deal, unlike most Western countries they are formally controlled by different bodies. The Ministry of Housing and Urban Development handles municipal wastes; the National Council for Environmental Protection and Nature Conservation (State Office) is responsible for hazardous industrial wastes, and the Ministry of Agriculture and Food for most agricultural wastes.

Solid municipal wastes amount nationally to 14 million m^3. Less than half of this was estimated in 1982 to be collected in an organized manner, the rest being "discharged into the environment in a manner impossible to control" [1]. How much of this waste is hazardous is unknown, but soil and water pollution is thought (from local monitoring) to be "considerable".

Liquid municipal wastes (i.e., mainly sewage) have to be transported to sewers when no sewage disposal systems exist, under fragmentary organizational control. No figures are available, but significant volumes of liquid industrial wastes are discharged into sewers.

The annual quantity of production wastes (from mining, building, and power production wastes) is estimated to be 19.5 mte, of which 7–8 mte are estimated to be bulk mining wastes, mainly harmless. About 5–6 mte are thought to be hazardous, based upon voluntary declarations by waste generators. Half of the latter figure consists of "red mud" waste from aluminum manufacture. Owing to the huge quantities involved, this fluoride-containing waste is managed by simply dumping at or around the site of production, a practice known to have caused pollution problems. Of the remaining 2–3 mte of hazardous industrial wastes, between 0.9–1.9 mte are regarded as sufficiently dilute or in such physical condition as to be ecologically harmless, even if in theory they arise from processes that may generate hazardous wastes. This leaves about 300 000 tonnes that are classified as genuinely hazardous industrial wastes, under the definitions in existing regulations – the 1981 Initial Act (see below).

However, this annual figure of industrial wastes that require special treatment and disposal facilities (T&D) is doubled by the generation of "infectious wastes" from the food industry, hospitals, etc. Ideally, these require expensive (and largely unavailable) treatment by incineration. The overall annual generation of about 1.2 mte of infectious wastes (mainly of animal origin) has been reduced by processing in the network of 9 factories under the National Enterprise for Producing Animal Protein Feeds. However, lack of full capacity and residual wastes leave about 300 000 tonnes of hazardous *infectious* wastes to be dealt with under the national hazardous waste regulations. Thus, the annual total of hazardous chemical and infectious biological wastes from production is about 600 000 tonnes.

Agricultural wastes overlap with food industry wastes (e.g., infectious animal carcasses) and chemical industry wastes (e.g., pesticides). About 20 mte of organic solid wastes and 42 mte of liquid manure are thought to be generated each year. Although these can produce highly toxic run-off liquors, neither is incorporated under regulation. Surplus pesticides that are "stored" by agricultural businesses and even at regional agricultural inspection stations have caused a control problem of "considerable quantity" [2], though there are no known figures for the volumes involved.

Of the 600 000 tonnes or so of hazardous industrial wastes, about 120 000 tonnes fall into the category designated as requiring high-temperature incineration, about 280 000 tonnes into that category requiring intermediate chemical treatment and/or controlled landfill, and the rest into the category for recycling (theoretical, but as yet not actual [3]). In practice, as we describe later, an infrastructure of proper T&D facilities – even of controlled, properly designed landfills – is not yet available, though it is now being created. At present it is estimated officially that about 15% of legally defined hazardous wastes are properly treated and disposed of according to the regulations, mostly by in-house incineration, e.g., at oil refineries. Because the landfill regulations, established in 1984 as part of the 1981 Initial Order, are stringent, no legal landfill sites yet exist in Hungary. Pending their construction, temporary storage sites (for up to 10 years) have been designated in relatively less sensitive areas (e.g., away from watercourses), to reduce the effects of uncontrolled landfill.

8.1.2. The formal institutional structure

In addition to these figures an outline of the formal organizational structure of regulation and related decision making is valuable, and given in *Figures 8.1* and *8.2*. *Figure 8.2* is an enlarged view to indicate a point, sometimes not appreciated, about centrally planned economies, namely the extensive institutional pluralism in the relevant decision-making field. Indeed, in many respects the problems caused by potential organizational fragmentation could be greater here than in more "decentralized" systems.

The two main regulatory bodies are the Ministry of Health and its regional inspectorates, and the State Office of the National Council for Environmental Protection and Nature Conservation (OKTH), which has a hazardous wastes department, with regional inspectorates.

The regional inspectorates are primarily responsible for checking compliance with the Initial Order, and naturally give their attention to the local environment and communities. In the case of hazardous waste regulation they monitor more directly the waste producers, rather than the environmental media. The Ministry of Health's longer-established provincial public health and epidemiology services are less directly authorized to check the

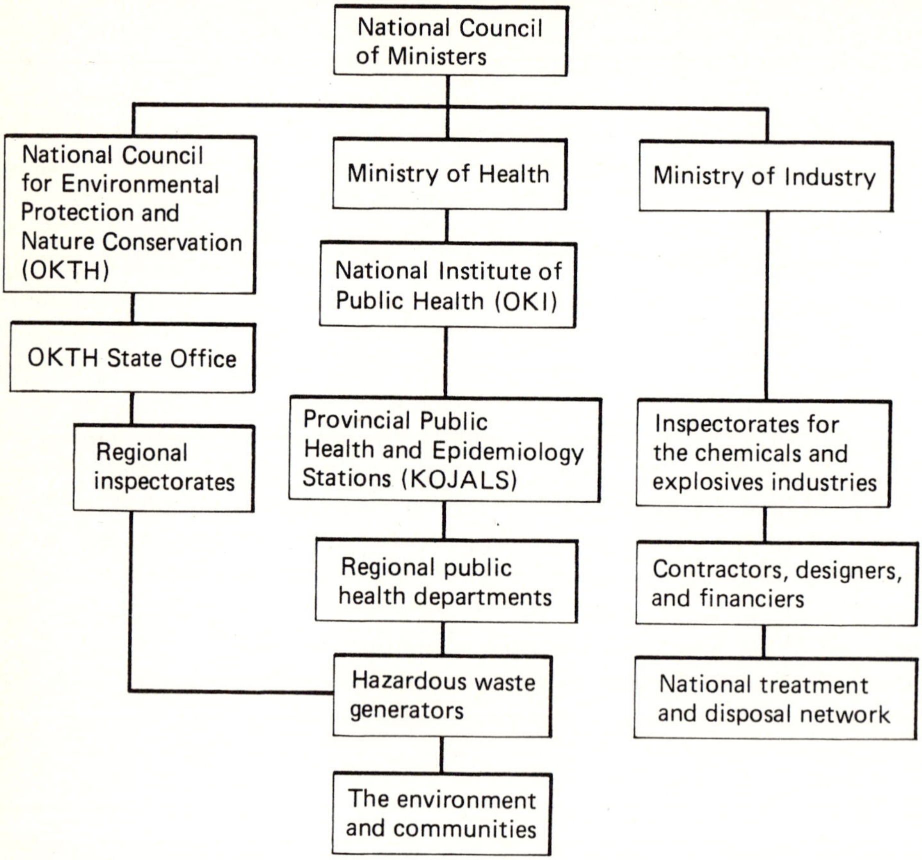

Figure 8.1. The directly relevant framework for the regulation of hazardous wastes in Hungary.

waste producers. Their primary task in relation to hazardous wastes is to conduct initial official investigations of environmental areas and communities found to be contaminated with hazardous wastes.

These two regional services usually cooperate, if not officially, in a concrete way at the local level. The effectiveness of this cooperation derives primarily from the personal relations between the local workers of the parallel regional services. There appears to be little conflict or competition between these regulatory agencies, so this loose-knit, informal collaboration apparently creates no extra problems for waste producers. At the national level a joint Council of Waste Coordination exists to promote the cooperation of all the relevant organizations and agencies.

The institutional picture is described in more detail in the rest of this chapter; the wider framework is shown in *Figure 8.2* to indicate the

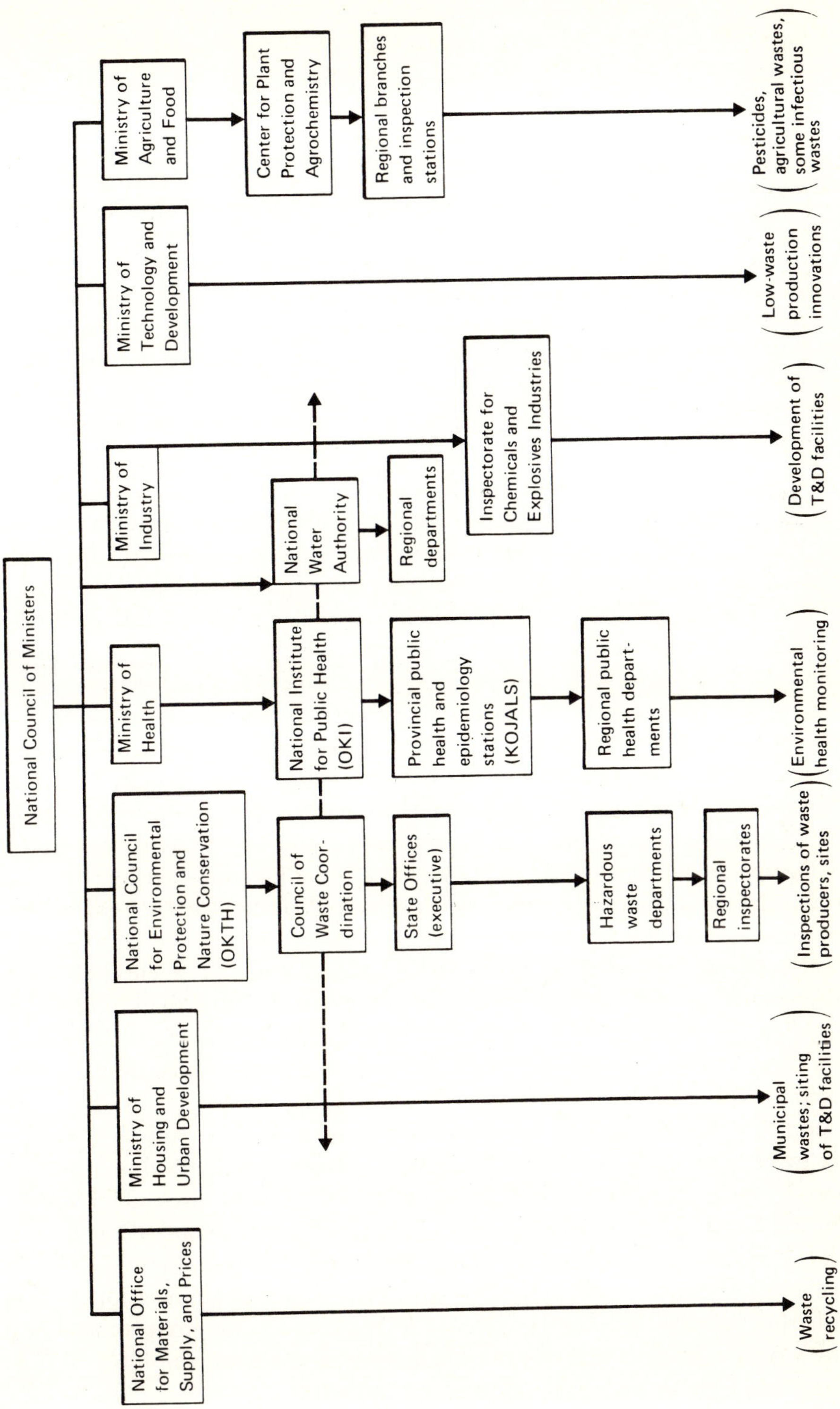

Figure 8.2. The wider institutional framework of waste regulation in Hungary.

distribution of responsibilities among the bodies at national level for, e.g., agricultural wastes and the creation of recycling and low-waste production technology innovations.

8.2. Historical Emergence of the Issue

A comprehensive program for modernizing industry was launched in Hungary after World War II, followed in the late 1960s by the intensive development of mechanized agriculture. As a result the industrial and agricultural infrastructure has developed considerably over the last 30 to 35 years. This extensive development of the economy included a tremendous growth of the chemical industry, which increased production almost 34-fold between 1950 and 1980. But this growth has imposed increasing burdens on the environment, including a considerable expansion in the types and amounts of hazardous wastes being produced. This output is expected to grow at about 3% per year for the foreseeable future, despite policy commitments to low-waste technologies. Since the late 1960s, the problem of environmental and health hazards due to hazardous wastes that emanate primarily, but not only, from the chemical industry has become an important issue.

Since the early 1970s, several major cases of environmental pollution from hazardous wastes have been discovered. Other cases of inadequate pollution control (especially nitrate contamination of drinking water, which has caused the use of bottled water for babies in some parts of Hungary) have amplified public sensitivity to waste management generally. The first case to cause concern was that of toxic waste sludge produced by an automobile and truck works. It was found that heavy metal salts and cyanide wastes had contaminated the subsoil and had spread uncontrolled beneath a waste dump. In another case, in Tatabánya, cyanide that leached from a waste dump into a nearby reservoir was detected by health authorities. In other cases, arsenic contamination was reported in areas of the Danube and Tisza Rivers, which resulted from the unsafe storage of a catalyst at a chemical plant. In addition to these and other domestic incidents, publicity surrounding studies on the potential dangers of hazardous wastes abroad, mainly from the USA and the FRG, helped to call the attention of policymakers in Hungary to this issue.

In 1971 the Hungarian National Institute of Public Health (OKI, a research-oriented division of the Ministry of Health) launched a study of industrial wastes. This study was carried out by ten provincial public health and epidemiology stations (KOJALs), with the cooperation of 140 industrial companies, selected as being representative (see *Figure 8.1*). In assessing the data, the Ministry of Health elaborated guidelines for storing and treating wastes that are considered hazardous to the environment and human health. At this time these guidelines did not have the force of law, but were intended as recommendations for industry.

Before the establishment of the executive State Office of OKTH in 1979, only the Ministry of Health had any responsibility relating to the risk management of hazardous wastes. However, its control was weak and indirect, since its main responsibility was general public health. The Ministry focused its monitoring and research on human populations and its control activities could not extend far beyond this point. Unless specific health problems could be associated causally with an industry's waste (as happened later at Vác, see below), it had no mandate to examine industrial waste-disposal practices. Nevertheless, perhaps because of the general political and cultural climate of industry–government collaboration in a noncompetitive economy; because it was not seen by the cooperating industries as a potential "policeman"; and perhaps because at this stage there was no public climate of concern and social uncertainty, OKI was given full access to "internal" industrial information. The "pilot" survey was followed up between 1978–1980 with a much enlarged exercise, covering 3 300 plants, involving 20 regional KOJALs. However, this later study was performed in the context of proposed formal legislation and control, and some have suggested that this new climate created a relationship that reduced the accuracy of the data supplied.

The larger waste survey was paralleled by surveys conducted by other Ministries, and led to regulations in 1981 which, among other things, defined further, more precise waste and production-process information requirements from industry. This is described more fully below.

At least among the upper reaches of society, well articulated awareness of environmental problems began as early as in other countries. In 1974 the Hungarian Popular Patriotic Front (PPF) held a well attended and widely publicized meeting on environmental problems, which led to several policy initiatives.

Subsequently, environmental problems in general were addressed by high-level, comprehensive policy initiatives on environmental protection. This resulted in the enactment of Law II/1976 on the protection of the human environment. This law provided the general framework for further legal regulations on environmental protection, but it was not then specific enough to act as a means for the effective regulation of hazardous waste disposal.

Even after 1976, however, producers could still deposit and "dispose" of wastes (some later classified as "hazardous") legally at their own discretion outside the premises of the company – unless the manner of disposal violated other public health regulations. Often such wastes were disposed of with municipal garbage or were "stored" indefinitely by the plant. Several pollution incidents were connected with such "storage", which effectively became long-term dumping. Even as late as 1982 an official joint paper from the Ministry of Health and OKTH [4] observed that:

> ... enterprises dispose of their own hazardous wastes at municipal waste dumps or their own dumps discharge them into public sewers or incinerate them under uncontrolled conditions, in the open air or in boilers not designed for the purpose. It is not unusual for wastes to be discharged in unknown places under conditions that cannot be controlled.

This lack of special treatment for hazardous wastes was due mainly to a lack of direct legislation that gave authority to inspect plants and waste movements. Fundamentally, there was little planning when designing industrial facilities for the treatment and reutilization of the wastes produced. The heavy political emphasis on the need to increase industrial output has discouraged consideration of waste problems in the industrial planning phase. Only a small number of companies have invested in their own waste disposal technologies; e.g., an oil refinery voluntarily installed an industrial incinerator. In other cases, incinerators of smaller capacity were built by the Chinoin Pharmaceutical Works and the Nitrokémia Chemical Works. But apart from these sporadic examples, industrial wastes of unknown quantity and composition appear to have been continuously deposited in the environment, i.e., in ditches, gullies, and other dumping sites, including untreated waste dumps at industrial sites.

In 1977, in response to unresolved environmental problems and rising public awareness, the Council of Ministers established OKTH as an advisory body, and the State Office as its executive body. The OKTH State Office was granted sole authority for developing regulations on air pollution and for nature conservation, and shared responsibility with the Ministry of Health for noise and hazardous wastes.

Law II/1976 on environmental protection, the basis for the foundation of OKTH and its State Office, only very gradually provided a statutory framework for the regulation of industrial wastes. One problem was that the traditional appreciation of nature in Hungary (reflected in the influence of the PPF) had not, in the 1970s, been translated into the creation of minimal pollution standards in the environmental media (except for drinking water). There was thus no *general* control framework to restrain toxic waste dumping. The dominant principle of concern reflected in "environmental" regulation policies was public health, which is one stage further downstream from the control of production and wastes than the installation of general environmental media standards.

Nevertheless, with the information obtained from its earlier survey, OKI called attention several times in the 1970s to the need for formal control of hazardous wastes. However, these demands foundered for several reasons: the relevant department of the Ministry of Health was being reorganized and anyway had only an indirect connection with the issue; it contained medical expertise, but no industrial chemistry expertise; and there was only fragmentary and circumstantial evidence for causal connections between toxic waste and local health problems.

OKTH, although being planned, was not yet in existence, and the Industry Ministry was more concerned about problems of increasing output. In the face of the very strong and coherently organized executive political concern to increase industrial production, environmentalist groups, though certainly active, were institutionally fragmented, dispersed, and had little scientific or other evidence to use as a resource to alter the balance. The ministry that guided and commissioned the second OKI survey, for example, was not a Health or Environmental Ministry, but Housing and Urban Development, which had responsibility for municipal (nonindustrial) wastes. However, in 1979 a disastrous situation, arising from improper waste disposal in the small town of Vác, stimulated urgent action not only to deal with that specific problem, but to produce more general remedies as well.

8.3. The Case of the Chinoin Plant in Vác

The Chinoin Pharmaceutical Works, near Vác, used one site to store raw materials and by-products, but from 1952 onward the company also used the same site to dispose of wastes, partly by open-air burning and partly by landfill. In 1977, after recording high levels of water and air pollution, the county council and the local water authority outlawed both disposal practices because of the contamination threats they posed to drinking water supplies. However, the accumulation of wastes in "storage" continued, in the absence of any practical alternative. Toward the end of the 1970s, in certain parts of Vác the drinking water became suddenly turbid, yellowish, and smelled of chemicals. Public concern was aroused throughout the town, and this was publicized in the media. According to the county public health department numerous illnesses with symptoms of vomiting and diarrhoea affected approximately 20% of the population. This situation continued for some time, since all initiatives to remedy it foundered in red tape. Finally, in several stages, a number of wells, primarily in the Vác-South water basin, were closed and have not been reopened. People began seeking other sources of drinking water; many local inhabitants who worked in Budapest brought home drinking water in cans for their families. Naturally, car owners had a distinct advantage, but car ownership was relatively low.

The local water and public health departments received many complaints about the quality of the drinking water but responded that it was safe. This was, in fact, true according to Hungarian standards; one county public health department official later admitted in a statement concerning the outbreak of illness that the reason for their reassuring replies to public concern was that they did not want "to create a panic"! Despite this frankness, they admitted that wells had been closed, not because they had definitely caused illness, but because one prevailing standard required that drinking water should not have any unusual taste or smell.

Because of the contamination, more than one third of the potable water supplies for Vác, Gödöllö, Dunakeszi, and 14 other small communities were temporarily cut off . The Institute for Water Quality Protection of the Water Management Research Center of the National Water Authority investigated the case (using mass spectrometry) and found that contaminants in the water included various solvents such as butanol, benzene, and toluene. Groundwater samples were taken from more than 20 observation wells drilled especially for this purpose. Knowledge of the Chinoin plant and other pollution sources in the area made it certain that most of the contaminants had originated from the plant. Experiments conducted with small mammal and human blood cell cultures from residents in the area confirmed detrimental health effects.

By spring 1980 the huge pit at the Chinoin works had been filled and was surrounded by approximately 12 000 barrels of various toxic, flammable, and explosive chemicals; many of the barrels were unsealed and unsuitable for transport. Through the intervention of the then Ministry of Industry, a license from the Ministry of Health was obtained for a one-off, open-air burning of the stored wastes, but the burning had to be discontinued because of high levels of air pollution detected by local monitoring instruments. The burning was later resumed, after some additional precautions had been taken. Chinoin has since been forced to export some of its more difficult wastes for T&D, using up limited foreign currency credits.

Many relevant authorities, institutions, and responsible individuals had been well aware for some time of the possibility that the water supply at Vác was being contaminated because the Chinoin waste site was large and well known; but no action was taken. Several weeks before the scandal broke, attempts had been made to clean up the Chinoin plant. The sudden move to dispose of the waste by burning had been made not only because of the fear of adverse publicity, but because the Council of Ministers decided to make a government committee and its chairman personally responsible for remedying the uncontrolled dumping and water supply pollution that had been going on. The public scandal engendered the creation of an *ad hoc* decision structure that was able to cut through the normal bureaucratic complexities.

8.4. Hazardous Waste Legislation

Partly as a result of the Vác scandal and using information from the enlarged 1978–1980 waste survey, in 1980 the OKTH State Office began to develop more precise legislation to address the problem of hazardous wastes, which at that time remained legally and technically undefined. This legislation was prepared by the OKTH State Office in cooperation with experts appointed by the Ministry of Health, which had conducted the two main waste surveys. The basis for this collaboration was provided by an order of the Council of Ministers, which conferred responsibility for dealing with toxic chemicals in the environment on the National Institute for Public Health (OKI) and on the local public health departments (KOJALs). In effect, the central body was responsible for planning and setting norms for waste disposal, whilst the local bodies were responsible for monitoring and enforcement. However, such matters as powers of inspection and analysis were left somewhat ambiguous. In November 1981 a law was passed by the Council of Ministers to deal specifically with hazardous wastes; this was the so-called "Initial Order".

The order began with an attempt to define hazardous wastes under legal regulation. It began with a general definition based upon waste types, drawing heavily upon the existing FRG waste catalog (see Chapter 6). The Hungarian version of this list was compiled by experts at the OKTH and the Ministry of Health. The decree enacting the 1981 Order by 1984 elaborated upon the general waste-type classification by adding a list of specific constituents, ranked in three hazard categories somewhat like the Dutch list. Then a five-fold set of characteristics tests was defined, which all doubtful or new wastes had to pass in order to be declared *non*hazardous. This latter "fail-safe" element is extra to the FRG approach, being more similar to the US one, although precise and standard laboratory protocols for sampling and testing could not be laid down. The enacting regulations also included specific norms for the proper T&D of given wastes, but because of a lack of actual facilities these norms have not been widely implemented as yet.

A draft of the Order was circulated among the ministries and state authorities concerned and a compromise order was finally worked out. There were some objections from the Ministry of Industry concerning the timing of the Order. As in other countries, they wanted to develop a T&D facility network before control legislation was imposed, but they were overruled. It was felt by a significant body of policy opinion that, especially in a situation of capital scarcity, investment in T&D facilities would only come about if "forced" by formal regulations that created a hazardous waste population to be managed by controlled T&D and if also supported by government R&D policies to develop industrial techniques.

In addition to governmental circulation, the initial draft was discussed in public, by industry, trade unions, scientific committees, committees of the Union of Technical and Scientific Associations, and by local branches of the PPF. However, only the opinions of government agencies had to be considered; the opinions of public interest groups did not.

An unexpected reaction of the companies concerned was a general sense of relief, because at least the Order eliminated a highly uncertain situation. They did complain, however, that their administrative burdens would increase with the compulsory reporting of their hazardous wastes, and that they would face additional costs in connection with the legally required controlled storage and neutralization of hazardous wastes. Nevertheless, the Initial Order was put into force on January 1, 1982, with further enacting regulations in 1984.

The Initial Order was supplemented by executive instructions from the Ministries of Industry, Agriculture, Housing, Transportation, Finance, and Defense. In addition, a regulation issued by the State Office gave a formula for calculating penalties for offenses under the hazardous wastes law. This formula took into consideration the nature of the violation, the amount of waste, and its hazard classification. The primary aim of the penalties – administered almost like normal taxes – was not to provide income for the budget (although the fines are paid into the account of the Central Environmental Protection Fund), but more to influence the attitudes of the companies involved. The increased expense incurred through improperly disposing of wastes and thereby polluting and endangering the environment, it was hoped, would act not only as an economic, but also as a social incentive to force the companies to comply with the regulations and to modify their processes or develop technologies for recycling wastes. The Order and its penalties were viewed not as punishments but as incentives.

The first fine actually imposed was in 1983. Most penalties are appealed by companies, and the legal procedure leading to actual payment of the penalty is quite lengthy. Based on present data fines totaling 160 million forints (approximately $4 million) have been assigned in 129 cases, but only Ft80 million have been collected. (For comparison, in 1982 a total of Ft210 million in fines for air pollution were paid.)

However, there are some legal loopholes that make implementation of the Order impossible in special situations:

(1) The Order does not apply to nonprofit organizations or to very small profit-making concerns.
(2) The sewage sludge from communal sewage plants is not regarded as a hazardous waste (although it is often hazardous).
(3) The Order stipulates that, in case of a direct and serious danger to the environment, the environmental protection authorities can close down the activity producing the hazardous waste. However, the ministry "controlling" (i.e., sponsoring and planning) the waste-producing company may override and cancel such an order for economic reasons.

Besides these legal loopholes, the small number of environmental protection staff also reduces the chances of implementing the Order. There are 16 000 plants that produce potentially hazardous waste in Hungary, and there are only about 30 to 35 inspectors. In the region of Budapest alone, for example, there are about 7 000 potentially hazardous waste-producing plants, but only three inspectors. With this ratio and an even inspection schedule, a given waste-producing site can be inspected personally roughly once every ten years!

In addition to the above problems, the correction and extension of the list of hazardous wastes is also an important question. It is obvious that structural changes in the economy and technological development are likely to produce additional hazardous wastes, not included previously in the initial list. Therefore, the Initial Order requires that each new industrial waste be regarded as hazardous until it can be declared non-hazardous. As in most other countries, the Hungarian regulations list properties and tests that define a waste as hazardous and thus formally "listed" for regulation. The waste producer is legally responsible for the risks from badly managed hazardous wastes, but the authorities are responsible for defining the waste as hazardous in the first place. In Hungary, a company that produces a new waste can apply for an exemption from the hazardous waste regulations. The Department of Public Health must then test the waste to either place it in one of the three categories of hazard in the list, or exempt it. Final decisions are made by the OKTH and the Ministry of Health. The characteristics tests include the usual measurements of pH, oxygen demand, flammability, leaching rates, heavy metal concentration, organic solvent content, microbiological tests, etc.

Before discussing the scope and problems of implementing regulations under the Initial Order framework, it is worth emphasizing that a central element of the Order is the information recording and analysis that it introduces. Unlike parallel exercises elsewhere, e.g., in the US, the UK, or the Netherlands, this does not focus upon waste arisings, subsequent movements, or final disposals. There is no "trip-ticket" registration system for wastes in Hungary, apart from the international conventions governing the general transport of dangerous goods. The information declaration system focuses instead upon industrial *production* processes and their waste arisings, and is part of a comprehensive national environmental information system that is being developed. The details of the information exercise are given later, but here it is worth stressing how this emphasis not only corresponds naturally with existing norms and administrative arrangements for gathering "internal" industrial data by government authorities (at least, documentary data; perhaps less so, direct inspection and analysis), but also corresponds with the ultimate aim of adapting production to low-waste technologies.

The Initial Order regulates four main issues: inspection of waste production; storage; treatment and disposal; and ways of directly preventing environmental pollution.

8.5. Inspecting and Recording Hazardous Waste Production

According to the 1981 Initial Order, each hazardous waste generator must complete a questionnaire that requests the daily and annual amount of wastes produced, their chemical types, and the method of collection, storage, pretreatment, and neutralization. The data required to be reported were selected so as to meet what the OKTH State Office defined as the "minimum" information needed to protect the environment. The aim behind the data definition is a complete materials balance of the processes involved. In practice, this is a very detailed data exercise if properly conducted, the questionnaire containing over 100 questions. For example, companies are required to detail which section of their plant gives rise to which particular wastes, and an associated "technology tree" describing the plant and processes must be submitted with the waste details.

The data will eventually be stored on computer at the Institute for Environmental Protection. The completed forms are "controlled" by the regional inspectorates of the State Office, by the regional stations of the Institute, and, when needed, by the local department of wastes. "Control" here means mainly a "desk evaluation" of the plausibility of the data provided. Regulators check the declared waste types and volumes against typical production coefficients for that industry, so that material inputs can be checked for discrepancies. This is a fairly crude form of initial control, but it allows for more detailed follow-up analysis. On occasion, inspectors conduct random on-site examinations, check the methods of storage, and ask companies to provide accounts of material balances and the full quantities of wastes reported. These inspectors are chemical engineers, all of whom have at least a general knowledge of the processes involved. Inaccuracies are often caused by the fact that the companies are not yet familiar with methods for conducting waste-oriented material balances, so that the data are distorted in many cases.

To date, some 60% of hazardous waste generators have returned more or less adequately completed forms, but many of these have had to be repeatedly queried and checked, since the whole process has proved more complex, uncertain, and time-consuming than was first imagined. So far, only about 50–60% of the reported data (i.e., 30–36% overall) can be considered complete and ready to be put on computer. The reliability of the data is influenced by several factors, the most important of which are:

(1) Within the companies that produce the majority (according to some estimates, about 70%) of hazardous wastes, the accounting of materials and technological descriptions are usually focused on end-products, so that the type and quantity of wastes produced in intermediate stages and in final waste streams (which are often combinations)

cannot be readily identified. Usually, only subjective estimates are provided.

(2) Sludges, which are a common physical form of wastes, contain a variable proportion of water and only the solid portion may be dangerous. This water content, which may be as high as 90–97%, cannot be determined accurately.

(3) Variations in production processes due to normal fluctuations in conditions, management, and worker practices, etc., cause waste volumes and compositions to vary uncontrollably within surprisingly wide limits.

(4) Producers do not always continuously record production levels. Often records are made on a quarterly basis, so that details of any short-term changes are concealed.

In order to increase the reliability of the data reported, the OKTH State Office intends to influence the Ministry of Finance to change its method of accounting for production credits, so that the types and quantity of wastes produced can also be recorded and penalized.

Records are kept not only by the OKTH State Office, but also by the Inspectorate for Chemicals and Explosives Industries commissioned by the Ministry of Industry. Between 1978 and 1980 the Inspectorate conducted a separate study of a sample of hazardous waste producers, which included 337 firms on 670 sites. The data included the origin of the waste, amount and type of waste, and a description of its chemical properties (solid content, water content, flammability, toxicity, solubility, etc.). By extrapolation to the national level this survey estimated that up to 220 000 tonnes of hazardous waste are disposed of annually by deposition and up to 140 000 tonnes are incinerated, mostly in an uncontrolled manner. Compared with the variability of waste arisings figures for most market economies (where production data are more secret), this is surprisingly consistent with estimates from other surveys. It is an interesting point of cultural comparison, however, that Hungarian regulators do not believe they have adequate knowledge of waste arisings and the production processes that create them, even though they appear to have far better knowledge, at the waste production end, than Western regulators. Whereas the latter tend to accept this situation and concentrate on waste movements and final T&D site licensing, the Hungarians are making more efforts to clarify their understanding of how wastes are produced, to influence upstream processes so as to facilitate downstream regulation.

Hazardous wastes that are produced in Hungary and that are not accurately covered in the various surveys can be divided into four categories:

(1) Hazardous wastes that are produced in such large quantities that they were simply omitted from the list in the Initial Order for pragmatic

reasons. These included such wastes as red mud produced during aluminum production and the smelting of aluminum and mineral sludges.

(2) Hazardous wastes that are processed by specialized companies, e.g., animal carcasses.

(3) Hazardous wastes that are burned in small-capacity in-house incinerators neutralized chemically, bound by special procedures (e.g., embedded in bitumen or ash), transported abroad for incineration, or recycled for further use in production.

(4) Very hazardous wastes "temporarily stored" by the waste generators (the distinction between "temporarily stored" and "deposited" is often extremely difficult to make).

In order to improve centralized control of hazardous waste disposal, the State Office of OKTH has submitted a proposal to the State Planning Committee recommending the establishment of a national network for hazardous wastes T&D facilities. This network is discussed below.

8.6. Treatment and Disposal Facilities

In 1980, the OKTH State Office commissioned the Institute for Environmental Protection to design a network of T&D based on the existing data on waste generation. The plans were completed in 1980 and included the construction of five incinerators and the designation of eight new controlled landfill locations. In discussions of the proposal the President of the OKTH State Office and the Minister of Industry agreed that, since the majority of hazardous wastes are produced by industry, the Ministry of Industry should be responsible for organizing and financing the national network. The Inspectorate for Chemicals and Explosives Industries was charged with this work. Hungarian law follows the same principle as in Western countries, that the producer is ultimately responsible for the wastes it generates, though this liability can, in some circumstances, formally pass to a transporter or handler. In 1982, the State Planning Committee approved the establishment of a country-wide T&D network, although the problem of who would be responsible for its financing remained to be negotiated. In 1983, the OKTH State Office defined the technical standards for treating and storing of hazardous wastes.

In establishing this network several difficulties have arisen:

(1) In Hungary, there are very few geologically and morphologically suitable areas for landfills; most of the country is covered by porous sedimentary rocks, with only a few underlain by impervious clay. In addition, there are no abandoned salt mines, such as those in the FRG,

which could be used as controlled dumping grounds. Of the 51 potential locations tested by geologists, very few were found to be suitable.

(2) The regions that were found to be suitable are now being subjected to detailed geological examination. If approved, the Ministry of Industry Inspectorate responsible for the investment must apply for county council permission to use the sites, but at this level serious difficulties have arisen. County officials try to avoid, whenever possible, the establishment of what have been called "poison cemeteries" in their area, for the simple reason that they are afraid of losing their popularity. People are afraid of "poison cemeteries" because of the previous media debate and official prevarication that created confusion in connection with the water contamination at Vác. "Dangerous" wastes have evoked public uncertainty and fear, and the issue is very prominent in public perceptions of environmental risks in Hungary.

(3) Present economic difficulties in Hungary do not favor the establishment of a capital-intensive network that will put a heavy burden on scarce capital resources. This situation is aggravated by additional circumstances. It has already become evident in the design phase currently underway that the main units of the high-temperature incinerators can be acquired only by import from Western countries. Funds for establishing what is essentially a pollution control network can only be provided from financial sources normally used for technological development. There is, therefore, direct competition in the same budget between funds for new production technologies, which might entail *reduced* wastes, and buying clean-up technologies to manage wastes from existing industrial processes. Furthermore, because of peculiarities of investment in Hungary, parts of the proposed network are relatively more expensive than in the case of a foreign establishment of similar capacity. For example, in the FRG a landfill with an annual capacity of 70 000 tonnes could be operated by a staff of five, while one designed in Hungary would require 50 to 60 persons.

(4) Management of the waste plants is to be carried out by consortia of waste-producing industries helping to finance the plant, customers, and local authorities. However, industries in Hungary cannot be forced to invest in the waste treatment network, so that there is no state-enforceable guarantee of industry participation and financial investment. Also, the state agencies involved have taken a long time to negotiate their proportional responsibilities for capital financial input. The overall uncertainty and resource limitations have led to cutbacks in the planned network, even though this will mean a shortfall of T&D facility supply in relation to demand.

Some of these difficulties are illustrated by the following case study.

8.6.1. The siting of a landfill in Zsámbék

The relevant county council agreed to geological investigations in the vicinity of the village of Zsámbék, in the county of Pest (near Budapest) in an area considered suitable for a hazardous waste landfill. A layer of clay was found in one section of the area, which, with respect to both its permeability and size, was regarded as geologically the most suitable site in Hungary.

Local residents were officially informed by the local council of the intention to establish a "poison cemetery" at the site. Fears were expressed (partially due to confusion surrounding the Vác scandal) that the drinking water supply would become contaminated, and protests were made against the establishment of the site. The protests continued despite further tests that have proved that the clay layer would act as a natural seal, and that the groundwater under the clay does not, in fact, flow toward the village. In addition, Zsámbék's water supply will soon be provided from the Karst Water Basin, in the course of a coal-mining program soon to be completed. The county council decided, however, to deny the license for the site because of the continuing protests.

When the president of the OKTH personally exerted pressure on the county council, they "mobilized" the parliamentary representative of Zsámbék district, who interceded with the Minister of Housing and Urban Development. The Minister argued successfully that since the proposal had resulted in such social unrest, a suitable site should be sought elsewhere. At that time, OKTH specialists were confident that they would find another clay-layered site and they did not push to keep the Zsámbék site open by further negotiation and lobbying through other government processes. However, an alternative site is yet to be located, so this decision is now regretted. The PPF was highly influential in defeating the proposal. The proposed site was located in woodlands used for recreation, including hunting by influential people. The PPF enjoys support from such people, and it has been said that was able to exert strong pressure in informal personal networks at high levels.

8.6.2. The current situation

Because of these difficulties, the planned size of the national network has been reduced. From five incinerators and eight landfills, the State Planning Committee now hopes to establish one incinerator and three landfills. In February 1984, the situation was as follows:

(1) The network is designed to contain an incinerator with an annual capacity of 30 000 tonnes and three landfills with a total capacity of 40 000 tonnes. These would be established and operated by an association of 178 companies concerned.
(2) Of the investment costs, 45% would be provided by the companies, 50% by loans, and the remainder from the Central Fund for Environmental Protection of the State Bank for Industrial Development.
(3) However, even these greatly reduced estimates of investment costs are twice as large as the available resources. The cost estimates are also highly uncertain, since no final decisions have been made as to the actual technology for the incinerator or the landfills.
(4) According to the most recent – but still unfinalized – plans, the first landfill would start operations in 1986–1987, the second in 1987, and the third in 1988, while the incinerator would start in 1988–1989.
(5) This reduced network will not be accepted as final by the OKTH State Office. It still plans eventually to establish the entire network (i.e., eight landfills, three incinerators) and interprets the present plan as only a first phase.

8.7. Discussion

From the perspective of the IIASA comparative project, the following points are most interesting about the Hungarian case.

When one examines the evolution of the public issue and the first phase of policymaking, namely legislation, one finds that Hungary is little different from Western countries – indeed, many hazard criteria and other norms have been more or less directly adopted, e.g., from the FRG, and used in the Hungarian legislation.

However, the national institutional relationships between industry and government generally have a direct effect upon the focus and implementability of regulatory legislation. Hungarian industry is centrally planned and financially supported in particular ways and, especially in the larger industries, there is no market competition. However, there is no central state *dictation* to industry, which is generally free to manage its own affairs. Thus, industry may, for example, choose not to participate in the planned waste T&D infrastructure, which is meant to be financed and run by consortia. It is even free to export wastes, without permission, so long as it has the collateral business to pay the necessary foreign currency. Note that this is a less restrictive framework than those of Hessen, Bavaria, or Denmark.

There are, of course, economic incentives that are designed to encourage participation and good waste-management practice generally; examples are higher charges for customers who do not invest in the planned network's T&D plants, and the fines system. In theory, excessive fines should make it optimal for an industry to decide instead to pay less money to be a part-financer of the infrastructure or, at least, to pay to be a customer of proper facilities. The fines and incentives do not appear to have been effective as a direct means of tightening up waste disposal. This is so because the lack of competition, established state support, and the national commitment to avoid unemployment mean that an industry might avoid paying fines for lax waste mangement to one state agency, while obtaining a countervailing subsidy from another. It has been difficult to gauge the true extent to which this happens, but existing institutional arrangements (which may, of course, be defensible on wider grounds) appear to allow it.

It is important to note that "central planning" does not at all mean *monolithic* planning. Centrally planned systems cannot *necessarily* overcome the institutional realities of interdepartmental fragmentation much better than others can. For example, for arbitrary reasons, some issues propagated by local bodies (such as the KOJALs of OKI who pursued hazardous wastes as a problem) often fall between the attention frameworks of higher level bodies that could place the issues more clearly on the policy agenda. The relevant department of the Ministry of Health was being reorganized when scientists and local inspectorates began demanding policy attention for hazardous waste control. The OKTH was only just coming into being – "everywhere there were new faces, new roles, new people" – and no effective attention. Interest and responsibility fell between, rather than squarely upon, the Health, Housing, and Industry Ministries and their various institutes, to the extent that they performed different surveys of waste arisings when they needed such data. In the process of developing regulations and other strategies, in surveys and in implementation, there is a pluralism of agencies and interests, and therefore of negotiation and diversity. Whether there is adequate *incentive* throughout the system – legal or economic – to solve the problems, is not a question we can answer, but there is an interesting relationship between regulation and attitudes, to which we return below.

Overall, therefore, there is the paradox that this centrally planned economy has little formal *institutional* resources to implement an urgently needed national T&D infrastructure. There is also no guarantee that industry will use and thus protect the viability of the infrastructural investment once in place. This institutional problem interacts with and exacerbates the already severe investment capital shortages.

Despite these institutional shortcomings for dealing with the *back end* of waste regulation, however, some institutional features of the Hungarian political economy are potentially superior to those of the "market economies" for achieving the ultimate goals of waste reduction.

The various potential points for regulation delineated in the schema discussed in Chapter 3 were: industrial production (P), waste generation (W), transport (T_p), treatment (T_t), and disposal (D). The Hungarian system of regulation places its weight upon the first two, since, as a centrally planned economy, it already has the institutional arrangements and accepted cultural practices that allow easy state agency access to "internal" industrial process data. As we have seen, this is being developed to take into account hazardous waste factors more systematically and comprehensively. Furthermore, the *idea* of influencing production to meet new social goals and values is quite normal in the Hungarian context, unlike for Western nations, with no such general traditions and whose dominant cultural values and entrenched interests maintain that in a competitive environment production information is private property. It is argued, largely successfully, that it is a sacred industrial right to maintain this autonomy. The comparative situation can therefore be depicted as in *Figure 8.3*.

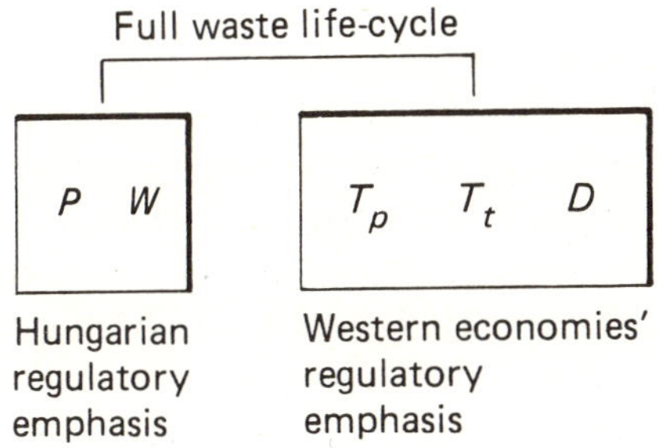

Figure 8.3. Comparison of emphasis in waste life-cycle, of Hungarian and most Western systems.

The disadvantage of the typical Western situation was discussed previously, namely that the most important, widely agreed as necessary mode of regulation (moving toward less waste-producing process technologies, or at least controlling the composition of unavoidable wastes, to encourage recycling) is the *least* accessible to regulation. The Hungarian institutional arrangements in principle overcome this problem, but in practice the financial resources needed to convert the information at this level into actual regulatory effects are in very short supply. Nevertheless, the system focuses attention at this more strategic point, even to the extent that there is no transport trip-ticket system in Hungary and, indeed, very few hazardous wastes are transported; most are "stored" or treated on-site.

The data compilation on production technology trees, material input–output balances, and waste outputs may, in the end, be less accurate than its elaborate framework suggests. Nevertheless, it appears that once available in computerized form it will, for the reasons already given, be much better than corresponding data from most if not all other countries. This is not only true of detail and accuracy, but also of the *focus* of the information – on production. Furthermore, a very real extra benefit may lie in the indirect effect of data gathering at the production process phase, in that it makes industrial managers more sensitive to the waste dimension in design, planning, and routine decisions, via the very exercise of producing such detailed data for the regulators. This attitudinal objective was often expressed to us during interviews.

The Hungarian case seems to substantiate the view that there are significantly different strategic definitions of hazardous waste management as a policy issue. In addition to the considerations above, in Hungary there seem to have been no scientific debates about specific risks of given compounds and wastes, and little formal risk analysis. The risk estimates implicit in other countries' hazard lists have been adopted, and government scientists have used informal judgments to decide whether waste descriptions from the industrial data surveys qualified the waste as a listed waste. These judgments have not been subject to formal review or justification. The problems of developing an industrial infrastructure have been far more pressing than the need to formalize and refine a risk assessment regulatory approach.

In order to make further advances much would now seem to depend upon the ability of the relatively young bureaucracy of OKTH to negotiate and bargain within and outside government, and upon the strength and clarity of public opinion on the need for national control (which means at least a national T&D network). It may well be that a scientific risk–benefit assessment approach will come to the fore to justify different positions in the severe competition for state resources that will figure centrally in the next phase of the issue in Hungary.

In Hungary as elsewhere, public concern and media debate about environmental risks is a very significant element in policy implementation, especially over siting. Although the economy may in some respects be less highly developed than those of many Western countries, public awareness is high, and the culture is sophisticated and articulate, with a strong tradition of independent journalism that sustains this.

The question of social attitudes and policy implementation is discussed in Chapters 11 and 12. It is clear from this discussion (particularly of the Vác situation) that in Hungary, as elsewhere, public outcry has had more effect and cut more bureaucratic knots more rapidly than the patient, private intragovernmental efforts of the Public Health Institute (OKI).

There is a broader, relevant political–cultural factor indicated in the references to attitudes and regulation; this cultural factor merits more attention. We noted earlier that in the socialist economies, the philosophy of directing production to reflect social goals and values is routinely accepted as part of the political–cultural landscape. Therefore, once the environmental value (and economic benefits) of preventive measures in hazardous waste management have gained a significant niche in government, the whole cultural context is already amenable to the principle of influencing industry's attitude.

The crucial element, that of a context of compliance and cooperation in production changes over and above formal regulations, is therefore already in existence, whereas the opposite is mostly true in Western market economies. The social assumptions underlying regulation in the latter situations (although there are variations) are that attitudes are autonomous and regulation involves setting up constraints against the free play of those attitudes. Even in the more "collaborative" setting of the UK, the philosophy of regulatory pragmatism dictates adaptation toward the regulated party's values, to try to find common ground from where "collaborative regulation" can begin. Marginal shifts of regulated parties' attitudes are the most that is conceivable.

In the Hungarian political culture, on the other hand, it is normal to assume that large-scale collective shifts of values and priorities are feasible as policy instruments and regulatory targets. Furthermore, it is clear that hazardous waste regulators see this not only as a shift in the culture and internal priorities of industrial production, but also of broader patterns of consumption. According to an official government report:

> ... neither the established technical–productive system, nor the predominant consumption patterns can be considered fundamentals. Every possible means available to us should be used with a view to reduce gradually the amount of materials becoming waste by transforming the above in a manner favorable to us ...
>
> There are two possibilities for reducing the quantity of wastes originating in consumption and commerce. One of them is the introduction of such new types of products whose basic materials after their use can be made fit for reutilization by simple means. The other is to influence the conscience, ways of thinking and attitude of the population so that as a result the speed of obsolescence of articles of consumption would slow down and only those materials would be discarded whose utilization is no more possible [5].

Such synoptic and ambitious "regulatory" goals could perhaps be more accurately described as cultural programs in the broadest sense. In Western market systems they would be rejected by most as paternalistic and authoritarian attempts at social engineering. Yet in terms of the expressed aims of policy everywhere – namely the revision of industrial innovation frameworks to include reduction of downstream hazardous wastes (as discarded products or as by-products of production) – these political cultural features and their institutional counterparts may be more realistic than those approaches that regard front-end production and technological innovation decisions as an autonomous sacred turf. It may be better to try to evolve systematic, measured ways of influencing this crucial area of decision than to have it influenced sporadically, inconsistently, and in private for the large companies who can afford the necessary R&D, through the unstable process of public reaction and threat.

To use the terms of existing discussion, Western regulatory systems seem to rely heavily in their public stance upon (formal) technical norms and languages, as if these will compensate for a receding and neglected (informal) moral and cultural context of shared values, and a will to compliance. The Hungarian system and collective planned systems generally seem to retain better the cultural climate of compliance, even if they do not have the practical technical and economic means to put it all into effect. It is probably easier to develop the latter than it is to recover the former.

Notes

Most of the information in this chapter was gathered by Erno Kiss through interviews and documentary research in Hungary. In addition, Brian Wynne and Joanne Linnerooth conducted a limited number of interviews with Hungarian policymakers. Attila Takacs, the Deputy Director of the Hazardous Wastes Office of the State Office for Environmental Protection and Nature Conservation, also visited IIASA for meetings, and subsequently communicated with Brian Wynne. Istvan Kiss, Pal Tamas, Anna Vari, and Janos Vescenyi also helped facilitate discussions and checked information for us. Because of language difficulties, an earlier draft by Ernö Kiss was revised by Brian Wynne and Mike Dowling, then agreed with Ernö Kiss. The discussion is by Brian Wynne.

[1] OKTH (1982), *The Conception and Stock of Requirements of Waste Management: Hungary*, p. 12 [National Authority for Environmental Protection and Nature Conservation (OKTH) and Ministry of Health, Budapest] (available from the Department of International Relations of OKTH, Box 33, Hungary 1531).

[2] *Ibid.*, p. 14.

[3] Takas, A. (1985), *Hazardous Waste Management Policy in Hungary*, paper presented to the UN Special World Commission on Environment and Development, Expert Meeting on Hazardous Wastes, Geneva.

[4] *Op. cit.* [1].

[5] *Op. cit.* [1], p. 6.

CHAPTER 9

Risk Assessment of Technological Systems – Dimensions of Uncertainty

Brian Wynne

9.1. Introduction

In all policy systems, "uncertainty" is a wild card that has come to undermine effectiveness and confidence in regulation. The characteristics of hazardous wastes discussed in Chapter 3 may make this issue more sensitive to uncertainties than others.

In the previous chapters we have analyzed concrete examples of the interactions between technical and institutional uncertainties, which have led regulation down many different avenues. We have also seen that underlying uncertainties are manifested at different levels and in different ways in different contexts. Some of the causative factors seem to be universal, but these may be given specific shape by more local factors within regulatory cultures and institutional arrangements. The emergence of uncertainty as a central issue also seems to have coincided with the rise of the problem of regulatory credibility, and recognition of the importance of implementation and its failings.

In this chapter we analyze some more universal factors embedded within the overall "problématique" of uncertainty and technological risks. Very roughly, we focus here upon analysis of uncertainties, while Chapter 10 deals more with institutional and scientific responses to them, especially in the light of the credibility problem.

To begin our analysis, we return to a question raised in Chapter 3, concerning ways of viewing technologies or risk-generating systems. We propose a general concept of technologies as organizational–technical *networks* of interdependent subarenas, not all of whose primary focus is the technological "output" or "system" in question. I argue that risk analysis and policymaking operate with the assumption that technology is well defined, with a unitary rationality. This may have been appropriate for highly structured engineering-risk systems, but the network concept corresponds more with the realities of risk generation, implementation, and social risk perception, in the diffuse risk systems that are becoming more typical. It is therefore more likely to provide a better basis for practically useful risk assessment.

Perceptual differences are often treated as rather exotic matters of public "irrationalities" only, having little or nothing to do with real technical knowledge. The present analysis argues, however, that "perceptual" differences also affect experts and frame their rigorous technical risk assessments as to:

(1) What a technology is.
(2) What are its significant components and connections.
(3) What are its boundaries and external context.

Yet these influences are usually unrecognized, and two pressures push risk analysts toward standardization of their model of "the" risk system:

(1) Normal scientific method, encouraged by the domination of risk analysis by relatively well structured, standardized engineering systems (e.g., chemical plant).
(2) The climate of demands for public justification, which tends to lead to the centralized projection of uniform scientific control and consistent cross-system decision rules.

These tendencies, unless recognized and countered, create problems for less well structured risk systems as they create a widening gap between expert analysis and implementation in actual risk situations. I suggest that although there is a need to recognize nonstandard, *situational* risks, the tendency to assume a single fundamental problem framework or risk-generating system still predominates, and obstructs clarification of the interactions between institutional and technical factors in regulation. We develop the argument that even expert risk analysis of highly structured risk systems can be seen to involve prior *framing* assumptions, yet conflicting risk analyses are attributed instead to technical imprecision and uncertainty due, for example, to lack of data. Framing commitments that underly technical analysis for regulation are not only *prior choices*; they are part of a tissue of informal judgments within science, which ultimately cannot be justified by more evidence, or by tightening, rules of inference, method, or logic. Yet the opposite model dominates public (and expert) attitudes and policymaking institutions [1]. As shown in Chapters 10–12, the pervasiveness of this intrinsic, informal dimension of science complicates the growing public requirements of formal accountability and standardization for authoritative regulation.

With this model of technologies as social–technical *networks*, we argue that in its different framing of "the same" risk-generating system, expert risk analysis incorporates informal and inadvertent judgments, for example about which parts of dispersed systems of risk are pertinent, the behavior of organizations or humans, or the credibility and boundary conditions of technical data, relationships, and models. The ensuing expert disagreements are not resolvable by more objective information because they reflect the latent conflict of institutional perspectives – *structural* or *institutional* uncertainty as outlined in the Introduction – and not technical uncertainty. If these uncertainties are resolvable at all it is by negotiation over prior frameworks and implicit assumptions, which accumulate to give knowledge (and uncertainty) its detailed shape. I show that this is true even for well-defined, extensively analyzed systems, such as nuclear reactors or LEG terminals, let alone for less well structured ones, such as hazardous wastes.

The connection with hazardous wastes is that disagreements about risks or dislocation between national systems over applied risk frameworks (hazard classifications) invariably invokes the response that more technical precision and rigor is the solution. This, we argue, may only artificially reduce uncertainty, whilst increasing systemic ignorance and vulnerability. The germane uncertainties are often not those "passive" ones to do with lack of precision in our measure of the objective world, but are "actively" created uncertainties that are functions of contending cultural backgrounds, institutional purposes, and needs *within* the risk system being defined and analyzed.

Although the strict impossibility to objectively define risk problems is stressed here, the aim is not to suggest that formal risk assessment is worthless; it is to lay bare the extreme fragility of decision processes based on such ideas of objectively attainable authority, especially once public skepticism begins to assert itself. This inherent vulnerability is multiplied by the large unknowns and the *behaviorally indeterminate*, ill-defined nature of the policy field in the hazardous waste case, properties that undermine attempts to discriminate and even rank with any scientific precision the risks associated with different chemicals, industries, wastes, and regulatory options. In these circumstances, those administrative cultures and regulatory institutions that do not realistically correlate the scope and flexibility of their risk assessment regulatory framework with the diversities and uncertainties of the actual risk-generating system, are more likely to find their policy analysis and implementation picked apart and undermined by the ensuing mismatches.

Risk assessment requires reliable estimates of the chances, processes and effects of exposure. This is a combination of intrinsic material properties and situational variations – how a waste is packaged, mixed, treated, confined, etc. Unfortunately, as discussed in Chapter 3, physical, chemical, and behavioral heterogeneity, as well as unpredictable multiple behavioral freedom in the system, mean that "downstream" unpredictables may swallow up putative "intrinsic" waste-stream risk differentials.

However underripe the field may be for it, increasingly formal risk assessment frameworks seem inevitable as demands for tighter control and reassurance multiply. It is therefore necessary to explore what the possibilities and implications are for using formal risk assessment approaches to hazardous waste management. There is currently a lively debate among policymakers in this area as to how standardized or situation-flexible risk analysis can and should be. As the intrinsic-situational risk management dilemma illustrates, the degree of elaboration and standardization of *analysis* depends upon institutional assumptions and commitments – at what points in the system, over which actors, are controls to be exercised, and for what purposes? How much can system actors who affect situational risks be trusted to act responsibly? These are matters of wide national variation, as shown in the previous chapters.

Even in the UK, the traditional stronghold of non-quantified, discretionary methods of decision making on issues involving risks, both a recent Royal Society Study Group and the Royal Commission on Environmental Pollution independently expressed strong support for more quantification of risk assessments [2]. The European Commission and the UK Hazardous Wastes Inspectorate have begun to explore what expert systems may have to offer through formal risk modeling, which has for many years been a US interest in the hazardous waste area, as in others. Formal risk frameworks (though not full probabilistic analyses) have been described in Chapters 4 and 5 for the Netherlands, the UK, the FRG, and Austria. Not only in the USA, therefore, but also in Europe there is growing pressure to adopt formal risk assessment methods in hazardous waste management. We therefore first briefly review present risk-analytic methods to show how they relate to our conceptual analysis of technological systems, expert knowledge, and the typical confusion of technical and institutional types of uncertainty.

9.2. Defining Risk

Risk analysis formalizes a model of a risk-generating process (or family of processes) so that quantitative methods of assessment can be applied to those risks. It is obvious that the more clearly structured is the real risk process, or regulatory problem, the more easily it will succumb to formalization. In this section I show how the analytical process is inevitably selective in choosing its focus, "freezing" the context of its selected focus, assuming that no contribution to risk arises from that "context", and thus limiting the kinds of conceivable risk mechanism. This selectivity creates more severe problems, the more the technology in question is dispersed and network-like.

The conventional definition of risk is the product of the degree of harm a given event would cause, and its probability of occurrence

$$R = P \times C$$

This would express a risk as, say, the estimated number of attributable deaths or other damage per unit time of operation of a given activity. But a chemical plant might accidentally emit lethal clouds of toxic gases every year in a remote region, and cause zero harm. Or a given chemical waste may be toxic and thus, in principle, of high hazard, but environmentally highly immobile and remote, so of low risk. *Hazard* may therefore describe the intrinsic "worst-case" damage a process or material could cause, whilst the above definition of *risk* incorporates variable situational qualifications, which reduce the probability of this worst-case damage [3]. The degree to which these "situational" risk factors are under the control of regulators, designers, or operators itself varies. Furthermore, risk analysts make different assumptions about their controllability, so that some of these factors are the focus of design analysis and choice, others are recognized but less structured, whilst still others may lie unrecognized, embedded in the structuring of the analytic problem, but excluded from the explicit analysis itself.

In the case of industrial plant, some of these situational qualifications are that: *properly* designed, constructed, and operated equipment has a low chance of failure; many parts of processes have fail-safe or redundancy built into the system in case of failure; and have monitoring systems that automatically react to early signals so as to prevent major failures. Other less explicit assumptions might be made about the professional expertise of operating staff, rigor of inspection, likely response to pressure to cut corners, the degree of design, construction, and operating experience, quality control of component manufacturers, management and organizational factors, etc.

In the case of hazardous chemicals, there are equivalent qualifying factors (though on the whole, less well defined), such as the physical form of a chemical (e.g., if it is an inhalation danger, is it in fine powder form or could it "easily" take such a form?), its mode of containment, and its local disposition (is it accessible to environmental pathways back to human populations?).

A typical schematic form of an *analytic* risk function would be as in *Figure 9.1.* On this formulation the same risk, R, can be given by different combinations of P and C: for example, $P_1C_1 = P_2C_2$. But these represent very different events, experiences, and associated uncertainties. Thus, a compelling criticism of the $R = PC$ approach has been that the universal dimensions so produced, takes no account at all of what may be major differences in the processes, kinds of damage, and human experiences involved – it does not compare like with like. "Risk" as conventionally defined and used in analysis and regulation is thus an artificially narrowed concept; it may not capture the features of an issue that its different participants define as its essential properties. There have been various attempts to overcome this limitation by factorizing technological risks into "attributes", but as argued in Chapter 11, these have failed to come to terms with the social–technical network nature of technologies.

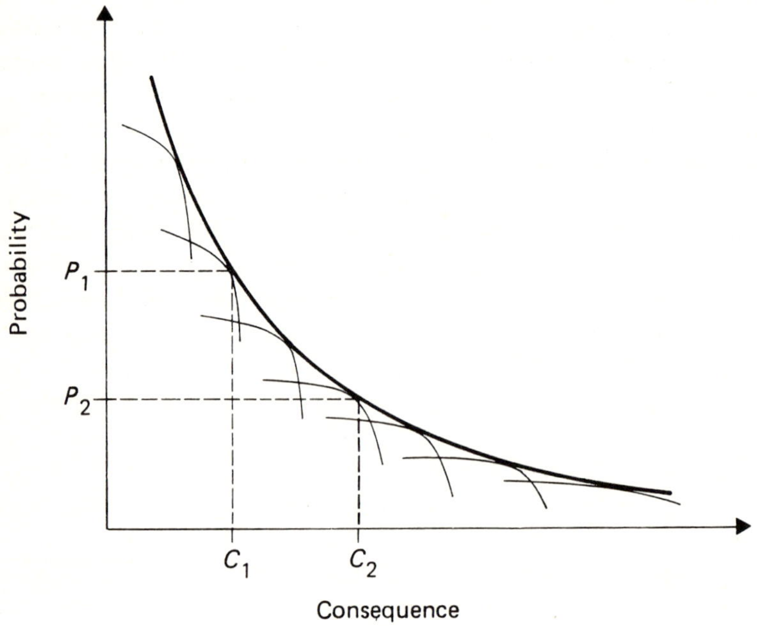

Figure 9.1. Typical overall risk function, smoothed over component events (small curves).

A related difficulty of compound risk approaches is that they may conceal value commitments in their definition. Risk expressed as the product *PC* may covertly incorporate different kinds of harm – mortality, morbidity, different kinds of economic loss – and these may be measured against different denominators. Thus, a comparison between two production processes in terms of risks per unit time, is very different from a comparison of the same processes by risks per unit of output or labor input if one process is more productive than another in terms of time or labor. A workforce may wish to know risks per unit of work time; a manager, per capital input or output; and a local resident, per unit of residence time. These often buried yardsticks can change the apparent scale and importance of risks very considerably [4]. Thus, the specific "technical" orientation of an analysis, even if not chosen on value grounds, may have value implications embedded within it, unknown to the analyst.

9.3. Event-Tree and Fault-Tree Analysis

The extra precision of risk assessment by the use of such techniques as event-tree or fault-tree analysis has been of great value. However, there is always a trade-off between precision and sensitivity with the artificial nature of analytical framing. The framing problem is more acute in poorly structured risk processes, but it is always present. It is worth outlining the analytic event-tree and fault-tree techniques, so as to demonstrate this.

Event trees and fault trees are complementary ways of modeling risk processes. Event-tree analysis starts with a postulated event, such as a pump- or valve-failure, and examines the cause–effect branching chains that may proliferate from the event into undesired end-points, such as the release of toxic chemicals or radioactivity. Fault-tree analysis works the other way about, by specifying a given end-point or "top-event" then analyzing all the possible chains of "upstream" cause and effect that could lead to the given end-point. A typical form is given in *Figure 9.2*.

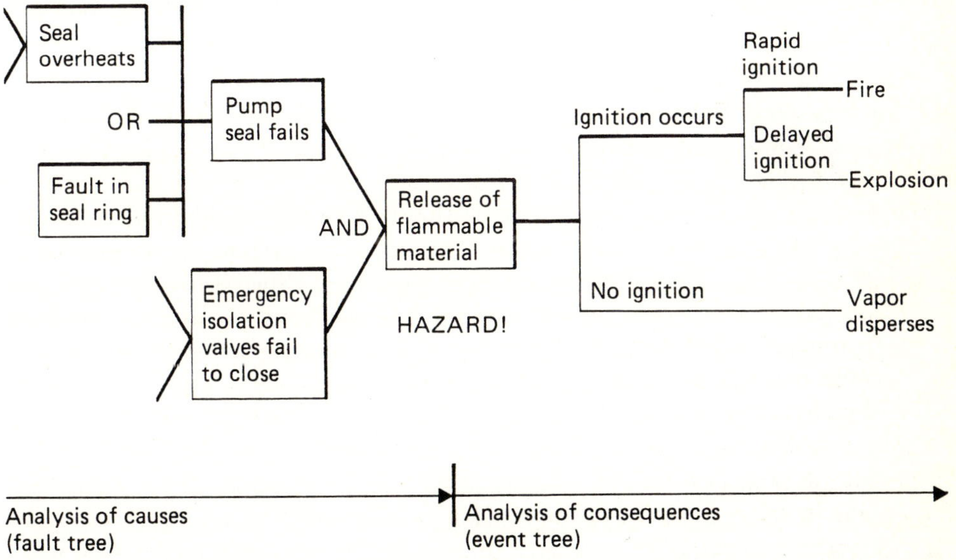

Figure 9.2. Complementary use of fault-tree and event-tree analysis.

The key property of such formal analyses is that they require specification of all the relevant end-points or initiating events, and they define cause–effect sequences in terms of YES–NO (either–or) gates. Along with other assumptions (see below) this allows computation of the estimated overall probabilities of different downstream consequences.

These techniques have several acknowledged problems:

(1) Initiating events still have to be identified and their probability P_I estimated as a starting point. Whether to treat this as a "black-box" estimate or break it down into a further fault tree is a matter of judgment. Completeness can never be guaranteed even in well defined, heavily analyzed systems. The extreme subjectivity and conflict in expert estimation of P_I is frequently obscured in statistical methods, for example by assigning probability weights to them, which represents the conflicting expert judgments as if they were random natural variation. Yet it is recognized that the effects of uncertainty in estimating initiating events – that is, uncertainties on or outside the framework of the formal risk analytic technique – carry through and dominate uncertainties in the overall conclusions.

(2) Formalization into trees suggests that all possible cause–effect sequences are "either–or" options, e.g., that a valve is either open or shut. Real processes are far more complex, so many significant system deviations and proliferating chains occur due to partial faults, e.g., of a pump working intermittently or below the designed power. This interacts with the problem of completeness of I. In a US Nuclear Regulatory Commission update of earlier reactor safety studies, for example, new initiating events had to be considered because those caused by *partial* loss of support systems, like power or cooling water (of higher P than complete loss), generated significant final risks. They could lead not only to shutdown, but also could undermine certain further safety systems.

(3) Related to the last point above is the well known problem of common-mode or common-cause failures. These may affect the probabilities of several branches in the event tree together, thus destroying the key assumption of independence that is necessary to be able to integrate multiple probabilities into overall risk estimates; in effectively fusing several otherwise separate sequences, common-mode failures can introduce large increases in risks.

External events, such as earthquakes and off-site power failures, are examples of previously underacknowledged initiating events and common-mode failure mechanisms. Another example is maintenance inadequacy. These may generate parallel failures in several sub-systems at once, markedly changing the overall risks in ways not easily captured in formal event tree analysis. They arise outside or in the interstices of the analytical framework.

To partly summarize, differences in risk analysis, even of well defined systems using formal models, are not only caused by incomplete data, which is where responsibility is usually placed. (Even data have to be selected from previous cases and applied to the "similar" case in hand, which involves choices of rule.) Analytical differences, or uncertainties, are also due to *framing differences* between different experts, which affect in fine detail the structure of the actual problem each is analyzing. The practical consequences for the divergence of overall risk estimates are very substantial – several orders of magnitude even for nuclear reactor analysis, which is probably the most heavily analyzed of all.

Further difficulties exist in using formal techniques in risk assessment, with responses typically concealing *framing* uncertainties as pure technical uncertainties.

(1) The technique of multiplying chained probabilities and then integrating convergent chains conceals the fact that different kinds of process and related uncertainty may be involved in the separate routes to given end-points. Some may involve only mechanical failures, yet others may embody human interactions of various sorts. Even within relatively well defined systems, aggregation is necessary. However, significant uncertainties can be inadvertently concealed, and there is no single optimal balance between overall compression and decomposition, because there are different needs that coexist within the same regulatory system.

(2) The selectivity of attention of formal analysis can be highlighted in the following way. Take a given set of events in a risk-analytic event tree. For each event we need to estimate its probability. One way of doing this is just to seek past performance data – e.g., of pump failure or boiler cracking. But any such event can be decomposed into subsidiary and conditioning processes. Storage tank rupture or boiler failure, for example, can be decomposed from a general class *inter alia* into materials fabrication and tank design processes, pressure dynamics, corrosion mechanisms, vibration processes, and crack propagation processes, *in the specific operating context envisaged.* Whether to decompose the formal analytic tree into such greater detail will depend on judgment about the overall sensitivity of outcomes to that part of the system, whether data or theoretical modeling can sustain more detailed analysis, etc. Not all parts of a system can be so decomposed; some have to remain aggregated and "black-boxed". Some may remain aggregated because the subsidiary processes they conceal have not even been recognized, e.g., human intervention may be buried in apparently technical parameters. The particular shape of selections made in analytical framing depends upon tacit evaluative (maybe inadvertent) judgments, and the specific methodological orientations

and skills of particular expert disciplines. Again, divergent expert analysis may be due to such subtly different constructions of problem definition based on legitimate commitments (not always expressly chosen) both outside of and within the domain of factual questions themselves.

(3) So far we have considered only the well defined part of risk processes, a typically complex plant with accidental releases as end-points. However, there are many different potential release events from a given plant, each of which has to be analyzed as to the estimated consequences, both for the workforce and externally. In order to reduce the estimations of "external" consequences to a feasible scale they are usually grouped into a smaller number of families. For example, hundreds of release possibilities were classified into 14 release categories in the US Rasmussen Reactor Safety Study WASH-1400, to give the inputs to analyses of external consequences. The external environmental dimension usually involves more complexity, uncertainty, and lack of structure, and thus more opportunity for different expert framing and structuring, than does "internal" plant risk analysis.

(4) Even before composite probabilities are estimated, the *description* of possible chains of events is in itself so complicated and open to interpretation that there is room, even after a real event such as the Three Mile Island accident, for dispute as to whether or not the real-world event sequence was actually described in the preceding analysis [5].

(5) The information basis of probability estimates for component failures and cascading sequences of events is highly variable. In some cases the empirical experience is good enough for reasonable statistical extrapolation; in other cases the applicability of historical data is questionable (e.g., what samples of past boiler failures should one use? Can data on conventional steam boilers, of smaller size, thicker or thinner metal, less quality control, etc., apply to nuclear pressure vessels?); should one take data on *all* like components, or those produced by the manufacturer for the plant in question? In other cases not amenable to even indirect "experimental" validation, theoretical estimation has to suffice. But in all cases, even where there are data to use, inference commitments are made. For example, using data on thin boiler failures at certain pressures for estimating failures in thicker boilers at a different pressure tacitly theorizes that these variables have no effect on failure rates. Expert commitments on such issues as these are equivalent to problem-framing commitments in that they are dependent on tacit evaluations of the meaning of information taken from other contexts of use. These evaluations are influenced by the expert's working context.

(6) When it is recognized that no usable data exist, recourse is usually taken to methods of eliciting and structuring explicitly subjective "expert" judgment, such as Bayesian statistics. Uncertainty so elicited is fundamentally different from that associated with random events or *imprecise* observations. For example, it depends critically upon the community of "experts" chosen to be consulted, the social structure of which affects the coherence of elicited judgments. Relatively close-knit scientific specialties can be expected to produce narrower "uncertainties" than less socially coherent research networks. This factor appears to affect Morgan's analysis of the aerial sulfur transport and damage question with regard to acid deposition [6]. The different scales of uncertainty found in the phases of aerial transport (narrow) and health effects (wide) could be interpreted as an artifact of the social cohesion of atmospheric chemistry as a discipline, and the opposite for the "community" of health effects experts.

Methods of representing subjective expert uncertainty also tend to obscure socially structured uncertainty as if it were objective technical imprecision. Use of log-normal degree-of-belief distributions for a parameter, such as a frequency, is usually presented as a median with an error factor. The error factor is the square root of the ratio of the 95th to the 5th percentile of a log-normal distribution. A typical subjective estimate would be given as 10^{-4}/yr with an error factor of 10. But the statistical quality of this kind of estimate is demonstrated by the fact that to obtain a comparable assessment based on a conventional statistical confidence interval would require data of only one occurrence in about 10 000 years! The unusably low quality of such estimates, and the rank ignorance behind them are concealed by the formal presentation.

Bayesian and related methods have thus contributed to the confusion of socially generated expert *framing* uncertainties and conflicts, with "objective" uncertainties to do with lack of data. Inadvertently, this has also suppressed pertinent questions about the authority of consulted experts to comprehend risk problems, and whether there are dimensions not just outside their domain of reliable information, but beyond their imagination and experience, as framed in their version of the risk system or analytical problem.

To summarize this section, expert definition and analysis of risk processes involves many dimensions of actively strategized, value-laden commitment (which is more extensive and subtle than *value* choice):

(1) In the overall choice of damage processes and risk units.
(2) Selection of some elements and not others for detailed attention.
(3) In the compression of some detailed process differences into unitary classes, yet distinguishing others.

(4) In the application of data generated in different, maybe unknown, contexts of use.
(5) In assumptions about behavioral–technical interactions, such as operator, maintenance, or other implementation reliability.
(6) In implicit assumptions about problem *boundaries* as to the stability of external contexts and of further relationships embedded in chosen functions or variables.

These various problem-framing and structuring commitments introduce divergent expert risk analyses that are not merely a result of inadequate data – they are uncertainties, or more accurately, conflicts introduced by active shaping of the analysis by those diverse commitments. Experts themselves are immersed in their own analytical frameworks, and work under scientific norms that encourage standardization in the pursuit of precision. The fact that they are perhaps the least able to appreciate the context-bound nature of their processes makes it difficult for them, naturally, to appreciate basic variances within a technological system and the importance of social contexts of implementation.

Bearing in mind our interest in regulation, one significant practical difference in risk problem-framing assumptions made by experts concerns the controllability or otherwise of elements of the risk system. This directly influences judgments about the scale and uncertainty of risks, and about what forms of regulatory influence are feasible. Thus, if a risk analysis is framed on the assumption that a particular element is either totally autonomous or already fully controlled (or even does not exist), it will highlight other elements for analysis, i.e., those that it already assumes to be (a) relevant and (b) regulable. Thus, a kind of circularity is possible where particular forms of risk analysis, when used in regulation, merely confirm their own value premises and implicit definitions of "the system" to be regulated.

The above analysis already indicates that "*intrinsic*" risk is strictly speaking a meaningless term, because risks are by definition *conditioned* by situations. "Intrinsic" or "objective" risk therefore implies freezing those assumed conditions, whatever they may be. Situational discrimination seems to represent an overall improvement in the clarity of risk definition, but the distinctions are not absolute. For example, if a chemical waste is treated (deliberately or not), thus reducing its hazardousness by making it, say, less soluble and therefore less environmentally mobile (as well as less gut-ingestible), is this an *intrinsic* change or a situational one (especially if it is a reversible change)? It depends upon a judgment about the regularity or uncertainty of the process or actors who could effect the change, deliberately or not. Containment or back-up devices for nuclear reactors or other hazardous installations may be regarded as "intrinsic" parts of plant design, because this change of design is virtually part and parcel of institutionally unified risk analysis, regulation, and design. There is no "natural" state of a material or technology by which to define its intrinsic risks, and which could act as a definitive basis for explicating all detailed or more macroscopic situational risk qualifiers. In this sense, by making assumptions – deliberate or inadvertent – about whether the components of an overall risk-generating system fall within the scope of risk analysis or are part of its given context, analysts are more deeply a part of the normative processes of regulation than is usually recognized. This is especially true for a *behaviorally structured* and diffuse risk system like hazardous wastes.

Even in more structured problems, it does not at all make sense to talk of experts merely discovering "facts", which are then distorted into divergent shapes by different policy interests. Nor does it make sense to reduce the complexity of the experts' prior framing to hidden value choices. We now turn to a range of examples to illustrate these points.

9.4. Liquid Energy Gas Facility Risks

LEG terminal facilities are an example of a well defined risk problem. The IIASA study of the risk analyses produced during four different national siting decisions identified 15 different risk analyses, but although the technologies and processes analyzed were very similar in all four cases [7], they reached very varied conclusions. In part, this was due to different analytical definitions of what was meant by risk – what kind of potential cost (e.g., population risk or critical group; risk per day, per job provided, etc.)? However, there were deeper incompatibilities than this. As Mandl and Lathrop note:

> ... several decisions must be made in the course of performing a risk assessment, such as how to characterize risk, what presentation formats to use, what gaps to fill with assumptions, what assumptions to adopt, which of several conflicting models to use, how to indicate the degree of confidence of the results, and which events simply to omit from the analysis. These decisions can push the results in any direction ... [8].

Some studies included shipping collisions or grounding and spills, others focused only on land storage tank rupture, others included transfer spills, but none analyzed potential sabotage. Even on specific events, estimates varied without any explanation. One study assumed that for a typical layout of six tanks surrounded by dikes, a valid estimate for a maximum credible spill size was 15% of the contents of one tank, whilst others took at least the full contents of one tank as a conservative estimate. The estimated probability of a spill at one site varied by a factor of 10^3 (10^{-3} to 10^{-6}) in three separate analyses. When the perspective is extended to cover *effects* of a release, the conflicting assumptions multiply. Different models of dispersion and ignition were used, different causes of damage were assumed – some took secondary blast effects to be the sole cause of deaths while others took thermal radiation.

It may be initially tempting to say that analysts chose their detailed problem definition to suit the conclusion they wanted; but not all such framing commitments are visible as choices to the analyst making them and, anyway, they do not always have identifiable effects on the conclusion. At least some of the separate commitments involved in problem definition are inadvertent and determined by social and methodological positions, such as the specific intellectual traditions of the analysts, etc. This has been found to occur in science generally [9]. Even when there *is* (to the external observer) a clear connection between an analyst's problem framing and her organizational position and interests, this may be a result of the framework having become naturalized within that organization (e.g., that its own operations and management are competent), which is fundamentally different from saying that there was deliberate choice of framework to suit the desired outcome.

As Mandl and Lathrop conclude:

> ... what is striking about the estimates is the magnitude of the differences. Societal risk, individual risk, and the risk of one or more fatalities vary over four orders of magnitude across sites, and the risk of ten or more fatalities varies over eight orders of magnitude across sites. It is hard to imagine another area of political concern where performance measures receiving as much attention as these did could vary over such a wide range. Yet even more striking are the differences between the three reports prepared for Point Conception. There is about a factor of ten difference in both societal and individual risk ... There is a difference of four orders of magnitude in the risk of ten or more fatalities. A policy maker faced with such variations could conclude that all three reports are based on very limited knowledge of the risks of LEG [10].

Furthermore,

> Each report poses as a representation of the current state of knowledge regarding LEG risks, but because that knowledge is incomplete, some of the reports represent it using probabilistic terms or error bounds. Yet each report is based on a different state of knowledge: different assumptions are made, models used, probabilities estimated, etc. No one report in fact represents a comprehensive representation of the current state of knowledge. When SAI gives a probability of 9.9×10^{-7}, and FERC [11] gives a probability of 8.1×10^{-3}, for the same event, the policy maker is likely to be somewhat at a loss as to the appropriate figure upon which to base his or her decisions ... each represents only a subset of the total state of knowledge. Yet neither report acknowledges that the other estimate exists!

The implication, not fully spelled out, is that through their inadvertent selectivity of detailed frameworks, formal risk analyses may appear to contain ignorance and "uncertainty" within apparently probabilistic bounds, as if they are all analytically manageable, therefore definable as "risk" [12]. The real uncertainties, however, may be better characterized by:

(1) Ignorance (there are factors and combinations of potential alternative frameworks that are not even identified, let alone "estimable").
(2) A wide scope of legitimate expert analytical choice in defining the relevant risk-generating system structure – *structural* "uncertainty".

Thus, the resulting knowledge is not characterized only by *passive* uncertainty, that due to the effects of *imprecisely known quantities*, but also by the *actively* (but not necessarily deliberately) shaped uncertainty of contradictory frames of reference. The fact that five of the 15 risk analyses did not even mention uncertainties only confirms Thompson's description of this as decision making and analysis under *contradictory certainties*, within different frameworks [13]. The conflict is *due to the cumulative effects of implicit analytical commitments even in defining what the "technology" is.* This point is seen better if technology is viewed as a social–organizational network, intersecting at several nodes with other networks [14]. The more heterogeneous and extensive the risk-generating system is, the more important this concept becomes, and the more misleading the conventional ethos of risk assessment.

9.5. From Technical Imprecision to Social Contradiction

Cox has identified unrecognized uncertainties that underly risk assessments caused by variation in the actual processes being evaluated, when fixed processes are being assumed in the risk analysis [15]. In order to simplify his example, Cox takes the evaluation of only workforce risks. Although he discusses risks of electricity production technologies, his point applies to all technological processes. The following is an outline of his argument.

Modern analysis can define technology as a chain of stages connected by input–output flows. A given stage is defined by its input–output structure, e.g., stages of mining, smelting, refining, manufacturing, and finishing in a typical metallurgical industry; waste arisings, "packaging", transport, storage, and treatment and disposal, in the case of hazardous wastes. More detailed models can be made of single stages.

The occupational risk per overall unit of output associated with a set of stages, J in the process, is

$$R_J = \sum_j^J \alpha_j L_j Q_j r_j \quad ,$$

where the set of stages J is defined as a technology (say, incineration), which is assumed to be well defined with a constant input–output structure; Q_j is the number of units of output from stage j per year; L_j is the number of man-hours of labor used in the production of one unit of output from stage j; r_j is the number of deaths per employee-hour in stage j; and α_j is the fraction of the annual output from stage j (e.g., x tonnes of enriched uranium from fuel reprocessing) needed to support whatever overall production unit is used as risk yardstick (e.g., per 1 GW of electricity produced or consumed).

Conventional uncertainties arise in the risk assessment and multiply in the usual multilinear combination of values. However, there are more basic uncertainties in defining the "system", "process", or "technology" in the first place. For example, every real-world process is an open system, with inputs from and outputs to an outside environment. Therefore, a process or *technology* has to be defined by placing limits on it, thereby also defining its environment. But then arises the thorny question of what is the appropriate system or problem boundary, or – put another way – the appropriate attribution of risk responsibility?

Should a nuclear *reactor* risk assessment include the proportional risks of reprocessing, transport, waste disposal, uranium mining, and even possible horizontal nuclear-weapons proliferation, since these are arguably associated with it as inevitable entailments? Should the risk assessment of fluoride-contaminated hazardous waste from aluminum smelting incorporate an element of the risks of coal mining or nuclear risks, because of the intensive use of electricity in aluminum production? In coal risks, does one include the risks involved in the manufacture of, say, the trucks found at mine heads, even though the same trucks would have been made, with the same risks, had there never been such a coal mine? Once begun, the possibilities of such network connections are limitless.

Inhaber's use [16] of essentially the same approach to analyzing energy system risks, for example, found high total risks for wind and solar power. But closer examination showed that these high risks resulted from an arbitrary assumption that dirty coal would be used as back up for these (intermittent) technologies used as base-load supply systems. Thus, Inhaber's definition of "solar technology" included dirty coal technology too! A normal definition of solar and wind technologies has them organized with storage systems or with clean back-up. This is a different definition of the technology as a social–organizational unit.

As Cox emphasizes, real economies and real technologies are far less simple than that implied in fixed internal structures, and thus fixed technological coefficients. Both the boundaries and internal structure of a technology can (a) vary in the real world, and (b) be defined variably by the risk analyst (and others) as "the" technology or "the" risk problem in question.

Another set of examples that enlarge the same point come from scientific disputes over the environmental risks of the proposed MacKenzie Valley pipeline from Arctic Canada to the USA [17]. Implicit, and eventually revealed, in the analysts' conflicting scientific conclusions were different social–behavioral judgments, which created tacitly different problem definitions. Thus, some scientists assumed that one pipeline could realistically be evaluated for its effects in isolation from further pipelines, roads, telegraph lines, airfields, residential service towns, and other developments (the "corridor"), which other scientists assumed would inevitably follow and should therefore, they believed, be a "natural" part of the system to be evaluated. If *socially* one initial pipeline would likely beget a whole corridor, who is to say which is more "objective"?

In another part of the same dispute, the damage to tundra from construction work was assumed by some analysts to be limited to that within *official* limitations of construction to winter months, when the tundra was hard-frozen. Other analysts assumed this was unrealistic because they believed the pressure of deadlines and huge investments would inevitably cause these limitations to be broken in practice, with summer-season construction leading to far greater damage. Different *behavioral* assumptions led to different "technical" risk problem definitions. In these cases, as before, there is no objective, singular problem definition or technological system that can be more and more precisely "revealed" by more analysis.

9.5.1. Nuclear technology

In the UK Windscale Inquiry in 1977 into a proposed oxide nuclear fuel reprocessing plant [18], the Inquiry chairman, the nuclear industry, and government agencies defined the risk assessment decision as that concerning a single reprocessing plant, and nothing more. Objectors on the other hand, assumed that the plant, which generated plutonium, and uranium for further rounds of nuclear power systems (and weapons) would create institutional momentum for more nuclear developments, including widespread fast breeder reactors and plutonium commerce. They defined the risk assessment question and the associated technological system as much larger and more diffuse. However, this was dismissed as "emotive" nonsense by the chairman.

Here was a conflicting choice of technology or problem definition which, despite the chairman's view, was not a "facts" versus "emotions" division. Although not recognized as such by the Inquiry, it was a conflict of founding problem definitions. The conflicting definitions were a symmetrical pair based upon different behavioral judgments and objective social experiences of the contending groups. To members of the establishment, it was rational to draw a decision boundary round the present plant, because they could objectively expect to influence and identify with the subsequent decisions whether or not to make further commitments. These future network decisions, the technologies and potential effects involved, could be logically fenced off and neglected. For outsiders to the decision-making establishment, however, an incompatible, but equally logical, view prevailed. From their social experience, it was rational to assume that they would have no real part in any of those subsequent decisions, as they had been excluded in the past. On past experience, to them they would not *be* decisions at all. It was therefore rational to condense all the possible foreseeable future network developments into the present "single plant" decision. The technology or risk system was thus defined to take these extensive further questions and probabilities into account. The relevant event-tree was enlarged in inverse proportion to the social access to (past and future) decision trees.

The important point is that *each* position, "expert" or otherwise, was based upon behavioral judgments and social experiences that were necessary to frame a problem at all. But each was equally defensible, or illogical, according to one's social position. No deeper, more objective definition of "the technology" existed. Nevertheless, the language of the Inquiry was totally that of an "objective" technology with "objective" effects, which could be *discovered* through the conflict by more rigorous analysis.

9.5.2. Pesticides

The official UK government scientific Advisory Committee on the Safety of Pesticides (PAC) evaluated the risks associated with 2,4,5-T in the late 1970s, when public suggestions about its pervasive harm were accumulating [19]. Having analyzed the latest scientific evidence, the Committee decreed that 2,4,5-T could continue in widespread use. After attempts to reopen the issue by the National Union of Agricultural and Allied Workers (NUAAW), the main labor union involved in spraying 2,4,5-T for farm and other employers (including many local authorities and government agencies), PAC reasserted the safety of 2,4,5-T, dismissing as unscientific the large NUAAW dossier of admittedly circumstantial clinical and other evidence of actual harm.

This rather patronizing scientific rebuttal only polarized the gathering conflict further and, eventually, in the face of further union action, PAC advanced the explicit qualification that its assertion of the safety of 2,4,5-T was conditional upon its proper manufacture, distribution, and use. These conditions were precisely where the farm workers' and others' direct experience and evidence was focused. For example, drums of 2,4,5-T often arrive with defaced or removed labels that are supposed to describe proper conditions of use. Even if these are known, the organizational realities of farm life often do not allow a farm worker to refuse to spray just because the climate is not correct, or because specified protective equipment is defective or nonexistent. Chemicals, called "adjuvants" that speed up the action of the main chemical are often added, despite the effects being unknown, and new spraying technologies designed to improve economic efficiency have had marked effects on exposures, e.g., by producing finer sprays. Also, the cultural reality of such a work life does not encourage a man to say he is concerned about the possible risks of such materials. In this *behavioral* reality of the technology of 2,4,5-T use, the workers were the experts and not the PAC scientists. To the workers the experts were "in cloud-cuckoo land behind a laboratory bench" [20]. The "objective" risk analysis of the PAC experts, which focused only upon the laboratory-controlled tests in the scientific literature, also required a certain set of concrete *situational* parameters to define it. In transforming these into a policymaking risk analysis of the process of use of 2,4,5-T, they were effectively assuming a whole standardized and *idealized* social world of use or implementation. They focused exclusively on only one part of the technological network, from one cognitive dimension. Note that their risk analysis was not *neutral* with respect to the real-world context, but would only have been valid if the real context of implementation accorded with the implicit ideal. The experts, entering with their highly selective frame, excluded *a priori* the realities of distribution and use of 2,4,5-T, which potentially radically altered its risks. There is a parallel here with the questions about "sensible" landfill as a hazardous waste policy, discussed in Chapter 7.

The point of this example is to demonstrate again the inadvertent selective analytical framing of the process or technology for risk assessment. Because of its insensitivity to the implementation context the narrow, unrealistic definition in this case eroded regulatory credibility beyond the specific issue.

A very similar example to the pesticides case concerns the conflict over beef-cattle growth hormones. In 1985, a scientific committee of the European Commission pronounced three nature-identical growth promoters to be safe if used under certain conditions (e.g., earlobe injection, dose threshold, 90-day waiting period before sale, etc). When the Council of Ministers rejected the scientific advice due to strong political pressure, Commission bureaucrats were outraged that their scientifically well prepared advice should be overturned by what they regarded as popular hysteria and superstition. However, it is arguable that popular feeling did reflect a question mark that the experts had ignored – are the crucial conditions of the risk analysis enforceable in social reality? These conditions define the risk system under which the scientists pronounced safety. The experts again appear to have automatically assumed that these would be met – an ideal social world of implementation. The implicit popular judgment, that they are most unlikely to be met, was equally legitimate, but not acknowledged and not impartially analyzed by the parties involved.

The importance of the network model for technology and risk processes is exemplified in the above cases. These networks are composed of *cross-cutting rationalities.* There are good reasons why farm workers spray herbicides in unsafe conditions and want to inject growth hormones under uncontrolled conditions, which are to do with the social and economic realities of farming. There are good reasons why transfer agents in hazardous wastes mix waste streams in ways not defined by regulators, to do with the feasibility and economics of waste treatment and transportation. In an urgent examination of arrangements for controlling methyl isocyanate (MIC) following the Bhopal disaster in 1985, a French inquiry found that MIC was being imported through the port of Marseilles to a plant in Beziers. Unloading of the MIC was being performed in a very unsafe manner, as if the barrels were bales of straw or grabs of coal. This apparently irresponsible activity, in the "periphery" of the technological network, was *locally* logical and necessary, because the economics of dock operation require a crane and its operator to fill a shift productively. This requires many loading and unloading operations, so that shifting attention, focus, and practice of the operator between different kinds of material is very difficult, especially if piece rates are being paid to increase productivity [21].

Thus, even when such network processes that are surreptitiously increasing a previously defined technological risk are discovered, they cannot necessarily be easily corrected by, as it were, flipping an operational switch, because the "deviant" action, as (perhaps) seen from the central risk management perspective of a given technology, is embedded and *normal* in a cross-cutting network of social, economic, technical, and other commitments. Even in relatively highly controlled risk situations, like nuclear plants, it is known that radiation workers take off their exposure-recording instruments in order to avoid being removed from radiation work (and the corresponding bonuses) because they have reached the official dose limit. When this can happen in "controlled" situations, truck drivers with hazardous cargoes can certainly try, with reasonable odds for success, to simply remove identifying hazard signs so as to cross frontiers at forbidden points or go through forbidden tunnels, and thereby perhaps save several hours of detour. "Central" risk management and unitary, standardized risk analyses may increase scientific precision only by excluding the so-called context, which is actually the substance in another framework.

It is arguable that the extensiveness and dependent intersections of modern industrial and technological processes are growing and becoming more acute, such as in the siting of modern chemicals industries in developing countries. The case of radiation exposure to unknown numbers of people when a cobalt-60 X-ray source was taken from a US hospital, allegedly with management acquiescence, and passed via scrap yards (in which children played) into steel for houses and furniture all over Mexico and parts of the USA, is a graphic illustration of what can happen. So too is the episode in Michigan in the 1970s when herbicide was stored with cattle food in a warehouse, leading to unknown amounts of contaminated beef on the market.

Once alert to the extent and complexity of network possibilities, one can see more clearly how expert risk analysis itself risks lack of effectiveness, and escalating loss of credibility, by its unreflective framework selectivity. In issues with strong network and behavioral properties, decentralized risk-analytic capability with institutional sensitivity is clearly necessary.

To summarize, different social assumptions influencing definitions of a "technological system" (including its *implementation*) can generate different expert problem definitions in "the same" risk analysis. There is no single objective definition of a technology or risk problem that supersedes all others. Risk analysts have to make commitments to a particular definition, *before and during* analysis – these commitments may differ. This point has been addressed at length, because it clarifies a key confusion frequently found between two quite different types of uncertainty in risk analysis.

The kind of *institutional* uncertainty we have identified *envelops* the system rather than being located within it, because "the system" or "the problem" is itself subject to conflicting definitions, in our case of "hazardous waste". This uncertainty is not *in* the data and definitions, but *surrounds them.*

9.6. Uncertainty by Strategic Design

Earlier we distinguished between "orthodox" technical uncertainty or imprecision (which may include real system indeterminacy) in risk analysis, and *structural* or *institutional* uncertainty, brought about by (frequently subtle) framing differences of analysts' problem definitions.

There is more room for conflict in the expert framing and definition of "the same" risk-analytic problem than has previously been recognized. The more that a regulatory problem area is behaviorally as well as technically diffuse and indeterminate, the more significant is this distinction. *Institutional* uncertainties are central in many processes of information-generation relevant to the hazardous wastes issue, because information is not unitary, but relative to the same frameworks that define technical terms in different ways.

In this section we examine the different kinds of uncertainty that underly the attempt to define terms and data normally regarded as absolutely central to a hazardous waste regulatory scheme. We progress from ordinary uncertainties or imprecision in measuring hazardous waste, via active, socially generated uncertainties in identifying them, to active, socially generated uncertainties in even *defining* hazardous wastes.

The act of defining a hazardous waste by regulatory authorities is tantamount to defining *actors*, and then defining their *actions* with that waste. The practical uncertainties are a product of two mutually amplifying factors:

(1) The inevitable imprecision in the regulatory definitions.
(2) The variable *interpretations* of these definitions by waste producers reporting wastes and by waste handlers, who may not even define them as wastes.

We have already seen the institutional and technical processes at work on these uncertainties in real cases, ranging from highly centralized management (at *Land* level in the FRG) to extremely decentralized management (UK). Now we attempt to clarify some of the different types of uncertainty, and their implications for rational regulation.

9.6.1. Conventional data uncertainties are softer than they appear

The General Problems

Even if the definitions of "hazardous" and "waste" were universally agreed, and there were also no intermediate interests that diffract "real" quantities and kinds of waste arisings into official regulatory data (see below), there would *still* be more problems than often recognized, simply in making accurate technical observations. A T&D company in the UK was criticized for not controlling the composition of the wastes delivered to it. The company contracted experts to help devise an accurate analytical sampling device for just one of its many consignments, an oil–water emulsion delivered in 4000 gallon tankers [22]. A standard vertically sectioned thief-tube was recommended. Trials found that even this simple two-phase physical sampling was impossible to perform except by very rough estimation. In reality, there were not two phases but at least four – water, sludge, oil, sludge – with very indistinct boundaries. Samples along the tanker (which was only baffled, not compartmentalized) showed variations of ±50%, though they should have been identical. This was the simplest physical sampling and analysis, yet it proved impossible to perform anywhere near accurately. This was also for only one load of only one type of consignment among many different sorts. It is not surprising in the light of such realities that even in the detailed data survey of waste arisings in Hungary (see Chapter 8), it was admitted that experts frequently had to resort to guesses to obtain figures at all.

This type of sampling and analysis variability exists even where the waste is a *laboratory* sample, as we discuss in Chapter 10, and it is not avoided by systems that use precise concentration limits as hazard definitions. If these uncertainties alone were extrapolated to bulk waste arisings and fully recognized, they would demonstrate the problems of depending upon precise hazardous waste arisings data for effective regulation. So, far from achieving regulation based at least on analysis and confined to the above orders of uncertainties, the situation in the UK, for example (which is probably fairly typical), is that:

> Only a minority of waste producers or waste disposal facilities are equipped with weighbridges. Waste arisings continue to be expressed in a variety of exotic units encompassing imperial and metric weights and volumes, together with the traditional and totally inadequate "guestimates" such as the number of lorry-loads, the number of bags or other containers. Compounding this confusion, the waste disposal authorities adopt non-uniform conversion factors to convert from volumetric measures to weight [23].

Thus, we can see how scientific sampling and analytical inconsistency, even in a well defined situation, overlaps with extremely loose recording, even of gross volumes, let alone composition.

This leads us to a case study of attempts to quantify hazardous waste arisings.

Determining the Amount of Hazardous Waste in Massachusetts

The state of Massachusetts has been relatively active and progressive in dealing with hazardous wastes. It has made several attempts to measure the volume of waste arisings that it has to control [24].

Attempt 1: The GCA study. In 1976, the Division of Water Pollution Control of the Commonwealth of Massachusetts commissioned a study from an environmental consulting firm, the GCA Corporation, to "survey the quantities, the geographic distribution, and the current practices of hazardous waste disposal in the commonwealth" [25]. As a first step, to determine the quantity of waste generated in the state, GCA reviewed the division's file of permit applications and monthly reports from licensed waste transporters. These reports were required under a Massachusetts law prior to the enactment of the federal RCRA regulation. The reports were supposed to include monthly summaries of where transporters picked up a waste, where it was sent, the waste type, and methods of treatment and disposal. However, GCA found this information incomplete and difficult to track or compile [26]. They therefore decided to conduct a telephone survey of a selected number of firms. Some 446 plants responded to their telephone requests for information, which usually represented the "best guess" of the plant manager or the plant's environmental engineer.

To yield statewide totals the waste figures reported were simply extrapolated on the basis of *number of employees* in the firms surveyed, compared with the total number of employees in the industries state-wide. The firms surveyed represented 36% of the state's manufacturing employees. This procedure assumed a linear relationship between waste generated and number of employees in a particular firm, which GCA admittedly had no evidence was correct. But they felt that the estimates of waste so generated were "probably accurate to within a factor of two". With this methodology GCA estimated that 37.57 million gallons of waste were being produced per year in the state [27].

Attempt 2: The New England Regional Commission study. In 1979, the New England Regional Commission employed Arthur D. Little (ADL) consultants to develop estimates of hazardous waste generation for the six-state New England region [28]. ADL performed no new analyses, but used the data of previous state studies, including the GCA report in Massachusetts. Taking GCA's raw data and performing the same extrapolation based on waste generated per employee ratios, ADL estimated that the total waste generated for Massachusetts in 1979 was 49.2 million gallons, an increase over GCA's total of approximately 30%, presumably due to changes in employee statistics [29].

The difficulty in using waste per employee ratios for extrapolation is shown by the wide range of ratios ADL found in New England (see *Table 9.1*). The report admitted that "variations between the states are not readily explained on the basis of industry differences [30]".

In addition to this estimate, ADL provided a "high sludge" estimate on the assumption that the introduction of planned wastewater treatment programs would lead to an increase in hazardous waste generation. Figures from Connecticut, which already had such a program, were used to estimate "high sludge" amounts for the other states. This amount for Massachusetts was reported as 84.9 million gallons per year [31]. It was this crudely estimated range of generated hazardous waste, 49.2–84.9 million gallons per year, that became the official state statistic for hazardous waste generation. It was published, however, in units of tonnes, by assuming a water density

Table 9.1. ADL ratios.

State	*Waste generated per employee per year*
Connecticut	255
Maine	258
Massachusetts	82
New Hampshire	156
Rhode Island	72
Vermont	155
Average	163
Standard deviations	81

of 240 gallons per tonne. This gave 200 000–350 000 tonnes of waste per year. Looking at one assumption alone, sludge or solid waste could be several times heavier than water, leading to *tonnage* figures several times greater than these estimates. With little reference to their uncertainties, the figures were used to argue for the enactment of a state hazardous waste control program modeled after RCRA.

Attempt 3: Department of Environmental Management. Obviously not satisfied with these attempts, the Massachusetts Department of Environmental Management (DEM) decided in 1981 to do its own survey [32]. They hired yet another consulting firm, Urban Systems Research and Engineering, to computerize and compile the information contained in the state's transporter reports (the same reports rejected as too incomplete by GCA). This study calculated 170 000 tonnes of hazardous waste produced in the state. In addition, DEM reviewed the EPA notification list of potential generators (compiled under RCRA) and an industrial directory in order to identify "potential" generators who were not reporting their wastes. Interviews were conducted on-site and by phone, and reviews of out-of-state manifest totals for waste from Massachusetts delivered to other states "revealed an additional 17 000 tonnes of hazardous waste not reflected in our totals." DEM's final estimate was 190 000 tonnes of hazardous waste per year [33], about six times the earlier estimate "accurate to a factor of two."

DEM identified sources of errors in this estimate, which were drawn from on-site interviews with 25 generators:

(1) Underreporting of waste from generators who appeared in the transporter reports.
(2) Generators who did not file transporter reports as required by law.

Finally, DEM admitted that these figures, "for the first time generated by hard data", took no account of waste being illegally dumped "by pouring it down sewers, incinerating it without approval ..., mixing it with conventional wastes, or using illegal disposal facilities". Quite honestly they concluded, "the extent of illegal disposal in Massachusetts is unknown [34]".

From a state to the nation. This case shows that estimates of hazardous waste generation, often presented as hard facts and used in developing policies for control, are inherently "soft" numbers. Even more sophisticated surveys are usually based on information voluntarily submitted by generators who face large incentives to underreport their wastes. Amounts of waste being illegally disposed of are simply unknown.

The problems of a state trying to determine such figures are even worse at the national level. In August 1983, the preliminary results of a US national survey of hazardous waste generators conducted for the EPA, again by a consulting firm, were released. This survey was again based on a telephone and mail questionnaire of approximately 10 000 generators who had identified themselves to the EPA as generators under the RCRA regulations. Yet in 1981 only 20% of the firms surveyed admitted to having generated any waste at all! The study concluded:

> The initial estimates are preliminary in nature and are subject to statistical uncertainties. Nonetheless, the study suggests that 150 mte of hazardous waste were generated across the US and its territories during 1981, in contrast to previous estimates of 40 mte [35].

This also contrasts with the 250 mte that the US Office of Technology Assessment estimated by simply adding up state estimates for the 50 states and territories [36]. All the elaborate effort, therefore, has produced figures for regulation that vary by a factor of about six.

In the end, one must conclude that regulatory resources might be better spent than subsidizing the environmental consulting-firm industry to produce numbers that inevitably have spurious authority conferred upon them simply because they exist. Donovan's summary of her experience tracking fuelwood consumption estimates in the Himalayas could be applied directly to the hazardous waste issue:

> All too often, consultants have neglected to explain the methods used to arrive at their expert opinions ... some estimates have been boldly quoted and requoted, often without citation, in ever more respectable documents, until a very casually contrived estimate has become the basis for policy formation and program planning.

Thus, despite qualifications and caveats, processing the very rough estimates of hazardous waste arisings through the regulatory bureaucracies and into public policy arenas has stripped away their *contextual* properties, so giving the impression that they are a measure of objective reality rather than a partly circular index of socially constructed and intrinsically open-ended definitions. A practical conclusion is that, even at the state level, it is very difficult to envisage clear guidelines for capital investment programs in T&D facilities, unless either the private market is allowed to use its own wisdom and take its own risks on a fully open basis, or the public authorities take comprehensive responsibility for regionally defined facilities, which focus less upon fine distinctions of risk as a qualifying criterion and more on producing an integrated *waste* management.

9.6.2. Institutional diffraction of key data and key terms

When the authorities began in 1945 to try to assemble data for risk assessment about people's precise exposure to the blast and radiation from A-bombs dropped on Japan, they relied upon self-reporting of where people were at the time of the blast, and calculated doses from this [38]. Many survivors apparently misreported themselves to have been out of town, because a strong social stigma had rapidly grown against association with having been exposed to radiation. Later, when free welfare services were offered to radiated bomb survivors, many of those who had been away suddenly recalled that they had, after all, been in town, so the figures for the exposed population expanded, with corresponding changes in risk estimates.

In some countries where production is centrally planned and fervently pursued, incentives such as bonuses are offered to plant managers and workers who reach or even exceed production targets. There is therefore a natural temptation to overrecord production levels. Systematic tendencies in this direction have been observed, involving large-scale, informal social coordination to avoid detection [39]. In some cases even entire factories have apparently been fictionally created for the benefit of central regulatory "data", so as to bolster "production" and bonuses. Ingenious supporting elaborations have had to be developed, such as massive routine wastage and breakages to explain why the extra production did not generate corresponding revenue. In another case, involving new equipment and funds for oil production:

> Reports submitted by the Ministry in the five-year period showed modernization proceeding according to schedule, and production figures barely short of plan indices. But an outside inspection in 1981 revealed virtually no attempt to introduce the new methods. Report data had simply been invented, and there was a shortfall of millions of tons of oil. Equipment had been left to rust and be pilfered. Inspectors found that thousands of barrels of imported reagents had been dumped beside railway tracks, thrown into abandoned river pits, and even encased in the foundations of railway embankments [40].

In this case, of course, inspection happened upon the target, but even so it took five years even for this scale of "data-diffraction" to be detected, by which time responsibility and remedy were extremely difficult, if not impossible.

When the Hungarian public health authorities were conducting their survey into hazardous waste generation, it was found that agricultural stations, of the government agriculture inspectorate, had been unofficially storing disused pesticides for many years, when they should have been declaring, properly treating, and disposing of them. The stations were found to have been systematically underreporting their inventories so as to "mask" their own inefficiency and negligence [41].

Reliable data about waste arisings – their types, properties, volumes, origins, movements, etc. – are regarded (along with a meaningful hazard classification or listing of wastes) as the main foundation of a hazardous waste regulatory scheme. Naturally, everyone expects such data to be fuzzy. When the Environmental Safety Group of Harwell, UK, gave evidence to the Gregson Committee Inquiry of the House of Lords, they gave a range of 2.5 to 4 million tonnes of estimated hazardous wastes arisings per year in the UK. Discussion with the Committee expanded the range to 6.6 million tonnes. From the Harwell experts' experience of the issue, and having themselves conducted surveys, they thought this range relatively narrow. The Gregson Committee, which contained some experienced environmental policymakers, but no hazardous waste specialists, was horrified that regulation should have to proceed in the face of such a large uncertainty as that reflected in the range of estimates [42].

The point about the examples above, however, is not so much the range of uncertainty, as if about a piece of reality which is physically difficult to observe, but the fact that the observation itself is *always* mediated by other social actors. Sitting between the observer (central regulator) and the reality he wishes to observe, these actors (whose behavior the observer wishes to control via the observed data) actively *diffract* the observation, often very substantially because they have different (and even changing, in the A-bomb survivors' case) interests that give the observations meaning and clothe them as "facts" or "data". They are data and uncertainties created by *actively strategizing agents.* Even when the strategy is not to deceive a regulator or researcher, the diffraction will still occur. It is uncertainty created by *conflicting* views of reality, not by the incompleteness of a single one.

Thompson and Warburton give a similar example of conflicting data on fuelwood consumption rates from research on Himalayan deforestation. Like hazardous waste arisings, this factor is apparently crucial in regulation, but the range of estimates differ by up to 70 times [43]. Such variations do not only fog, they obliterate the "discoverable" differences between factual effects of policy options on which normal policymaking subsists. The observer *inevitably* depends upon the regulated for data by which to regulate them. As we see below the less controllable are the regulated, the more it is tempting to tighten the data and definitions. But if this becomes detached from supporting institutional measures, and too extreme, it simply becomes a framework of delusion on a grand scale.

The same kind of *actively created* uncertainties, or conflict of social meanings and frameworks, pervades the creation of central data in the hazardous waste management issue, but this problem has a further wicked twist to it.

For hazardous wastes, not only is there a strong diffraction caused by cross-cutting observation and reporting frameworks, but there are even deeper uncertainties as to *what* (and who) is being observed and regulated anyway. As we have seen in the previous chapters, different basic definitions of what is "hazardous" and what is "waste" are used by different actors in the system, and by different regulatory bodies in different state or national systems. There is no *intrinsic* risk of a given waste, only risks depending upon situations. How tightly regulatory systems try to specify or control these varies according to local institutional processes embedded in wider cultural structures. Even the most elaborate attempts to specify tight criteria and definitions still fail to eradicate generous interpretational flexibility in several dimensions. Thus, even within a single system, fluctuating definitions of the key terms "hazardous" and "waste" sharply affect the estimated volumes of hazardous waste arisings, even before one takes into account more deliberate diffraction. In this case, therefore, the underlying "solid reality" becomes a mosaic subject to social definition, even before the observation phase brings more uncertainties into play.

9.6.3. Hazards in waste

The context-dependent character of "hazard" has been a theme of the whole book. It was specifically discussed in Chapters 3 and 5, and earlier in this chapter; we also return to a further aspect of this in Chapter 10. Let us now, therefore, look at "waste".

A very common and serious ambiguity lies in the storage or even sale of toxic waste against the possible future recovery of constituents, when it is then defined as a (nonregulable) resource. For example, a liter of silver- or mercury-contaminated oil or solvent may be a (hazardous) waste if taken as such. But if chosen to be defined as one of a thousand different liters, to be collected and combined, it may legitimately be an unregulated resource. Speculation against *future* market prices of contaminants allows a large area of license in deciding whether or not to subject a material to regulation. Furthermore, because this kind of judgment may be made "downstream" in the life-cycle of a waste, say at a transfer station where the transfer agent decides to sell an incoming waste as a resource to a recovery operation, it will then disappear from the regulatory records. No one will know, without specific checks on that individual consignment, whether it was illegally disposed of or its character legally changed from waste to good.

Many industrial economies have thriving, if insecure, T&D sectors based on entrepreneurs who buy and sell wastes or recycleables, and profit from their knowledge of where a "waste" (for whose disposal they have been paid) can be sold as a net good. The uncertainties are amplified by the cultural style of risk taking in this commercial area. Brokers go all out to sell their T&D services, often ignorant of precisely what materials they will receive once deals have been struck. They *exploit* the uncertainty, hoping that they will more than compensate for unexpectedly nasty consignments by "windfall" double-profits when a "waste" can be sold as a good. The recovery and recycling industry is also well known for its "instability", with many companies starting and quickly disappearing from sight. This, of course, makes control all the more difficult. The US EPA will attempt to control "speculative storage" in future, but this adds an unknown number of extra agents to be policed.

An intriguing, but important, case of the definitional–perceptual anomaly occurs when tankers or drums of pressurized or frozen liquid energy gas are "emptied". Because they are no longer carrying their usual load (of goods), they are treated as containing *nothing*, so they are considered nonhazardous. In fact, they now contain a residual "good" that has become a hazardous waste – gas *vapor* at ambient pressure. Many accidents have occurred because of this perceptual anomaly.

The silt that is carried naturally down the Rhine and deposited as a great economic nuisance in Rotterdam harbor used to be converted into a resource by dredging all 40 million m^3 of it every year and spreading it on agricultural land, where its high organic content was a benefit [44]. But in recent years, the toxic contamination of the Rhine has made this silt a hazardous waste, useless for agricultural land, and it has even been disqualified from being dumped at sea by the London–Oslo Conventions. (Notice that if the *Rhine* had dumped it in the North Sea, it would not have been disqualified!) However, the imaginative Dutch have turned it back into a resource by using it to landfill a new polder and, suitably insulated, to create new residential or industrial land. While the material concerned has remained true to itself, it has had various truths *invested in it* by the surrounding institutions and actors dealing over it.

9.7. Ignorance, Uncertainty, and Expertise

Our examples have shown qualitatively different kinds of uncertainty. We began by showing that even for well defined risk-generating problems, uncertainty and disagreement blamed upon observational imprecision and lack of data usually conceal deeper uncertainties that are actually more subtle conflicts of expert framing of the problem. We have also shown how, when we view technologies as technical–organizational networks of intersecting and interdependent processes, the problem of framing becomes even more acute, though it ought to be more visible. The network concept of technology allows us to see more clearly:

(1) The significance of "downstream" or "peripheral" network variants and intersections.

(2) The multiple social relationships that people can have with a technological system and the cross-cutting rationalities of diverse actions and perceptions relating to the technology. This becomes especially relevant when we consider public reactions (see Chapter 11).

(3) The vulnerability and brittleness of regulation based upon risk-analytic expertise which focuses too precisely on only one part of the overall network. The normal context-bounding processes of scientific analysis, the unconscious framing processes described earlier, and the natural preoccupation of scientists with precision, all tend to create realms of *ignorance*, which underly the arenas of defined uncertainty and risk that are formally structured by scientific methods. Thus, the validity, effectiveness, and credibility of regulation is vulnerable if elements of the network and its diverse rationalities and variations are excluded. This exclusion in *analysis* has a practical counterpart in the neglect of implementation worlds when risk analysis moves into design of optimal regulations.

Full appreciation of the practical importance of this point requires a brief digression into the philosophy of science. It is usually taken for granted that science *embraces* uncertainty, feeds on it, and reduces it painstakingly to more reliable knowledge. Partly this is true, but the part that is false is more important, because it is not recognized. The uncertainties embraced and highlighted by science are already a reduced population, selected by prior cognitive framing commitments which *a priori* exclude other frameworks from the explicit problem domain. This prior framing is culturally ingrained, by the cumulative theoretical and methodological commitment of a technical specialty. In gradually increasing precision within the defined domain, questions about the context and framing are suspended, and die away. Unexplained anomalies that arise within the domain of well defined scientific "puzzles" are passed over into the background, and either written off as freaks or shelved in the faith that one day the existing framework will be able to incorporate them. (The parallels with risk management are striking. There are interventions from "outsiders", to be written off as irrational, and there are quietly admitted problems, such as more at-risk subpopulations or synergy effects, which are for another day.) Furthermore, the difficulties involved in overcoming this intellectual reductionism are more than "merely" intellectual, because the processes are a main constituent of *social* processes of order, cohesion, and progress in scientific communities; thus, they are more deeply rooted.

Effectively, therefore, science decreases uncertainty (solves its defined puzzles) whilst increasing ignorance (suppressing the excluded context). Whilst this may be functional up to a point within science, it is not necessarily an appropriate mode for science in public – for example, dealing with uncertainty in risk management. In public contexts this mode invites regulatory commitments that are vulnerable to dislocation arising from the external "context" (or network dimensions), of which it has created its own ignorance and inadvertently concealed. Furthermore, these artificially excluded dimensions are not only physical, but social and perceptual.

Thus, one could say that uncertainty is inversely proportional to ignorance. In other words, decreasing *acknowledged* uncertainty by using standardized risk assessments in heterogeneous risk arenas is increasing ignorance, and hence regulatory vulnerability. Equally, when ignorance is reduced, e.g., by the radical surprise of an unimagined "external" network intervention (the 2,4,5–T experience or the Love Canal discovery), uncertainty is increased. But uncertainty as seen from a single point can also be *deliberately* increased, e.g., by decreasing the standardization or centralization of risk assessments and their institutional use, as a means of reducing *overall* ignorance and system vulnerability to surprise.

The practical distinction and interaction between uncertainty and ignorance is illustrated in a brief discussion by a UK review committee of the issue of potential synergistic effects of chemicals in mixed wastes. The local authorities' association representative had advocated revision and tightening of the regulatory definitions so as to include this possible problem. However, the committee's overall reaction was skeptical:

> The Committee accepts that, with tens of thousands of chemicals in commercial use, synergistic couples may occur in mixed wastes. It is not however convinced that the present state of knowledge would permit them even to be properly identified, or the definition of special waste to be amended to take account of their existence. Given that one of the most persistent criticisms of the definition of special waste has been the reputed scarcity of reliable toxicity data, a fortiori it would be impracticable to seek to qualify the definition by reference to a genuinely obscure and poorly documented phenomenon [45].

In other words, the committee seems to be saying that such risks exist, but since they cannot even be estimated, regulation must be structured as if they did not exist. Note that in this case, at least the excluded question was recognized before being re-buried. In other cases of the scientific framing of problems, they are not – recall the examples of 2,4,5–T.

The point is not whether the committee was correct or otherwise in ignoring synergy, but whether the implications of leaving it out of technical regulatory definitions are picked up and compensated for by institutional correlates. Thus, it would be realistic to exclude this dimension from regulatory definitions if, instead, one relied upon local authority experts to know what in their area is being mixed or might be mixed, and what the most risky combinations are likely to be (of course, this would require intensive investment in local regulatory institutions). The overall point, however, is that reliance upon unitary formal regulatory risk definitions and analyses, and the assumption that their advancing precision (i.e., standard situations) will adequately match ill-defined behavioral–technical contexts of the sort that exist for hazardous wastes, only makes for greater ignorance. This is because the analytical–regulatory framework, if it remains within a unitary institutional orientation, is less able to recognize and respond to network pluralism. In appearing to manage uncertainty, it fosters ignorance and the potential erosion of public credibility.

9.8. Conclusions

Regulation is the crucial arena in which universal principles and ideal methods or starting assumptions must be converted into concrete effects. It is here that the concepts embodied in science and in policy rhetoric have to grapple with real, varying social experiences, as well as organizational and physical constraints, uncertainties, and contradictions [46]. Science, here risk analysis, is one of the few generally valued resources in making environmental policies, and in making them credible. In the risk assessment of chemicals, scientific uncertainty and ignorance are being relentlessly expanded by institutional driving forces, because of the escalating understanding and control demanded of science [47]. This topic is discussed in Chapter 10.

The dominant model of science in policy and risk management portrays a given pool of facts defocused by the surrounding imprecision or technical uncertainty. This fuzzy area stands as a "feasibility space" for credible, conflicting technical views of risks, etc. [48]. When the facts are uncertain enough, different groups may even, with equal credibility, define spaces that do not overlap. More objective analysis or disciplined debate (or both), so this model goes, will tighten reality's constraints on conflicting claims, and thus approach a singular feasibility or *credibility* space within which all policy options must be located.

Conventional approaches to policy decision-making under uncertainty depend upon the idea that the factual domain, though uncertain, can be treated as prior to and independent of values, and that decision-evaluation processes (e.g., into standards for regulation) optimize within the mathematically formalized uncertainties. Even uncertainties in values are treatable in this framework, using multicriteria methods. However, if the factual domain and its uncertainties are actually *reflections* of values, and complex ones at that, the whole framework of thinking about science, risk assessment, and regulation changes. Ignorance may surround the framework of *explicit* uncertainties; this ignorance is more significant when different frameworks exist in the very system being analyzed. The aim in this chapter is to show that "factual" technical uncertainty, including that developed in highly controlled expert analysis, already embodies multiple institutional commitments of a value kind, the important ones of which are often tacit, incremental, and inadvertent. This is not recognized at all by the dominant model, which portrays expert disagreement as due to residual data imprecision or to *deliberate* external policy interests illegitimately concealed in the science.

We have used examples from technological risk analysis to show that intrinsically different definitions of the relevant problem system always exist, even in well defined systems. (This does *not* mean that *any* definition can be made up.)

The essence of formal risk analysis is to advance precision, but this may reduce uncertainty only to increase ignorance. Institutionally, this correlates with an assumption of, or need for, risk situations that are standardized in behavioral and technical detail; this in turn corresponds with centralized uniform regulation. Yet the features of the hazardous wastes issue emphasize the severe limits of this approach. The cross-cutting, multiple rationalities and the data diffracting processes outlined earlier in this chapter, show how unrealistic is regulation based upon uniform, precisely equal life-cycles and risk situations for the same waste, wherever it arises. Actors *in the system* can define even apparently unambiguous technical terms in unforseen ways. Only in 1984 did it come to the notice of UK regulators that liquid wastes banned from landfill were being redefined and legally landfilled as solid waste by putting them in drums – which were immediately burst open by the landfill-compressing caterpillar tractor.

The problems of creating coherent frames of reference for expert risk analysis were seen to exist even for nuclear reactors; but the level of flexibility, the indeterminacy, and the lack of structure in the case of hazardous wastes places the goal of such a single-frame coherence beyond reasonable limits of credibility. Even at the level of national systems, if we could imagine that these were internally coherent, we show in Chapter 5 that incompatible overall approaches, criteria, and technical frameworks were developed for regulatory use. Assumptions as to the underlying risk situations (including their degrees of standardization) vary enormously, depending upon institutional structures of regulation.

In using risk analysis for regulation, we have first to define the structure of the risk-generating system, to decide which factors are important risk creators, which of these can be controlled or mitigated, and how. If it is very badly structured, this could be due to lack of information (it is better structured than we know); or to *genuine* variability and indeterminacy; or to both. Hazardous wastes suffers both, but the diverse behavioral freedom within the system looms large. If every actor were perfectly trustworthy and infallible, no overall structure would need to be conceived. Central policy could issue environmental protection norms and let the system run itself. Each actor would conduct her own risk analysis according to her own situational activities, aims, and constraints. From a central point of view this would be amorphous and insecure (except that the actors are *trusted*), but still an optimal sytem. At the opposite extreme, if *no one* could be trusted, then their freedoms would be totally removed, and central control of all wastes would be necessary, with a single organizational structure for the whole life-cycle. A single risk analysis framework (akin to the WET model) could define "best practicable environmental" (i.e., T&D) options for every waste, and these could be enacted by the controlling agency.

The whole point, of course, is that neither of these scenarios is realistic, for different reasons. The strategic question is how much freedom to give to actors within the system – economic agents and local regulators – for example, whether they should be given tight, direct behavioral prescriptions ("all solvents to high-temperature incineration"); removed altogether (e.g., replaced by a monopoly public corperation); or allowed to perform their own risk analysis against indirect norms (e.g. "if a leach test shows potential toxicity above a threshold, landfill is not allowed"). There are many detailed options, (licensing, inspection, mandatory rules, codes or practice, etc., etc.).

The overall point is that risk-analytic uncertanties that appear as *technical* uncertainties have underlying *institutional* determinants, and thus *potential* institutional solutions.

In Chapters 6 and 7 we examined institutional approaches that have very different implications for the way in which the underlying tension is managed between standardized, more precise risk assessments that are less rooted in diverse situational realities (high ignorance, low uncertainty), and less standardized, less precise but more situationally connected ones (low ignorance, high uncertainty). The centralized management systems of Bavaria and Hessen effectively collapsed this tension by institutionally reducing life-cycle variability to within the control of one management framework. The UK has implicitly accepted the need to keep risk assessment more situationally located, dispersed and flexible (but without the *institutionally* distributed regulatory resources to make it work well). The USA has adopted the most standardized, precise, and centralized technical approach of all, yet for an *institutionally* dispersed system. This high ignorance, low uncertainty US approach is possibly the most prone to uneven results – dramatic failures and perhaps dramatic successes. These different approaches not only correlate the different technical approaches of the countries with their characteristic institutional structures, but also correspond with broader political cultural properties.

Note the radically different view of the role of science in policy that emerges from this analysis, as compared with the conventional one. In the latter, it is scientific lack of resolution that leaves room for policy values to play on and produce versions of the facts to suit conflicting partisan positions. Two implications are:

(1) That the values are *chosen*.
(2) That more rigorous scientific analysis will produce less uncertainty, therefore less opportunity for concealed value conflict. This, indeed, is how Weinberg's notion of transscience structures the relationship [49], except that he identified areas of science that, for the lack of intrinsic resolution, could not reduce uncertainty beyond a certain point.

This conventional stance corresponds with the policy belief that if regulation is incomplete and uncertainty is high, then the crucial need is for greater technical rigor in defining "hazardous waste".

Our perspective shows that since expertise is always fundamentally context-bound, intensifying formal analytic inputs can actually *increase* uncertainty and incoherence, as each elaborates its own incommensurable underlying institutional framework. This is exemplified by many modern cases of scientific experts in policy debates [50].

The implication is that the cognitive *framing* of risk analysis – what is taken as the idealized context, what degrees of aggregation at what points, what kinds of uncertainty to try to analyze and control, what to ignore – does not merely flow from "the objective" problem, "nature", or "scientific method". But neither is it reducible to chosen values. Furthermore, it has direct *practical* institutional dimensions, because the relevant risk-generating processes interpenetrate technical and human behavioral actions. Thus, the way a risk analysis is framed implies that certain *behavioral* relationships in the regulated system are unproblematic, and that others are problematic.

The effect of our argument that scientific uncertainty is more intractable and, indeed, qualitatively more fundamental than is recognized is to cast doubt upon the viability of any regulatory system, national or international, that depends upon technical precision and standardization of risk analysis, unless it can first "standardize" *institutionally*. If this is deemed impossible or unacceptable, it may imply decentralizing analysis and regulation of risk situations, but there are two key caveats:

(1) The institutional *resources* must be available in distributed form.

(2) It is unrealistic to expect to *fully* match the extreme heterogeneity and indeterminacy of hazardous waste life-cycles by institutional dispersion of regulation, because (a) the uncertainties are deeper and the degree to which they can be formalized for analysis and control is more limited than recognized (see also Chapter 10); (b) the desired decentralization comes into direct conflict beyond a certain point with the limits of (1) above; and (c) public credibility demands a degree of unification in the overall view of the system, to justify its operating rules.

There are severe limits to the ability of formal risk analysis to analyze and control technical uncertainties when these, and the risk-analytic framing, are driven by underlying *institutional* uncertainties. A logical response to these dilemmas would therefore seem to be either: to limit the *institutional* uncertainties and reduce the need for refined analysis, by redefining hazardous waste life-cycles into single management systems on a partly decentralized (e.g., regional or state) level; or to accept the intractability of the uncertainties by attempting to strengthen self-regulation mechanisms. A problem with the latter is that one of the main methods of doing this has been to lay out precise normative risk-analytic criteria (e.g., testing methods), but these only address one point in the life-cycle, so the problems of relevance and overstandardization still apply. Also, the question of ensuring enforcement remains, even if the risk analysis is carried out.

Whatever one's preferences, the overall conclusion of the analysis is that the transaction costs of controlling back-end uncertainties in hazardous waste management are realistically far higher than presently recognized. Thus, means must be found to reflect this in the relevant upstream decisions.

A main theme of this chapter has been to argue that scientific risk analysis always has to frame a risk problem, which means defining a "model" risk-generating process. This involves tacit judgments, which may vary among equally competent experts. Science progresses by freezing a defined context and assuming an ideal world. *Progressive* science "pulls itself up by its own bootstraps", by responding to signals that indicate deviations from reality in the set of assumptions. The world becomes more rich and complex, the science more interesting and useful. Nevertheless, a basic and necessary tendency also pulls in the opposite direction, to cling to existing themes and their primary framing assumptions until they are "really" worked out. *Progressive* conflict clarifies the nature of these commitments, so that their validity, implications, and alternatives can be reexamined. Sometimes, however, scientific paradigms may manage to insulate themselves from critical evidence and alternative premises, so that their primary assumptions are not challenged and developed. The prevailing analytical framework does not enter into constructive interaction with its surroundings, and so develops its own ideal-world assumptions.

We have shown in this chapter how the risk analyses of several systems – pesticides, growth hormones, the McKenzie Valley oil pipeline, the Windscale nuclear fuel reprocessing inquiry – were oblivious to the centrality of the social assumptions that they had to make, even to define a coherent *technical* risk-analytic problem. They were therefore in no position to explore the implications and validity of those naive, ideal-world framing premises and system definitions, nor to interact constructively with alternative assumptions or evidence. They keep uncertainty low and ignorance high. Our suggestion is that risk analysis and regulation must nurture *institutional* methods for identifying their founding value premises and social assumptions, and to interact constructively with alternative experiences and premises. Especially for the more dispersed, *network*-like technological risk systems, intensifying the conventional, "context-free" framework only risks decreasing public credibility. Yet there are strong pressures toward this identification.

In Chapter 10 we examine this tension by moving from the risk analysis of technological systems to that of toxic chemicals and environmental processes.

Notes

[1] Wynne, B. (1981), Institutional mythologies and dual societies in the management of risk, in Kunreuther, H. and Ley, E. (Eds), *The Risk Analysis Controversy: An Institutional Perspective*, pp. 127–143 (Springer, Berlin); see also Barnes, S.B. and Edge, D.O. (Eds) (1982), *Science in Context*, part V (Open University Press, London). For the origins of the "informalist" view of science, see Polanyi, M. (1959), *Tacit Knowledge* (Routledge and Kegan Paul, London); Ravetz, J.R. (1974), *Scientific Knowledge and its Social Problems* (Penguin, Harmondsworth, UK); and Hesse, M.B. (1974), *The Structure of Scientific Inference* (Cambridge University Press, London).

[2] Warner, F. (Chairman) (1983), *Risk Assessment: Study Group Report* (Royal Society, London); Southwood, R. (Chairman) (1984), *Tackling Pollution: Experience and Prospects*, Royal Commission on Environmental Pollution, 10th Report (HMSO, London).

[3] This section provides a fairly standard treatment, though a reader new to the field must be aware that terms like "hazard" and "risk" often vary in meaning, in the way outlined, without being defined. The literature on risk is huge. See the journal *Risk Analysis* and *Risk Abstracts*.

[4] Wynne, B. (1982), Nuclear power — Is the health risk too great? *Journal of Medical Ethics*, **8**, 77–85.

[5] Kemeny, J.G. (Chairman) (1979), *The Accident at Three Mile Island*, Report of a Presidential Commission (US Government Printing Office, Washington, DC). See also the Royal Society Report, *op. cit.* [2].

[6] Morgan, M.G. (1984), Technical uncertainty in quantitative policy analysis – A sulfur air pollution example, *Risk Analysis*, **4**(3), 201–216. See also the reply by Ravetz, J.R. and Funtowicz, S.O. (1984), *Risk Analysis*, **4**(3), 219–222.

[7] The major difference was that three of the facilities were for liquefied natural gas, methane, which requires very low temperatures (−161.5°C), the other was for liquefied petroleum gas, mainly propane and butane, which are less volatile and can be stored at nearly ambient temperature and pressure. In fact, this technical difference was dominated by other differences introduced by the analysts in the studies.

[8] Mandl, C. and Lathrop, J. (1983), LEG risk assessments: experts disagree, in Kunreuther, H. and Linnerooth, J., *Risk Analysis and Decision Processes: Liquid Energy Gas Facility Siting in Four Countries*, pp 148–177 (Springer, Berlin).

[9] Trenn, T. and Merton, R.K. (Eds) (1979), *The Genesis and Development of a Scientific Fact* (University of Chicago Press, Chicago, IL); Knorr, K. *et al.* (Eds) (1980), The social process of scientific investigation, *Sociology of the Sciences Yearbook, 1980*, No. IV (Reidel, Dordrecht); Barnes and Edge, *op. cit.* [1].

[10] *Op. cit.* [1] p 166.

[11] SAI and FERC are abbreviations of the names of different research firms offering risk analysis for the California siting decision.

[12] Schon, D.A. (1982), The fear of innovation, in Barnes and Edge, *op. cit.* [1].

[13] Thompson, M. (1983), Postscript: A cultural basis for comparison, in Kunreuther and Linnerooth, *op. cit.* [8].

[14] Wynne, B. (1983), Redefining the issues of risk and public acceptance: The social viability of technology, *Futures*, **15**, 1–31; also Wynne, *op. cit.* [1], and Wynne, B. (1985), From public perception of risk to technology as cultural process, in Covello, V., Mumpower, J., Stallen, P., and Uppuluri, V. (Eds), *Environmental Impact Assessment Technology Assessment and Risk Analysis* (Springer, Berlin); Johnston, R. (1984), Controlling technology: An issue for the social studies of science, *Social Studies of Science*, **14**, 97–114.

[15] Cox, L.A., Jr., (1982), Artifactual uncertainty in risk analysis, *Risk Analysis*, **2**, 121–135.

[16] Inhaber, H. (1978), *Risk of Energy Production*, AECD-1119/Rev-1 (Atomic Energy Control Board of Canada); and Holdren, J.P. *et al.* (1979), *Risk of Renewable Energy Sources: A Critique of the Inhaber Report*, ERG 79-3 (Energy and Resources Group, University of California, Berkeley, CA).

[17] Campbell, B. (1981), *Disagreement between Experts: Scientific Disputes in the MacKenzie Valley Pipeline Inquiry*, unpublished PhD Thesis (McMaster University, Hamilton, Canada).

[18] Wynne, B. (1982), *Rationality and Ritual: The Windscale Inquiry and Nuclear Decisions in Britain* (British Society for the History of Science, Chalfont St. Giles, UK).

[19] Cook, J. and Kaufman, G. (1982), *Portrait of a Poison: The 2,4,5-T Story* (Pluto, London); Wynne, *op. cit.* [14].

[20] Clutterbuck, C. (1980), Pesticides risks, *Landworker*, **June**, 8–10.

[21] I am grateful to Henri Smets and Patric Lagadec for this example.

[22] This account is taken from an interview with E. Finnecy, Environmental Safety Group, Harwell, UK, who was involved in the case.

[23] UK Hazardous Waste Inspectorate (1985) *Hazardous Waste Management: An Overview*, p. 27, First Report of the HWI, DOE, London.

[24] This section is taken from a case study by Michael Dowling, who was employed by the Boston Municipal Waste Management Decision and took part in the survey.

[25] Fennely, P. *et al*, (1976), *The Generation and Disposal of Hazardous Wastes in Massachusetts* (GCA Corp., Bedford, MA).

[26] *Ibid.*, p. 13.

[27] *Ibid.*, p. 19. It is interesting to note that a figure "accurate to a factor of two" was given to two decimal places.

[28] Arthur D. Little (1979), *A Plan for Development of Hazardous Waste Management Facilities in the New England Region* (Arthur D. Little, Cambridge, MA).

[29] *Ibid.*, p. B-10.

[30] *Ibid.*, p. B-9.

[31] *Ibid.*, p. B-12.

[32] Department of Environment Management (1982), *Hazardous Waste in Massachusetts, 1982 Statewide Environmental Impact Report* (Department of Environmental Management, Boston, MA).

[33] *Ibid.*, p. i.

[34] *Ibid.*, p. ii.

[35] US EPA (1983), *National Survey of Hazardous Waste Generators and Treatment, Storage and Disposal Facilities Regulated under RCRA in 1981* (EPA, Washington, DC).

[36] US Congress, Office of Technology Assessment, (1983), *Technologies and Management Strategies for Hazardous Waste Control*, OTA-M-196 (OTA, Washington, DC).

[37] Donovan, D.G. (1981), Fuelwood: How much do we need?, *Newsletter*, DGD 9 (Institute of Current World Affairs, Hanover, NH), quoted by Thompson, M. and Warburton, M. (1984), *Knowing Where to Hit It: A Conceptual Framework for the Sustainable Development of the Himalayas*, Ch 3, Working Paper WP-84-30 (International Institute for Applied Systems Analysis, Laxenburg, Austria).

[38] This information was given in interviews with radiation research experts A. Stewart, E.P. Radford, and J. Rotblat. How much difference it makes to risk estimates is, of course, along with most other parts of the issue, a matter of scientific disagreement.

[39] Clarke, M. (Ed) (1983), *Corruption: Causes, Consequences, and Control*, Chs. 13–16 (Frances Pinter, London).

[40] Shenfield, S. (1983), *Pripiski*: False statistical reporting in Soviet-type economies, in Clarke, *op. cit.* [39], Ch 14, p. 243. Of course, such systematic data diffraction is not at all unique to such economies, but is a tendency of all bureaucratic systems.

[41] E. Kiss, personal communication from Hungarian case study research.

[42] Lord Gregson (Chairman) (1981), *Hazardous Waste Management*, Vol. 2, pp. 246–280, UK House of Lords, Select Committee on Science and Technology (HMSO, London).

[43] Thompson and Warburton, *op. cit.* [37].

[44] This case is taken from research by Dutch case study collaborators.

[45] Department of the Environment (1985) *Report of a Review of the Control of Pollution (Special Wastes) Regulations 1980*, p.74, DOE, London.

[46] Wilson, J.Q. (Ed) (1981), *The Politics of Regulation* (Basic Books, New York, NY).

[47] Nicholson, W.J. (Ed) (1981), Management of Assessed Risk for Carcinogens, *Annals of the New York Academy of Sciences*, **361**.

[48] Wynne, *op. cit.* [1].

[49] Weinberg, A. (1973), Science and transcience, *Minerva*, **XI**, 261–279.

[50] In the chlorofluorocarbons-ozone dispute, for example, the US government supported attempts to develop one-dimensional models, which would create data that could be compared between studies and thus (it was hoped) create a consensus. British approaches were dominated by two-dimensional models, because there was less anxiety abut policy consensus. But these overall "interests" were shaped by cascading elaborations of methodological, cognitive, and other framing commitments, with correspondingly different detailed profiles of observation, key variables, and perceived uncertainties. This same view of uncertainty and cognitive conflict can be derived from the social analysis of science. See [1], [9], [18], [39] and Whittemore, A. (1983) Facts and values in risk analysis for environmental toxicants, *Risk Analysis*, **3** 23–34. Typical of communications from the White House Office for Scientific Technology Policy is that issued by D. Calkins *et al.* (1980), Identification, characterization, and control of potential human carcinogens: A framework for federal government decision making, *Journal of the National Cancer Institute,* **64**; 169–176. When a new area of "uncertainty" is identified, such as chronic toxic chemical effects, the different entrants bring with them their different conceptual and methodological resources already shaped by previous problem definitions, explanatory interests, and research traditions. These scientific positions evolve in different ways through the unfolding policy issue so that at any one time what is regarded as known, as uncertain, and as problematic by each school of thought is already shaped by these institutionalized (social and cognitive) commitments. As Rip has summarized it,

> ... this way of looking at the socio-cognitive dynamics of a controversy makes it impossible to speak of areas of uncertainty that leave room for different interpretations which are guided by the differing values and interests of the parties in the controversy. These values and interests have already been at work, through problem definitions and research agendas, in determining what the lie of the land will be. And the controversy is often not about the interpretation of a given area of uncertainty, but about which areas are to be considered certain and which areas of uncertainty are sufficiently irrelevant to remain uncertain.

Thus, it becomes vacuous to suggest that there would be no controversy if there were no uncertainty. But this does seem to be an assumption in current controversy studies: because there is a space left open by uncertainty, "controversies over science and technology develop over competing political, economic, or ethical values." This might be the case in NIMBY-type conflicts, where people do not want a new plant, a powerline, or some other technological project "in their backyard." Even then, however, socio-cognitive dynamics appear as soon as the assessment of hazards and other science-related aspects of decision making are drawn into the debate and subjected to further negotiation and articulation.

Rip, A. (1983), *Comparative Study of Science-related Controversies: Avoiding Blind Spots*, paper presented to the Annual Meeting of the Society for the Social Study of Science, Blackberg, VA (mimeo, University of Leiden, Holland).

CHAPTER 10

Risk Assessment and Regulation for Hazardous Wastes

Brian Wynne

10.1. Introduction

This chapter examines the application of risk analysis to toxic chemicals, especially to hazardous chemical wastes. It focuses the general perspective of Chapter 9 on the more specific problems of toxic chemical wastes and their environmental distributions. The overall area is usually differentiated into three foci – risk source (e.g., chemical plant) at the point of *emissions*; environmental movements and concentrations, etc., or *exposures*; and *effects* on humans. The conception in Chapter 9 of technological systems as networks complicates this simple framework. For hazardous wastes, the complications deepen because the first two categories interpenetrate, since environmental movements and transformations of wastes occur by *human* design as well as natural processes, creating many different emission situations and mechanisms, even for the same type of waste.

In this chapter we first outline the emergence of chemical–environmental risk assessment and indicate the degree to which its analytical capability falls short of comprehending the actual processes of risk generation. Recalling the heterogeneity and lack of structure, the multiple points of partial regulation, and the analytical confusion of natural and human behavioral factors in chemical waste life-cycles, we return to the problem of how regulation tackles the yawning gulf between the state of risk-analytic knowledge and actual risk processes and situations. Informal expert judgment is found in two related areas: the relatively controlled scientific field of toxicology; and risk analysis of *ad hoc* chemical risk situations. The existence of informal, subjective judgments within science is not in itself remarkable, but the extent of its *domination* of attempts to formulate accountable scientific analysis in regulation whilst attempting to retain a language of formal expert control, suggests that regulatory bureaucracies are engaged in the artificial conversion of rank ignorance and implicit conflicts of perspective into "manageable uncertainty". The evidence suggests that the increasing concern with credibility and reassurance that is inspiring the development of risk analysis also encourages further development of the split within science, between a public language of order and control and informal disorder. We conclude this chapter with a discussion of this predicament.

10.2. Environmental Risk Assessment: The Poverty Trap

The institutionalization of public health concern in legislation and associated regulation has meant the natural extension of risk assessment into areas of environmental pathways and human or environmental end-effects [1]. Although some disciplines relevant to the latter – for example, toxicology – have long traditions, their previous ethos has often been clinical and individual, and so badly suited to questions of *collective* public health effects [2].

Although formal risk analysis was dominated in the 1950s and 1960s by risk source, or "mechanical system" analysis, this period also saw the beginnings of more systematic environmental exposure and dose–effect analysis. Indeed, attempts systematically to gather evidence and define the risk associated with *radioactive* exposures had begun with the establishment of the International Commission on Radiological Protection (ICRP) in 1927 [3]. However, this originated in the concern of clinical radiologists about individual risks to themselves and patients in clinical X-ray therapy and diagnosis. The effort took on new impetus and was reorganized in the 1950s, following the industrialization of nuclear energy and a shift toward collective public and workforce exposures and risks. Early work in the field was dominated by pharmacology and experimental pathology (as well as the earlier clinical traditions), from which chemical toxicology also developed. The established paradigm was built around the simple concept of a threshold so as to reduce acute doses and associated acute effects (e.g., gross tissue damage as measured by histology). Thus, approaches to risk and regulation standards setting involved short-term experiments to establish "no observable effects levels" (NOELs) for such gross effects. These were then converted straightforwardly into maximum allowed exposure limits by the application of a safety factor, typically 10 for workforces or 100 for public dose limits. The implication of this approach was that such exposure limits involved zero risk.

A whole institutional framework was established on these conceptual and methodological commitments. Initially, even radiation-induced carcinogenesis was also thought to be associated with only gross tissue damage. However, with the observation in genetics of radiomutagenesis and the somatic cell mutation theory of carcinogenesis (an early version of the "one-hit" model of carcinogenesis), the alternative idea slowly became established amongst a new sector that the origins, at least of cancer, lie in more microscale damage to far more sensitive entities, such as genetic material; this would emerge as recognizable effects only over the long term. This new idea suggested that there may well be no dose threshold for health damage, including mortality, so that no "zero-risk" standard could ever exist. However, long latency periods and multiple causes became associated with the observation of effects and entirely different scientific end-points, methods of observation, and supporting interpretive frameworks became involved.

It was through the gradual though contested establishment of this no-threshold idea that risk analysis in its presently recognized form developed [4]. In this new form, standards setting took on a concern for risk–benefit balancing, on the grounds that if no zero-risk levels of exposure could be found, and zero exposure to most agents was impracticable, a level of exposure and corresponding risk would have to be set that made an acceptable trade-off between risk levels and the costs of reducing exposures. Thus, all the now-familiar procedures of optimization, evaluation of "best practicable" control technologies, elaborate analysis of low dose–effect relationships for various agents, and concerns for public distributions and acceptance of risk, entered into mainstream regulatory agendas and processes.

As compared with radioactive risk assessment, two further factors complicated risk assessment for chemicals.

First, exposure is to relatively few different kinds of radiation (e.g., α-emitters, β-emitters, γ-emitters) from *relatively* few sources or events. Chemicals and their forms of environmental release are so multivariate and diffuse that they present different orders of magnitude too complex even to identify exposure pathways and populations, let alone to begin to attach quantitative estimates to them.

Second, radiation penetrates tissue physically and delivers its damaging energy in relatively well understood ways, so that specific organ or tissue doses can be reasonably calculated from external exposure effects. Although biological processes are important too, e.g., metabolic uptake and movement of particulate radiation (especially α-emitters, which deliver highly localized doses), well defined physical models are available to show specific local damage. The same is far from true for most chemicals, so that specific internal organ or tissue doses cannot be calculated even if *exposure* levels are known.

When we consider exposures to chemical wastes, this ignorance and complexity is expanded even more, because the chemical compositions, points of generation and disposal, and subsequent mixing of wastes are frequently badly known at best.

Thus, in both dimensions of "external" risk analysis – exposure pathways (including points and focus of emission) and dose–effect estimations (including the judgment of which possible effects to explore) – hazardous waste management is severely underdeveloped [5]. Probabilistic estimations often conceal downright inestimability or ignorance. Some regulatory purposes can avoid some of these lacunae – for example, the intrinsic hazards of chemicals can be ranked without having to analyze situational exposure routes and magnitudes; and different disposal options and exposures for the same chemical can be evaluated assuming the same dose–effect relationship. On the other hand, ignorance within each may multiply. For example, if an existing inhalation pathway for a given chemical is not identified and inhalation creates more damage than ingestion, then the mistake begun by ignorance in one dimension is multiplied by ignorance in the other.

One can put the state of development of chemicals risk analysis in perspective by observing that far less attention has been devoted to it than to nuclear radiation, and yet it is a far more diverse and ill-defined area [6]; but even with the relatively huge attention devoted to it, for decades, the low-dose radiation risk issue still defies scientific definition and consensus. Thus, in its review between 1976 and 1980 [7] the US National Academy of Sciences' Biological Effects of Ionizing Radiation (BEIR III) Committee in the end split irreconcilably over carcinogenic risks, and had to conclude that ignorance was still so deep that no single dose–effect inference was scientifically warranted at doses beneath 10 rad, the very area of public health and regulatory concern.

For perspective it is worth thinking of event-tree analysis as a method for examining the movements of environmental chemicals:

(1) Even in localized situations the processes are characterized more by continuities and gradations than are mechanical systems (though recall from Chapter 9 the problem of partial failure of components or of grades of human intervention). Thus, more complex, continuous and nonlinear functions and models are needed.
(2) In most broader situations, especially taking cross-interactions into account, we are simply too ignorant to construct models.
(3) Even where process parameters or relationships exist, we have little or no information that would allow a normal risk-analytic estimation of *probabilities* of exposure, or of mechanisms of possible exposure.

Furthermore, emissions risk analysis for a complex plant either stops at release categories, or has single-phase dynamics, such as vapor- or radioactivity-dispersion, to model. Toxic chemicals are often mixed and, unlike accidental releases from a plant failure, the *starting* dispositions may

Table 10.1 The range of potential human health end-points to examine in chemicals risk assessment (nonexhaustive).[a]

REPRODUCTIVE	NEUROTOXIC
Sexual dysfunction:	Various tissue toxic attack – axons, neurons, myeline, glin, blood vessels
decreased libido	Distal axonopathy – various sites
impotence	Vibration sensation
Sperm abnormalities:	Motor nerve conduction
decreased number	Peripheral neuropathy
decreased motility	Sensory conduction
abnormal morphology	Cranial, spinal, lemniscal and thalamo-cortical nerve activity
Subfecundity:	Toxic encephalopathy – convulsions, hallucinations
abnormal gonads, ducts, or external genitalia	Inflammations
abnormal pubertal development	Tremor
infertility (of male or female origin)	(Pigmentation changes)
amenorrhea	GENETIC
anovulatory cycles	Biochemical:
delay in conception	altered protein electrophoretic mobility, altered enzyme function
Illness during pregnancy and parturition:	Chromosomal:
toxemia	sister chromatid exchanges (different cells and parts of cell cycle)
hemorrhage	chromosomal aberrations and breaks, micronuclei formation
Early fetal loss (to 28 weeks) and stillbirth:	Sperm:
intrapartum death	abnormal morphology
death in first week	lactic dehydrogenaze isozymes
Decreased birthweight	CARCINOGENESIS
Gestional age at delivery:	Various sites and kinds of tumour
prematurity	"ASYMPTOTIC" DISEASES
postmaturity	Reduced performance of normal functions – perceptual, memory, motor skills, reflexes, balance, intelligence, problem solving, attention levels, sleep, etc., etc.
Altered sex ratio	
Multiple births	
Birth defects:	
major/minor	
chromosome abnormalities (detected in early fetuses, through amniocentesis, in perinatal deaths, in live births)	
Infant mortality	
Childhood morbidity	
Childhood malignancies	
Age at menopause	

[a] Adapted from Lowrance (ed.), ref [5].

Many of these categories interact in different ways, e.g., observations at one level may be symptoms of 'observations' at another.

be very poorly known. There is also a dauntingly wide array of different health effects to have to try to identify, evaluate, and ultimately model (see *Table 10.1*). Although some of these may be near-enough related as to be under the domain of one expert specialty, most imply separate commitments from scientific traditions of observation, etc., for which a given end-point has been naturalized to the exclusion of others. Thus, risks cannot be evaluated comprehensively by choosing between and then comparing or synthesizing different kinds of health end-point. Each one represents a whole way of life of an expert discipline, and may be brought into high focus (or out again) by cultural values and attention cycles. To say, as do some decision models (see Chapter 11), that risk management can analyze evenly for the range of effects of *Table 10.1*, then "choose" among them by weighting for values, is to ignore the different "institutional" traditions and commitments that constitute each mode of analysis.

The environmental–biological side of chemical risk analysis is therefore in its infancy. Indeed, in important respects it becomes more infant as time passes, because the rate of new chemicals and mixtures arising, combined with the rate of "new" exposures from "old" chemicals (e.g., past dumps) is greater than the rate of toxicity research. Recall from Chapter 3 the arrival of up to 1000 *new* chemicals on the scene every year and the extreme costs of fully testing these, not to mention the timescale of approximately three years. This is for new chemicals only. At the Love Canal waste dump, after a colossal analytical program involving over half a million data points [8], over 400 different chemicals were eventually identified, with a range of characteristics (*Table 10.2*) estimated from a cursory scientific literature survey on animal and *in vitro* testing.

Table 10.2. Characteristics of the 400 chemicals found at the Love Canal site.

Characteristics	*% of total*
Mutagens	11
Carcinogens	13
Embryotoxicants (teratogens)	7
Hepatotoxicants (liver)	10
Neurotoxicants (brain)	16
Renal toxicants (kidney)	10
Pulmonary toxicants (lung)	9

However, *over 50% of the chemicals identified had no research record at all. Their properties were simply unknown, even though they had been produced and disposed of up to 30–40 years ago.* Even the data on the rest were so uncertain that the EPA's Carcinogen Assessment Group judged that only 4% (15 of 400) of the chemicals found could be given reasonably secure individual risk estimates (under unrealistically well defined dose conditions).

The pressures upon scientific knowledge as a regulatory (and antiregulatory) resource have strained the system beyond breaking point in recent years. Thus, in 1984 three top executives of a giant US company, Industrial BioTest (IBT), which it is estimated had been responsible for about one third of all toxicity tests for new chemicals in the entire world, were jailed "in one of the biggest scientific frauds ever seen in America" [9]. Thousands of chemicals in products (and wastes) were accepted worldwide on the basis of IBT's "tests", without independent repetition because of the enormous costs. Thousands of IBT's studies, performed for "a Who's-Who of the chemical, food and drugs world", were utterly worthless, with directly fabricated results. In others, nearly half of the exposed population of animals had died through lack of care, then been disposed of and either recorded as zero-damage results or replaced by unexposed animals.

Since IBT's exposure, US agencies have found that this was not an isolated case. Two executives at a New Jersey laboratory were convicted for falsifying data, and two more laboratories closed down following government investigations. Six more had government contracts canceled and four others sanctioned for lesser misdemeanors. The economic implications of these regulatory holes are indicated by just one case, that of the artificial sweetener Aspartame. World sales are estimated at $600 million per year, or two thirds of the total turnover of the producer, Searle Pharmaceuticals. According to an investigation for TV [10], when the US Food and Drugs Administration (FDA) examined Searle's own tests on Aspartame and six other products, they found them to be so shoddy and inadequate that they recommended a Grand Jury be convened so as to investigate possible criminal violations [11]. Tumors had been removed unreported from animals, and chemicals fed to the animals were not ground up and fully mixed into the animals' diet, but were given as lumps of the chemicals, which the animals could easily avoid, thus receiving a zero dose.

The credibility of these tests was defended on the grounds that "scientists in regulatory agencies around the world, some 36 different countries have affirmed that Aspartame is safe. They've all reviewed the same data". Yet this argument only underlines two important points:

(1) That the data reviewed were still Searle's own data; to have obtained independent data (which is what the usual reassuring image of scientific method would suggest) would have been unrealistic.

(2) That even if independent data were too costly to have obtained, the scientists "of 36 countries" demonstrated a normal property of science, that is *credulity*, contrary to the public image of unflinching skepticism and independent testing of claims.

In chemical risk regulation, at least, the costs of this false reassurance provided by a misleading public image of science far outweigh any value it may once (if ever) have had. The misconception encourages the idea that safety claims have been separately tested under formal reproducible scientific conditions, even when this has not been done. Any short-term public reassurance on specific questions is often repaid with interest in a later loss of general credibility by the institutions involved.

Chemical risk analysis is in a kind of intellectual "poverty trap". Because it is not well developed and the demands for knowledge are nearly always created under conditions of urgency, the kind of "research" performed is closely tied to urgent, retrospective regulatory demands. Even if this did not generate sloppy science and corruption, it draws resources from more carefully conceived fundamental research, which, simply because it is pitched at a more general conceptual level, when it does advance may solve problems of larger classes – it is more *anticipatory*. There is also a natural tendency for regulatory bodies (and regulated bodies for that matter) to be concerned only with the "science" (such as immediate monitoring) that corresponds with their minimal legal and regulatory needs. Both these factors create a certain framework for regulatory science in this area which we examine in greater detail later, in the context of public reactions. Contrary to the myth that science proceeds independently of its institutional setting, we see that it is substantively shaped by it.

In the next section we examine one response to the severe shortfall of regulatory resources and attention, namely how priority is given to which materials to select for risk analysis.

10.3. Ignorance, Judgment, and Stepwise Testing

As indicated in Chapter 3, even where it is most elaborate, regulatory attention for hazardous chemicals focuses upon only a small part of the available problem domain. The growing excess of chemicals and their potential effects over the available analytical resources combine to reduce the area of rigorous attention relative to "demand". Also, as analytical techniques become more powerful and refined, e.g., chromosome damage techniques and various neurological field techniques [12], they ironically expand uncertainty because they move further ahead of the ability to interpret the meaning of their new observations [13].

In the risk analysis of chemicals and chemical wastes, therefore, stepwise or tier-testing schemes have become normal currency, although their structure is far from fully defined [14]. *Tier testing* is designed to identify the priority in which smaller populations of material should be subjected to more refined, more expensive, and time-consuming risk analysis. Stepwise *hazard ranking* attempts to create a practical hierarchy of increasingly strict regulatory control for increasingly hazardous materials, or combinations of materials and situations.

In principle, these two phases are distinct because the first is an analytical process, which ought to be only a preliminary to the second, practical regulatory approach. In reality, the scientific poverty trap outlined above means that regulations are made directly on the basis of tier testing, which, being based on standardized laboratory protocols, does not address the situation variance problem in waste life-cycles.

10.3.1. Tier testing

Given that only a small fraction of all potentially harmful chemicals can be tested, the aim of tier testing is to first perform cheaper, more rapid, and cruder screening tests on a wider range, in order to try to find out what materials to test in more detail. One very important pragmatic limitation of such schemes is that they usually only apply to newly introduced commercial chemicals [15]. Therefore, those already in circulation or created anew in wastes are exempted unless particular ones are arbitrarily "chosen" and brought to attention dramatically, as with PCBs or with Love Canal, Gouderak (the Netherlands), and other dumps. Tier testing therefore bypasses a first, greater level of ignorance, particularly true of hazardous wastes, which involves not even knowing what chemicals, where, and in what form, to even screen for more detailed risk testing. In the hazardous waste area, tier testing has been used most in the analysis of past waste dump problems.

As already noted, there is a daunting range of potential adverse health effects to test for, and there is no reasonable scientific connection between rapid, acute effects and chronic, low-level effects. Thus, the ordering of tests is scientifically rather arbitrary. Whereas regulators see it as a rational means of optimizing regulatory knowledge within very limited resources, industry sees it as a rational means of cutting down regulation and thus the costs of introducing new products [16]. Many different schemes have been proposed or used. Typical is that proposed for the 1976 US Toxic Substances Control Act by a "consensus group" of industry and professional environmental bodies, convened by the Conservation Foundation [17]. This scheme is organized into four tiers, the first being a review of primary chemical and physical properties, such as chemical structure, volatility, purity, solubility, adsorption/desorption, etc. The second tier begins the biological tests, e.g., for acute toxicity in mammals, birds, and fish. Given that each LD 50 value (the mean lethal dose of 50% of subjects tested) is a statistical determination needing many data points, this already implies many laboratory experiments. This tier also entails plant and seed toxicity tests, analysis of transformation and degradation processes, and short-term mutagenicity and carcinogenicity tests, such as *in vitro* cell transformations. Already, in addition to escalating costs, it is notable that strict requirements for standardized experimental designs and reproducible methods are involved. These are by no means easy to develop and establish in practice, even between technically experienced practitioners, and require elaborate development involving interlaboratory comparisons and exchanges, thus further multiplying the costs and time scales. The laboratory conditions also, by definition, simulate a real-world risk process, and may or may not be a realistic model thereof. We return to this important question later.

If the earlier tests indicate the need, a third tier goes into greater detail, with "longer term", typically 90-day tests, for subacute, chronic, and teratogenic effects *in vivo* and *in vitro*, in a range of species and cell cultures. Biodegradability, bioaccumulation, and other environmental movement and/or transformation tests are also required at this stage, so as to develop a picture of potential exposures under different conditions.

If indications from the third tier suggest it, the final tier involves lifetime chronic toxicity tests in at least two mammalian species, with associated histopathological examinations and further mutagenicity tests. Neurological tests in various species, life-cycle plant growth and reproduction tests are also involved. Metabolic and environmental transportation studies are required to include more examination of the by-products of such transformations, since these may be toxic even if the parent chemical is not.

Some kind of escalating set of tests as outlined above is a pragmatic necessity. However, necessity should not be confused with fulfillment of rationality (and risk control) in any stronger sense. Merely listing the escalating tests in this way gives no sense at all of the vast array of large-scale and detailed experimental and interpretive uncertainties at every point. Furthermore, the structuring of the burden of proof in the stepwise screening process means that a chemical is exonerated from further tests unless it gives some positive indication in the earlier tier. But, for example, the acute tests and effects that occur earlier in the testing scheme may not relate to chronic effects, which occur by entirely different metabolic pathways, and act upon different entities and functions from acute effects. Thus, formally tiered risk analysis schemes and experimental protocols are not a substitute for experience and judgment. They are usually thought to replace these less accessible, less "controllable" decision processes with more transparent and controllable ones; however, the more elaborate formalization of risk analysis there is, the *more* informal judgment may be needed to fill it out and work it. It is important to realize that we are not dealing with a trade-off between formal or informal decision processes. Rather, the required scope of formalized analysis is being expanded by an ever-increased scope of *required* management and control, of perceived unknowns in the environment. It is entirely possible for uncertainty and informal judgment to expand along with formalization, as the required area of "control" expands.

10.3.2. Problems of using these models

Tier tests are framed as numerical scoring systems using risk assessment models. These systems fit bureaucratic needs by assimilating results of different test batteries into single, weighted aggregate scales, which are automatically calculated. One such model [18], attempting to distinguish degrees of hazard for wastes, has

> ... a single value represent all types of effects from different types of cancers to different kinds of graded responses such as liver damage. The conceptual link ... is the probability of an incident per unit dose. A score of "2" on our scale, for example, is intended to represent roughly a 1% risk of either contracting cancer or having an adverse effect from consuming 1 mg of pollutant per 1 kg of body weight per day ... The model assumes ... that most reported MEDs, or minimum effective doses, correspond to a risk of about 10%.

In the model one finds that the "roughly" 1% risk referred to is actually an order of magnitude estimate. This is only one link in chains of such multiplying uncertainties.

One can understand the reasons for attempting to develop such formal models, but their use is not without cost. The model synthesizes not only qualitatively different effects and processes, but also multiplies qualitatively different *uncertainties* into single dimensions. Acknowledging some of the uncertainties, the modelers then proceed to frame them:

> There are two general problems in scoring risks to health posed by hazardous wastes. First, most waste streams are complex mixtures, and many are poorly characterized. Second, even if adequate information were available on the composition of the wastes, for most chemicals, there is insufficient information on inherent hazard and environmental behavior to permit precise estimates of risks. For these reasons, scoring for risks to health is subject to large uncertainties and in many cases depends on scientific judgment. The scoring scheme we developed does, however, base application of scientific judgment on a formal scientific underpinning. By providing rigorous definitions of the factors to be scored and by specifying numerical equivalents for the scoring categories, it is possible to assign fairly precise scores to a number of well-studied chemicals. These scores can then guide the assignment of scores to less well-studied chemicals and to mixtures ... [19].

This confidence in the controlling effects of the method of formalizing the scores, for 140 different chemicals in diverse wastes, could only be justified if the "relatively well-studied" chemicals did not suffer significant extra uncertainties by dint of being *wastes* (as the authors themselves acknowledge they do) and if the waste and environmental processes for all the chemicals were the same, so that they could be assumed constant on a relative risk basis. These are unrealistic assumptions, yet the model simply combines all the uncertainties as if qualitatively the same into a log-normal probability distribution. The "guidance" provided by existing scores might give a false sense of solidity, especially in "analogous" case analysis.

The four major sources of uncertainty identified are:

(1) Lack of information (e.g., of inherent hazard).
(2) Site to site variability (e.g., quantity of materials released).
(3) Environment variability (e.g., population susceptibility).
(4) Chance events (e.g., leakage).

They indicate nothing of the real life-cycle system transformations and indeterminacies due to organizational–behavioral factors, and account only (and incompletely) for physical, natural system uncertainties. In addition, the kind of assumptions made about environmental movements in the model involves "simple dispersion relations and some subjective judgments based on common sense to arrive at a scoring system for exposure" [20].

Despite these very crude means of encoding processes that are riddled with uncertainties and ignorance into an elaborate computer model, a reviewer was very frank about its ultimate role. In addition to providing guidance about priorities for more rigorous regulation (note the inadvertent switch from tier *testing* to direct hazard ranking), its main property is its *manipulability*, and its main role in life is *persuasion*:

> Certainly the Model can be manipulated to reach almost any desired outcome provided all of its ramifications are clearly understood and can be put in perspective. The key will be to utilize this knowledge for appropriate policy and economic purposes. Since the framework will accommodate most outcomes, manipulation of the Model will allow for flexible control of hazardous waste issues without requiring major policy departures ...
>
> The key issue will be to convince the public that the Model represents reality with respect to the perceived risks [21].

What hardly needs to be pointed out is the glaring contradiction between the rather dubious basis for credibility of regulatory policies provided by the model and the central public persuasion role envisaged for it. Even where the predominance of informal judgments throughout the formal modeling is admitted, as it is here, the implications of this pervasive subjectivity are disregarded. Indeed, it seems that it is the need for public persuasion using "science" that drives this production of formal caricatures of knowledge.

It is important to emphasize that the point of this account of a formal risk model is not to deny the worth of trying to structure relative risks and options; it is to emphasize the inevitable dilemma that accompanies such attempts. This dilemma is that the synthetic processes of modeling such extreme levels of aggregation of phenomena using extremely artificial parameters and indices does not *clarify* research questions and issues for further resolution; it *conceals* them in the combined processes of aggregation and negotiation for social credibility in which the "knowledge" is utterly entangled.

The bureaucratization of diverse basic processes, values, and interacting uncertainties into one simple numerical dimension is encouraged by the felt need for the "objective" authority that is supposed to emanate from formal models [22]. Powerful forces press for their development and use, and they may even serve as a ritual framework in which opposing parties share in negotiating policy consensus. However, a more realistic possibility is that the gaps and inconsistencies, apparent behind the pretensions, will fuel, not dampen, continuing dispute and lack of credibility. On one item alone, for example, expert estimates of chemical toxicities vary by over two orders of magnitude, while the *aggregate* uncertainty allowed in the model indices of hazard is one. Under these circumstances risk analysis may actually obstruct rather than encourage the development of credibility (not to mention more mature and discriminating science) in regulation. In environmental–chemical risk analysis it is typical to find that many of the supposedly precise, experimental tests, let alone interpretations of results, have to be "done in the hands of a master" to mean anything at all [23]. Another widely acknowledged expert, McKay, has criticized the whole framework of stepwise testing and ranking in chemical risk analysis as a misbegotten attempt to substitute for experience and judgment:

> Undoubtedly, "scoring," or "rating," or "prioritizing" numerical schemes will be developed, allegedly to assist in identifying the most hazardous substances. It is the author's opinion that such schemes are usually misleading and are often pursued only because of intellectual laziness. There is no substitute for the careful gathering and assimilation of reliable, physical–chemical, biological, and industrial data by a broadly experienced group of well-informed and well-intentioned individuals who can then make a balanced judgment in which all the issues have been weighed subjectively [24].

This defense of science as a craft skill incorporating tacit experience, intuitive judgment, and subtly negotiated intraexpert agreements is often counterposed to formal models and methods. Yet craft skills may still be *organized* and rooted in useful practice. The main problem is that the kind of science McKay is defending costs more, takes longer and makes explicit the need for public trust in expert institutions. Regulation's use of formal caricatures as if they were science falsely implies the existence of an adequate knowledge for regulation, and attempts to side-step the issue of public trust.

Stepwise ranking schemes for hazardous chemicals, including wastes, extend in two directions from the kind of tier-testing scheme outlined before.

First, even to pass the entry gate, selection has to be made. Only about 6% of chemicals in the marketplace have been tested for animal carcinogenicity. Thus, for example, the Interagency Testing Committee (ITC) for the US Toxic Substances Control Act was required in 1979 to provide a scoring system to meet a six-month congressional deadline for the full repertoire of 70 000–100 000 or so chemicals already in commercial circulation. This was to determine which of these should enter into a more careful scheme akin to that outlined before. The expert group that conducted the scoring was itself forced to use a series of crude gates and selection principles established *ad hoc* [25]. The key one was simply to borrow lists from existing regulations in the hope that these had more definitive origins – a familiar resort. Furthermore, for all its heroic attempts to be definitive and explicit, the group repeatedly had to bridge huge gaps of ignorance and uncertainty with subjective judgments.

Second, the starting point for the ranking of hazardous chemical goods is *relatively* well defined compared with that for hazardous wastes. The quantitative ranking process for hazardous goods either takes the approximately 500 new chemicals per year, cuts them down first to about 40, then down to the three to six per year that can actually be subjected to full STAR (Scientific and Technological Assessment and Review) testing at a cost of $500 000; or, as in the exercise above, it selects 900 from the original 70 000 or more, and works these down in a similar way. Hazardous chemical wastes, on the other hand, are much less well known because, being socially defined as wastes, their chemical compositions and purity have no positive economic value; therefore, they may also be mixed and vary indiscriminately with unknown chemicals that result from uncontrolled interactions. As Finnecy has noted:

> Firstly there are many thousands of possible constituents in a waste, and waste is *very* rarely even an approximately pure substance. Secondly waste is rarely ever completely analyzed by anyone. Rather, analyses (or estimates of one sort or another) are made of a few 'significant' components while the rest of the waste is largely unidentified in anything other than general terms [26].

Even leaving aside for the moment this extra dimension of ignorance of hazardous wastes, it is worth quoting at length the reflections of a scientist involved in the ITC ranking exercise:

> I would draw several conclusions from this exercise ... that the scheme that was used incorporated a very significant degree of scientific judgment at every step. At each screening step, the chemicals screened out were examined manually by experts and the scoring involved at least as much scientific judgment as the use of objective data ... that the scoring system which considered 15 factors and had a range of 0 to 3 or 0 to 4, was at least as complicated for each of the factors as was justified by the available data. It may in fact have been too complicated. It may have tried to divide the chemicals more finely than our knowledge would justify. Frankly I believe that any reasonable scoring system can handle hundreds of chemicals with no finer subdivision than 0 to 3, or zero, low, medium and high ... that the screening process was limited primarily and very severely by lack of data on most chemicals. In fact, I would say that, if for each of the 900 chemicals we had placed information on the seven factors into separate boxes, at least two thirds of the boxes would have remained empty. My conclusion is that the scoring process was not limited by scoring methodology or by any other screening methodology. It was limited by the lack of data. I would therefore suggest that elaborate scoring systems are not justified at the present state of our knowledge. Until we have such more extensive data on most of the factors for which we have to screen, there is no point in developing elaborate scoring systems because they cannot be used [27].

This expert not only recognized that schemes and rankings apparently controlled by formal criteria actually required *more* informal judgments, rather than less, but also that risk-analytic ranking systems may *conceal* ignorance. The kind of streamlined "scientific knowledge" that such classifications require is easily confused with more fundamental scientific knowledge, which embodies open gaps and conflicts, conditional limitations, situational adjustments, and so on. Because it converts ignorance into apparently manageable uncertainty, this regulatory science cannot easily develop strategies to cope with surprise and *irreducible* conflict or ignorance. Regulatory systems *may* have to act *as if* knowledge existed and ignorance were strictly bounded. But the institutional processes involved confuse this regulatory "as if" confidence and associated "knowledge" with deeper scientific knowledge.

We have shown in Chapter 9 that a central dimension of expert informal judgment lies in framing a model of the risk-generating *process* for analysis; but in this chapter we have shown already that this is not the only source of informal judgments in risk analysis. We develop this point further, but first we need to look at the regulatory contexts in which tacit expert definitions of risk-generating processes meet other kinds of informal judgment.

10.4. Degrees of Hazard and Situational Risk Optimization

Although risk analysis begins with the estimation of "intrinsic" hazards, of engineering plant or chemicals or both, various situation-specific factors enter into regulatory risk management. The basic aim is simply to refine regulation to correspond with degrees of hazard in real cases, rather than with blanket worst-case scenarios. This analytic refinement is also an economic one; it parallels the regulatory attempt to use risk-benefit optimization [28] in that in trying to make regulation more "efficient", by tailoring defined risks to varying specific situations, margins for control costs (but also for allowable uncertainty and error and thus of safety) are also naturally reduced. This, therefore, implies the need for a greater regulatory ability to define the specific controlling properties of such varying situations. This assumption is highly problematic.

The "intrinsic" risks of a chemical plant might be defined as the probabilities of given release rates of harmful agents. Situational factors – such as environmental attenuations, siting (proximity of populations and other sensitive entities, such as drinking water), typical particle size, chemical form, etc., of releases – affect estimates of the actual situational risks. In the case of large-scale plants, such as liquid gas terminals, the releases and environmental pathways to be considered as situational risk qualifiers may be very few, usually to do with atmospheric dispersion characteristics of vapor clouds (liquid gas plant) or particles (chemical and nuclear plant). These are complex enough, but simple when compared with waste life-cycle situations.

The definition of any specific risk situation is basically identical to the problem discussed in Chapter 9, of defining the relevant structural characteristics of a technological risk system. For hazardous wastes the risks of understating uncertainties and variations are greater because of deeper scientific ignorance and greater situational variation.

Various possible approaches exist to define situational risk qualifiers and to organize these into regulatory degree-of-hazard schemes [29]. The usual assumption is that the more situation-specific risk definition identifies risk-*reducing* features, which therefore allows the severity of centralized regulation to be relaxed to a minimal base line, supplementing it with the risk-qualifying factors introduced by more autonomous, situation-specific *operator experience* and good management practices. The former may be "directly" regulated with specific rules, etc., while the latter may be only indirectly regulated (e.g., by economic incentives), or merely assumed to take place.

However, as the 2,4,5–T case illustrates, note that it is not necessarily the case that situational realities *reduce* risks. Whether laboratory experiments with individual pure components capture the "real, intrinsic" hazards of chemicals in environmental circulation is questionable and, as mentioned earlier, the distinction between the "intrinsic" or "natural" hazard of a substance and its situational hazard is not an objective distinction free from social determination and variability. Often, what is defined by one party as an *intrinsic* hazard already contains unrecognized assumptions about standard situations and institutional factors, which appear as natural to that party, but which others see as questionable or false.

10.4.1. Regulating degrees of hazard

Different regulatory systems construct different distributions of direct constraints and autonomous responsibility when balancing "universal" risk characteristics and flexible situation-specific realities [30]. For example:

(1) Listing for universal control (e.g., registration) those hazardous wastes with intrinsic hazardous characteristics; then taking account of, say, volumes produced, physical form, typical concentrations of hazardous constituents, professionalism of industrial management, siting, etc., only in separate *local* decisions (e.g., plant licensing) about specific T&D practices.

(2) Exempting, even from primary listing, those wastes, even if intrinsically hazardous, that are not produced in large volumes or concentrations, produced perhaps only by large technically well-endowed companies, or arise in aqueous form if they are mainly an inhalation risk [31], etc.

The regulatory arguments in favor of the latter approach are that it optimizes more flexibly to variable situations, and that it slims down primary regulatory lists, thus allowing more attention to focus on the more intense hazards [32]. Thus, situational optimization can coincide with risk ranking. This also coincides with industrial interest in greater autonomy from regulation. Thus, for example, Dow Chemicals' main criticism of the EPA hazard listings was that they were based upon the estimated intrinsic hazards of constituents only [33], effectively hypothesizing that waste would be improperly managed. Dow proposed, instead, a stepwise degree-of-hazard scheme, which placed estimated real exposures (including production volumes) at the front end, thus reducing the apparent risks in an "equivalent toxicity" measure. Likewise, the European Chemicals' Industry Toxicological Research Institute (ECETOC) has advanced a three-tiered degree-of-hazard approach, which combines intrinsic toxicity criteria with "exposure conditions which correspond to those in man or where the relevance of the exposure conditions can be deduced" [34]. Thus, again, situation-specific, socially determined factors are tacitly incorporated into the same standard degree-of-hazard regulatory ranking scheme as if they were natural, *intrinsic* factors.

This is generally typical of European institutional processes, which compound "intrinsic" scientific factors with extrinsic factors via informal procedural mechanisms and imprecise, flexible "technical" criteria [35]. These institutional mechanisms may involve joint advisory committees with interest group representatives, not only scientists, who deliberate in private and negotiate the particular weaving together of scientific and extrinsic ("intrinsic" and "situational") factors.

In the UK, situational risk-qualifying factors for a given waste have been identified, such as [36]:

(1) Physical location.
(2) Quantity.
(3) Concentration of hazardous components.
(4) Physical form.
(5) Detailed environmental disposition (e.g., if a fine powder, is it open to wind dispersion; is it adsorbed by clay surroundings).
(6) Sensitivity and number of targets exposed to it.
(7) "How damaging the hazardous effects are to the target," i.e., the intrinsic hazards of the chemical constituents of the waste.

The apparently "intrinsic hazard" list for regulatory control in the UK actually already incorporates informal judgments of several such situational factors; furthermore, the decision structure provides for automatic *delisting* of specific wastes on this list unless they *also* meet the regulatory criteria for toxicity, etc. [37]. The US system, on the other hand, specifies a list of controlled wastes and *adds*, as a catch-all, several tests that all *unlisted* wastes must also pass in order to be exempt from control [38]. Whereas the US system is designed to automatically *include unlisted* hazardous wastes, the UK system is designed to automatically *exclude listed* wastes that *in their specific circumstances* are not deemed hazardous. This decision is an *ad hoc* local one. Each system has a combination of intrinsic hazard definitions and methods of defining situational variants, but these are very different in their institutional and technical roles, and in the degree of standardization that they imply.

10.4.2. Situational variables – natural and social

In addition to the extra uncertainties and narrowed error margins of situational variation, a further complication with systems that attempt to incorporate situation-specific factors in more flexible actual hazard rankings is that they risk confusing *naturalistic* risk factors with *socially determined* ones. This point is crucial and deserves close examination.

We are now familiar with the idea of the "life-cycle" of a hazardous waste as it passes from generation, via various phases including transformation, to "final" disposal (see Chapter 3). In emphasizing the need to incorporate situational risk-qualifying factors, for example, it is observed that:

> ... hazardous waste may be hazardous in only one phase of its life cycle. Acid waste for example, may be hazardous only up to the point where it is neutralized before final disposal. Organic solvent waste may pose a hazard only until it is burnt in an incinerator [39].

Notice, however, that these are examples of "life-cycle changes"; situational risk reductions are not *natural* transformations, such as biodegradation with time into harmless products (some natural transformations *increase* risks). They are the results of deliberate human interventions. It is precisely these for which regulatory controls are required, first, to define as necessary and, second, to enforce. We cannot assume that benign behavioral interventions will always occur naturally; otherwise regulation would be unnecessary. Systems vary in their dominating value judgments as to how far the net of trust in the behavior of actors in the system should (or, perhaps inevitably, has to) stretch. This affects how much precision and standardization is built into regulatory test rules and definitions, and how tight they are. There is, as shown above, a tendency to confuse these behavioral dimensions with genuinely natural factors, which may in some situations reduce exposures and risks from intrinsic "worst-case" levels by attenuation or benign transformations. Even many of these "natural" processes, however, must also be activated by deliberate human action (e.g., to site a landfill on thick clay and to allow only certain compounds to be filled; or to maintain pumps properly), and are thus *not* natural in the sense that they cannot be assumed to take place automatically without regulatory control to ensure good management.

Risk modeling often conceals these behavioral assumptions in "technical" parameters or coefficients, making the results – and ensuing regulation – brittle to ignorance of the variabilities in the concealed behavioral factors. The discussion in Chapter 7 of UK dependence on landfill codisposal regulated by site selection, licensing, and inspection, is a good example. There, the technical policy depends on *behavioral* assumptions about decentralized *situation* control, which may not be realistic.

As the accumulating empirical research on actual environmental regulatory implementation is showing, such proper management is not even ensured when direct regulatory statutes and bodies exist to "enforce" it [40]. Yet degree-of-hazard schemes incorporating situational factors often mix questionable assumptions about the ideal behavior of the complex sets of actors involved in hazardous waste, together with genuine physical environmental factors that may reduce real exposures (and thus damage) to below theoretical worst-case possibilities.

Much hazardous waste policy analysis argues that environmental situational factors, which (it is assumed) reduce exposures from "worst-case" possibilities, should be given more prominence in risk assessments for regulation [41]. Underlying this argument is a feeling that widespread publication of disquieting *laboratory* evidence of toxic damage from a whole host of chemicals in the last decade or so has encouraged exaggeration of the real risks and a consequent overregulation based upon "worst-case" assumptions that neglect exposure-reducing realities. The following arguments are typical:

> Perhaps the most direct approach [is] to assemble a list of toxic effects, along with a list of tests that establish the presence or absence of those effects ... Such an exhaustive approach has been commonly required for food additives, pesticides, new human drugs and animal drugs ... While such an approach may define the biological effects due to a substance, it is necessarily incomplete, since such information is only a part of what we need to know to define the *actual* hazard to health or the environment
>
> ... it is generally less expensive, more technically appropriate, and equally protective to evaluate the potential environmental impact to the degree necessary to make decisions on the degree of containment control based upon actual expected impact, rather than treat routinely for worst case conditions in which it is assumed that the toxicity associated with a particular source extends for considerable time and distance [42].

Reflecting this concern to modify an intrinsic risk approach with environmental risk factors that attenuate and/or dilute, data frameworks stress criteria such as:

(1) Production volumes, places, and durations.
(2) Modes of dispersion.
(3) Physical form and containment.
(4) Environmental transformation, absorption, partitioning, etc.
(5) Proximity of populations (including eating and habits of environmental usage).
(6) Food chains and other possible exposure routes and limiting factors.
(7) Likely exposure levels and durations.

Again, however, we can see that many of these criteria beg the question of controlling the *behavior* of multiple, unidentified actors in waste life-cycles, because many of the apparently physical parameters are affected by such behavioral processes. Corresponding risk analysis models and data frameworks combine exposure-related hypotheses and calculations with estimates of intrinsic effects. At the same time, strong arguments are made to incorporate such "situational" dimensions in formal regulation. These are supported by optimistic claims that:

> ... there is emerging a capacity to predict the environmental fate of newly introduced chemicals by means of techniques such as ... evaluative models. Such techniques predict the likely compartments of the environment into which the contaminants will flow and accumulate, thus exposing biota and humans to toxic effects [43].

However, the insecure foundations of such claims were outlined before. They are also exposed by the same author's admission that this same area "contains a vast number of species with varying and poorly understood interdependencies" in which, yet again, "scientific knowledge severely lags behind regulatory needs" [44].

Even considering the physical–chemical and biological unknowns in the domain of environmental movements and exposures, therefore, there are severe difficulties and *risks* involved in trying to optimize risk analysis and regulation to varying situations, without destroying or overstepping safety margins. We have seen that scientific analysts confuse natural and behavioral factors that affect risks, and do not appreciate the behavioral or social assumptions on which their proposed technical refinements of risk management are based. This is true even for relatively pure, well analyzed chemicals. For hazardous waste one must add the extra complications of badly known and more variable waste compositions, and semi-autonomous, ill defined behavioral factors that sharply affect their physical dispositions, releases, and thus environmental conditions. The risks are especially sensitive to some of these behavioral factors. The intellectual difficulties of distributing the constraints properly between intrinsic, standardized risk situations and varying situational ones is paralleled by the *institutional* question of how to distribute *discretion* in defining such situations and thus controlling actual regulation at the sharp end of implementation.

In the next section we illustrate that, even in the highly controlled and artificially simplified context of scientific testing, ignorance and indeterminacy of situational variations are too great to allow credible standardization.

10.5. Standard Tests and Situational Risk Assessments

10.5.1. Case study of the US elutriate procedure (EP) test

Conventional science routinely has to reconcile the variability of specific cases with the search for underlying universals. Regulation must manage a similar reconciliation, but the principles and purposes are different. Conflicting pressures and contradictory strategic aims are exposed at the heart of regulatory science.

If regulations embody key definitions of materials for control, such as "hazardous", "special", or "controlled", these are defined either by lists or properties or both. We can call these definitions decision rules as to whether a waste must be included in a further set of behavioral rules (regulations). In order to define whether a given waste has a given property, clear criteria must also be defined for each property, and standard, reliably reproducible tests established. We can call these *observation rules* that support the decision rules. For properties such as flammability, corrosivity, etc., relatively simple indicators can be defined, e.g., flash point or pH (though even these are not without ambiguity), but it is for characteristics such as toxicity or carcinogenicity that the major difficulties arise.

The first area of difficulty is identical to that discussed above for all chemicals, even those with well defined compositions. This is that any standardized hazard criterion:

> ... is not technically valid for assessing the environmental hazard associated with solid wastes primarily because the transport and transformation (environmental chemistry) of the solids-associated contaminants can be markedly different for each specific environment, and usually play a dominant role in determining the hazard associated with the introduction of solid into the environment. The environmental chemistry of contaminants is influenced by many factors ... and must therefore be evaluated on a site-specific and solid waste-specific basis [45].

It is important to note that these situation-specific arguments apply not only to the particular environmental conditions that surround a waste in any given site, but also to the variable condition of wastes themselves. Wastes given the same name in a hazard classification or list, and which are thus identical in regulatory terms, usually vary in composition and physical form between plants performing the same process and even from the same plant over time (and over their subsequent life-cycle, as already discussed). These variations may cause significant changes in risk characteristics, e.g., leaching properties for hazardous constituents. Even if the *chemical* composition of different wastes is identical, changes in the *physical* form, such as particle size distribution, degree of aeration, etc., may radically affect the release of hazardous constituents (and thus risks) in the same environmental site.

Despite the force of these arguments, regulatory bodies have understandably been tempted by the opposite appeal of a single universal criterion and a corresponding standard test, at least for the initial characterization of wastes as hazardous before going on to consider situation-specific variations. An apparently simple example of such standard criteria are concentration thresholds for listed constituents of wastes, with "standard" analytical methods for tests. A regulatory test for toxicity in the USA is the elutriate procedure, or EP test, which is formally designated as a standard, statutory test to bring wastes that fail it under regulatory control [46].

The EP test is designed to test the leaching rates of potentially hazardous waste constituents into water, so as to simulate releases from landfill sites. A "representative sample" of a waste is mixed with a solution (pH 5 acetic acid) that is supposed to represent typical landfill conditions, and the leachate separated from this mixing is then analyzed for certain listed constituents. If these are present above a specified concentration, the waste is legally hazardous. So far the listed constituents are 14 chemicals or elements taken from water quality regulations under the Clean Water Act. The concentration thresholds in the leachate are set at the standards for acceptable drinking water quality, multiplied by an arbitrary factor of 100. This large "situational" factor is to allow for assumed further attenuation or dilution between leaching and escape from a waste site, and possible eventual contamination of drinking water.

As one expert has remarked:

> The primary requirement of a method to be used in making such an inherently expensive decision as whether or not the leaching from a waste is hazardous or whether a disposal method is safe, is repeatability. Not just that one technician in one laboratory can run three replicates, and get the same answer. Ten technicians in ten laboratories must be able to [47].

Interlaboratory precision and repeatability is absolutely vital to regulatory use of such standardized hazard classification tests, especially since very strong economic and political interests are at stake in the results. The regulations therefore stipulate detailed methods, which are obligatory for conducting such tests. Even so, the possibilities for methodological variations and disparate measurements, *even on the same laboratory sample taken from a waste*, are very large and still undefined.

Even with the standardized method, an interlaboratory evaluation program in 1979 found poor reproducibility from standard laboratory samples. (In other words, all the severe variations in *sampling* a real waste have been excluded.) Thus, for example, the EP test on a sample of fly ash from the coal-burn waste in arsenic leaching found a mean concentration of 0.227 ppm, but with a standard deviation in the results of $\pm$ 0.226, i.e., $\pm$ 100% uncertainty [48]! Chromium fared somewhat better at $\pm$ 50%. Other standard tests, all involving apparently precisely controlled statements of method, also showed such poor reproducibility that *even for the same laboratory sample*, it has been concluded that the EP test is highly unreliable as a regulatory instrument. Sampling variations introduce an even further dimension of variability and uncertainty into the attempt to define the "intrinsic" hazard of a waste. These are uncertainties quite extra to the question of whether the standardized test criteria (i.e., definition of the notional risk situation for wastes) reflect real risk situations.

The experimental method for the EP test is thought to be well-defined and controlled. Yet some of the factors that produce variable results are very subtle, and may not even be consciously recognized as elements of method by the practitioners themselves. These factors include [49]:

(1) Leachate:

  (a) Precise purity and composition.
  (b) Redox conditions (e.g., dissolved oxygen).
  (c) Temperature.
  (d) pH, including buffering.
  (e) Method of preparation and storage.

(2) Batch or column (continuous) test.
(3) Volume of sample.
(4) Leachate–solid waste ratio.
(5) Method, vigor, and duration of agitation.
(6) Method of addition of solution to solid waste sample.
(7) Material, design, and even exact position of mixing vessel.
(8) Contact–agitation time.
(9) Sample preparation, e.g., grinding, homogenization.
(10) Organic contents of waste sample.
(11) Particle-size distribution, porosity, etc.
(12) Number of elutions performed per waste sample.
(13) Leachate–solid separation method – centrifuge, settling, filtering – and time.
(14) Preservation of samples before leaching.
(15) Preservation of leachates after mixing and before analysis – freezing, drying, light exposure, time, etc.

This is not an exhaustive list, but it already presents formidable difficulties for precise repeatability. Some of these experimental factors have been specified in the standard test protocol, others have not. It is therefore important to understand the several levels of variability that accumulate:

(1) (a) Is the "standard laboratory sample" a representative sample of a real waste in a real situation?
  (b) Is the "standard sample" taken from a waste by any given laboratory the same as the "standard sample" taken by another laboratory?
  (c) Is the "standard sample" taken by a given laboratory the same as that taken at another time by the same laboratory?

(2) Laboratory methods and detailed practices that affect leaching – the same three questions apply as in (1).

(3) Chemical analyses of a standard leachate from the experimental leach test are also highly variable, especially for the low levels of concentration that are relevant. The same interlaboratory program mentioned above [50] found that, with the same leachate sample and the same analytical technique, *analytical* results could vary by ±100% in this concentration range.

These kinds of recalcitrant analytical variation are not unusual in standardization programs. As Collins and others have documented, experimental reproducibility in science, even for "simple" experiments, involves many tacit, craft aspects of detailed laboratory practice, which are barely specifiable, if at all [51]. Some practical procedures correspond with theoretical commitments and problem definitions, so that variations of practice cannot easily be ironed out even if they *can* all be identified. Thus, reliable standardization for consistent and secure regulation is far more difficult than is usually recognized, and would anyway in practice be prohibitively expensive, even if it were achievable in principle.

The potential variations outlined above incorporate two separate kinds of uncertainty. First, there is that resulting from variation in detailed laboratory practices, including sampling and analysis. Second, there is the uncertainty as to whether the standard *laboratory* test conditions laid down in the attempt to gain clarity and reproducibility bear any resemblance to the *real conditions* of a waste. This is not only a question of representative sampling, but also of whether, e.g., a specific agitation method and duration mandated for the test is a realistic simulation of solid–liquid mixing and leaching in a real landfill, or whether pH 5 leachate is always realistic, etc., etc.

Clearly, for a laboratory test protocol to be practical, it must simplify conditions from its real-world counterpart; and to be valid it must identify the factors and detailed experimental conditions that match the key actual factors that affect leaching of hazardous constituents from a landfill. Yet these are poorly understood even in general, let alone in the fantastic variety of specific sites, their particular properties (which vary within one landfill and over time), waste inventories, and management practices. Nor do the possible differences between laboratory tests and real conditions always produce overconservative regulatory results. The leachate pH specified in the EP test, for example, is 5, to represent what is thought to be a worst-case scenario of disposal of an industrial waste with municipal waste. Yet compounds of some toxic metals (such as arsenic, selenium, and chromium) are more soluble and therefore subject to greater leaching in the alkaline conditions of some landfills. Other landfills are much more acid than pH 5. The range of pH found in a study of US landfills was 1.5–9.5 [52]. A variety of other uncontrolled and poorly understood factors in different waste site conditions could similarly lead to under-estimations of real risks by the standard testing protocol.

To summarize, attempts to standardize central regulatory observation rules fail because of situational variations, even in the highly controlled process of laboratory scientific testing. Yet this is already an artificially simplified sample, and science is supposed to be definitive and controlled by clear and precise rules. If situational variation is a problem even here, we must multiply the problems manifold for the real world of wastes and situations.

10.5.2. Formal science, informal science, and the allocation of authority

This brief review of the EP test illustrates a fundamental dilemma in the use of science for this kind of regulatory instrument. Efficient, optimal, and defensible risk assessment requires that methods for measuring risks match real risk situations. The wide variability of such situations, even for the major single treatment and disposal method – controlled landfill – would thus require the local modification of test methods and experimental parameters to match particular cases. This has been widely and vigorously advocated by industry and other bodies, but such diverse, *ad hoc* modifications destroy the basis of uniform and accountable regulatory management. Furthermore, it is inherently impossible to distinguish between legitimate situation-specific test adjustments and those adjustments that affect the consistency and reproducibility and, thus, the external credibility of the tests.

It has been estimated that developing just one of the several candidate extraction tests from which the EP test was chosen cost one central research body $1.5 million [53]. The EP test is an example of a formalistic caricature of science. Its regulatory establishment changed the orientation of research away from developing an adequate predictive understanding of leaching mechanisms under different conditions, toward the more mechanical and relatively superficial problems of experimental reproducibility and classification under artificially defined standard laboratory conditions. The origins of this inflexibility lie in an extreme institutional need to regulate, using science of a particularly formal, transparent, and uniform kind [54].

The test represents an expensive attempt to use science to reconcile a deeper institutional conflict. The technical conflict between a fictional standard situation and widely variable real situations embodies an *institutional* conflict between central, standardized control, and industrial (or local regulator) autonomy. Using standard, precise test protocols, like using laboratory science to define real-life pesticides risks, effectively requires that the world be reorganized wholly in the image of those laboratory tests. They are a code for implicitly required behavioral standardization. Luckily, this is utterly unrealistic [55]. Central standardization implies that regulators do not trust industries to devise and do their own tests and allow for local factors in a responsible way. Perhaps a more accurate way of putting this is that regulators are forced by their own social relationships to show themselves (to courts, Congress, environmentalists, and other industries) to be regulating everyone with total equality (even if this does not lead to equal effects). Apparently, no intermediate position is conceivable in which *guidelines* are centrally issued, and uncontrolled industrial autonomy in their use is restrained by more substantial local regulatory institutions. These could have a better chance of understanding local realities, but still retain and actively interpret the overall central regulatory philosophy, goals, and guidelines. The level of uncertainty and error in the EP test, and the extent of its shortfall with respect to reality, is a function not of the inadequacies of scientists, but of the resolving power and social standardization *expected* of such a test. If it is so brittle and unreal, more robust criteria and tests could retreat from the attempted scientific precision and place the burden presently overloading the scientific domain upon more distributed institutional processes. But this in itself would require adequate resources at that level, a central trust in state or other local authorities, and a political culture that generally allows such devolution in issues like this. Analysis of the *technical* problems, therefore, reveals questions about the institutional structure of regulation and, ultimately, about political cultures.

For comparison it is instructive to recall the equivalent "test" for toxicity used in the UK to define specially controlled wastes. As described in Chapter 5, the test is "simply" that if any 5 cm^3 sample of a waste would cause acute toxicity ("death or serious tissue damage") if ingested by a 20 kg child, it is to be designated special waste [56]. Although the weak technical basis and ambiguity of this test has been criticized, and it has been described as offering a potential field day for lawyers [57], its institutional role and surroundings in the UK system render it relatively uncomplicated to administer. (Whether it has the desired effect is a different matter.) Responsibility for case-by-case interpretation in practical regulation is allocated to the discretionary wisdom of those with statutory responsibility for control, namely local authorities, with occasional advice from government experts and even the waste producer. Although explanatory guidelines are also offered for advice, these are still rich in the need and opportunity for situation-specific judgment, which in principle allows more flexible tailoring of regulatory classifications to real-world variations.

In lacking universal, clear decision rules, as we saw in Chapter 7, the UK approach thereby offers less opportunity for third-party access, review, and accountability, but it is impressively strong on institutional confidence and required collaboration, and correspondingly economical in its demands on formal public science. Informal expertise buttressed by institutional arrangements and broad rituals, which stress this elitist "craft" image of scientific expertise in the UK, replace the elaborate public use of formal precision and "public" science in the USA. When appeals to formal science for authority begin to escalate beyond a certain ill-defined but strongly influential (and rather low) threshold, these are superseded by normative appeals to accept *institutional* authority (e.g., "Competent industries and local authorities acting in good faith should reach a socially negotiated consensus which weighs technical judgment with economic and other factors. Third parties should accept such consensus as legitimate"). It is when these fail that major rituals (such as public or parliamentary select committee inquiries) to reemphasize such norms are performed [58].

10.6. Institutions, Formal Decision Rules, and Informal Judgment

In Chapter 9 we made the important claim that scientific framing processes shape the content of analysis by a complex involvement of social assumptions, value judgments, technical judgments with value implications, and technical judgments with no clear value implications. These framing commitments go on during the detailed development of analysis as well as prior to it. Many informal judgments buttress formal methods, definitions, and reasoning. We argued that this is true even for well defined technological risk analysis, and this led to a novel perspective on uncertainty, expert disagreement, and credibility in policy.

Earlier in this chapter we indicated that risk analysis of toxic chemicals – especially aggregated risk-ranking and environmental modeling – was being overstretched into formal frameworks. These embody extremely crude subjective judgments, but have an envisaged central role in policy persuasion and conflict management that plays down this informal side. Yet, because of the strong role of subjective factors in the models, the decision rules are eventually exposed as being too manipulable to be credible. This is all part of a general process that we called the bureaucratization of uncertainty. By the same token, specification of the EP test observation rules appeared to be too *inflexible*, yet its production of results still too uncertain, to be credible in broad-based regulatory use. We pointed out that the bureaucratization of uncertainty is driven by a felt need for formal, aggregated, and standardized systems of regulatory knowledge, as part of public legitimation.

10.6.1. Observation rules

A further aspect of the growing specification of precise observation rules like the EP test as part of regulatory decision rules is the reshaping and control of science that it embodies. Research is increasingly constrained by standardized and carefully specified rules that reflect narrow external objectives. This is evident in the toxic chemicals field especially, as illustrated by the international propagation of toxicity testing codes and the external control of scientific inferences when these have policy consequences but no clear scientific warrant for any one of the options. We now examine both of these areas so as to develop our perspective on uncertainty and institutional responses to it. This will prepare for the wider discussion of issues of public credibility given in Chapter 11.

The proposals of the European Chemicals Industry Ecology and Toxicology Centre (ECETOC) on Codes of Good Laboratory Practice [59] were a direct result of the legislation of such testing codes by the US FDA, following the episodes described in Section 10.2. The ECETOC proposals were careful to try to defend scope for expert craft judgment, but even they listed nearly 100 separate rules that a laboratory performing such toxicity tests would have to follow. These ranged from agreed definitions, through organizational requirements, such as how to replace a Study Director, to planning, operating, and reporting procedures. To illustrate, the standard operating procedures to be specified in a test protocol are given below:

SUB-PART E: TESTING FACILITY OPERATION

E.1. STANDARD OPERATING PROCEDURES

A testing facility shall have written and approved standard operating procedures that are adequate to ensure the quality and integrity of the data generated in the course of the study. Deviation from these standard operating procedures shall be authorized by a competent person as designated by management, and documented as raw data in the study.

Each separate laboratory area shall have immediately available standard operating procedures relevant to the activities being performed there. Published text-books and articles, manuals and directions may be used as supplements to these standard operating procedures.

A historical file of standard operating procedures and revisions thereof, including the dates of such changes, shall be maintained as in sub-part J.3.

Standard operating procedures, where applicable, shall describe but not be limited to the following:

a) Test, control and reference substances (henceforth called substances)

i) Receipt, identification, characterization, handling, formulation and storage of substances.
ii) Testing the homogeneity and stability of substances, and concentration of substances in mixtures with carriers.
iii) Administration of substances.

b) Test system
   i) Room preparation and external conditions for the test system as appropriate.
   ii) Procedures for receipt, transfer, proper placement, characterization, identification and care of the test system.
   iii) Test system observations and examinations.
   iv) Laboratory tests and analyses.
   v) Handling of individuals within the test system found moribund or dead during the study, where applicable.
   vi) Termination of an experimental study and/or necropsy of the test system.
   vii) Collection, identification and handling of specimens.
   viii) Histopathology.

c) Equipment
   i) Use of equipment.
   ii) Maintenance, cleaning, calibration and/or standardization.

d) Documentation, evaluation and reporting
   i) Data collection, handling, storage and retrieval.
   ii) Preparation of reports.

Handling of test system animals required a whole separate section of rules and principles. Note, too, that these are *principles* to be applied in *specifying* test protocols. Any actual test protocol would be more detailed still.

The purposes of this detailed managerial control of scientific research are to exclude discretion and even minor variations, to "mechanize" research into programmable observation and decision rules. The laboratory test codes, or observation rules, apply to the *generation* of data. The inference rules being increasingly dug out, specified, and controlled (as shown below) apply to the *interpretation* of data into regulations. Both aspects combine to produce a comprehensive network of growing rigidification of the risk analysis. The problem is that the necessary purpose of reducing incompetent or otherwise dubious analysis, and even merely variable competent analysis, conflicts with the need to allow discretion to define test parameters so as to match diverse real conditions of exposure.

By so standardizing specified risk situations into those embodied in more and more elaborate, rigid, and precise test protocols, this approach actually draws risk analysis away from an appreciation of the network and situational variants that affect the real creation and delivery of risks. It therefore becomes inflexible to diverse situational factors, less real from the experience of people actually in the system, and it thus lays itself open to irrelevance and lack of credibility. Ironically, the standardization born of concern to establish credibility of regulations in the hazardous waste area becomes too rigid to retain credibility.

10.6.2. Interpretation rules

Given that so much regulatory dependence is now placed upon toxicological risk analysis, there is pressure to clarify the "internal" decision rules of the science so as to render the consequent regulatory decisions transparent and open to detailed review. The US White House Office of Science and Technology Policy proposed in 1980 that the observed infusion of scientific risk analysis by policy values could be overcome by a "return" to strict separation of facts and values. A more sophisticated approach was developed by the US National Research Council, Committee on Risk Assessment in the Federal Government [60]. This study proposed a distinction between scientific risk assessment, risk assessment policy, and risk management. Risk management would consider the conventional economic, social, and other factors extrinsic to science, which are weighed in decisions upon "acceptable" risk standards. Scientific risk assessment means the conventional fields of science unsullied by policy considerations.

The interesting dimension is the middle one – risk assessment policy. Here questions of a scientific nature are nevertheless strictly unanswerable by science, either because of uncertainty due to gaps in science, or to *inherently* trans-scientific properties of the issue. Any one of several scientific inference bridges or decision rules could be legitimately used to reach across the gaps and allow the construction of policy-relevant scientific knowledge – each might be consistent with, but not determined by, existing scientific knowledge. Yet each may have its own policy implications, so that the choice of a "scientific" decision rule is inevitably partly a policy matter. A good example is the choice of an extrapolation rule for low-dose toxicity or carcinogenicity effects in humans, when what empirical data there are rest upon high doses, in animals. Choice of a linear, quadratic, linear-quadratic, or threshold low-dose–effect relationship is more or less equally legitimate according to available high-dose data, but the choice often dramatically affects the estimated effects, e.g., excess cancers, depending upon the constants employed. Whittemore gives an example of three equally credible alternative models of dose–response curve for nasal tumors in rats exposed to the defumigant, ethylene dibromide, showing that the choice of model changes the estimated tumor rate at low doses by over two orders of magnitude [61]. In other cases the change can be as high as six orders of magnitude. It therefore seems to be necessary for policy to make an inference bridge, but which *scientific* inference rule to choose as a "risk assessment policy" is legitimately a matter of *policy* choice.

Leaving aside more recognizable value inputs, such as discounting future risks, various ways in which values more subtly infuse the *factual* analysis in toxicological risk assessment have now been recognized. Familiar examples, in addition to the high–low dose–response function question, are:

(1) How should the relative significance of apparently divergent results in different species, different organs, different genders, etc., be weighed?
(2) Should body weight, surface area, or metabolic system correction factors be used to extrapolate from animal data to human estimates?
(3) What are the effects of different means of administering doses, and different dose rates?
(4) Should "benign" tumors be counted (and how can they be distinguished from malignant ones)?
(5) How can the balance be weighed between statistical probability of false positive results, on the one hand, and weakening statistical power to detect true carcinogens, on the other, by the choice of statistical confidence limits?
(6) How can this balance be weighed when separate studies give opposite results at the same confidence limit?
(7) Should maximum-likelihood effect functions or upper confidence limits be used?

These inference choices are usually visible and clear in their effects upon the policy conclusion. However, they are continuous with a further range of finer and deeper inference options within scientific knowledge. An important general category of these concerns the *range of validity* of observed data and relationships. Some of the above are in this class; for example, whether an observed tissue or organ sensitivity to some chemical is species-specific; or whether a given dose administration method produces artificially high results because of some idiosyncratic factor, or is of wider validity.

10.6.3. Inference rules and plural scientific cultures

It is important to realize that many of these scientific commitments are made without any choice or realization that an inference option exists. They are part of the given scientific specialty's culture. For example, some data in carcinogenicity or toxicity testing are generated from experiments on species chosen for their known, exceptionally high yield of tumors, because their purpose was the examination of the mechanisms of tumor development. This "special case" knowledge may be lost in the reporting of bare data on excess tumors for a given dose, and the risk data unwittingly used as a basis for wider extrapolation. In other cases, structure–activity relationships of a new chemical with a known one may suggest certain potential toxic effects in the new chemical, even though the experimental evidence is ambiguous. What *kind* of structural parallels are regarded as significant or insignificant, and how much weight to give such relationships compared with experimental data, may well affect risk estimates substantially, but these commitments may be already established in the "natural" state of accepted knowledge and practices of a given scientific specialty.

The question of how much weight to place on animal studies compared with human data (which are usually more sparse, less controlled, but more direct) is also a case in point. Epidemiology as a specialty naturally gives more weight to human data, and its intellectual diet is precisely in making the fine distinctions and judgments necessary to evaluate badly structured human data. Conversely, animal pathologists condense all of those uncertainties into one aggregate uncertainty surrounding "uncontrolled" human data and epidemiology generally; their discipline thrives on the uncertainties that surround distinctions between forms and sites of damage in animal experiments, etc. Their *natural* values are to place credence, suitably refined and discriminating as befits their own special expertise, in animal studies.

Even the dose–response function question provides an example of this same cultural divergence within science. Before the influence of epidemiology and genetic-damage theories of carcinogenesis, pathologists (who relied upon tissue- and cell-damage techniques to examine damage processes) and pharmacologists (who observed physiological effects) both saw dose–effect *thresholds* through these well established methods of observation. These were the natural frameworks when these disciplines turned toward new problems, such as modern toxic chemicals. They were not even *seen* as "inference choices", as if alternatives existed, until challenged by other approaches that had not been socialized within that taken-for-granted disciplinary paradigm. They were the *framework* within which further scientific puzzles were shaped and evidence interpreted, not the *subject of* such questions.

It is worth noting that the natural *methodological* commitments outlined above are integrated with their own theoretical commitments too. For example, whether one treats benign tumors as significant depends on beliefs about the exact mechanism of cancer, the variability of metabolic or immune system processes, and so on. There is a certain circularity about this – premises define observations, which confirm premises. Each specific scientific commitment relies upon a network of supplementary commitments, some supported by evidence, others suppositions that have gradually become buried in the taken-for-granted context. When such a scientific commitment is challenged because it is seen to have policy implications (e.g., about risk levels) those supplementary inferences and suppositions can be identified and examined for their scientific warrant. An example of what happens is given by Jasanoff in an analysis of the formaldehyde risk debate in the USA [62]. Formaldehyde is acknowledged to produce excess cancers in rats at doses similar to known human doses. However, there have been several null results and also the distribution of cancer sites in rats is known to differ from those observed in exposed humans. In particular, the observed preponderance of nasal cancers in rats does not occur in humans, even though formaldehyde is mainly inhaled as a gas. To proponents of stricter regulation, these bodies of evidence encourage certain new balances of significance in supplementary issues – cancer-site correspondence for known carcinogens is not always found between animals and humans; and humans may breathe more through the mouth when irritants are around. Industrial proponents argued instead that the null results indicated a threshold dose, and that the absence of human nasal cancers was very significant. Combined with the ambivalent mouse data, this was used to argue that nasal cancer in rats is species-specific, and thus nongeneralizable – a result of cell damage in the nasal tissues of rats, which, it was suggested, must be hypersensitive.

This argument developed into a conflict over the proper interpretation of the very limited mouse data. These showed some excess nasal cancers at high doses, but not in a statistically significant number. However, the opposing sides differed on what value of statistical significance is to be used when the natural background rate of a cancer is very small, as it is for nasal cancers in mice. The US EPA argued that significance standards should be relaxed in such cases, and again raised new inferences that challenged taken-for-granted ideas in the existing knowledge structure – they suggested that mice may reduce their volumetric breathing rate when subjected to inhalation irritants, thus effectively protecting themselves.

Each exchange in this continuing debate threw up new challenges to previously taken-for-granted and "closed" scientific assumptions and commitments, which supplemented the *explicit* commitments. There is no reason to suppose that each process of deconstruction could not continue, digging deeper and deeper into the established body of scientific knowledge held by different parties.

In a working paper for the aforementioned US NRC Risk Assessment study, for example, McCray explains how, having begun with the "naive" conventional wisdom that values and facts could be cleanly separated, he was surprised to find how many value choices had to be made in making necessary inferences within toxicology [63]. He first identified 36 such inference choices, including those listed before, but this soon grew to over 50. Moreover, some of these were recognized to be *composite* commitments, which could be broken down into multiple component inferences. For example, the question was raised in the formaldehyde case of what significance to attach to small numbers of excess cancers in mice, which have a very low natural "background" rate. To set the statistical confidence limits lower for low-background rate species or sites – as the EPA experts argued one should – implies a commitment to one particular alternative in the available theories of the cancer mechanism. The use of absolute or relative risk models in extrapolating limited-time results into full-term damage estimates also implies similar alternative commitments. How to evaluate apparent "saturation effects" in metabolic processes or toxic effects depends upon commitments to different assumptions about metabolic mechanisms. The list proliferates. As a general matter, McCray concluded that so many observations and relationships in toxicology were specific only to a given experimental context that it was impossible to develop generic decision rules within the scientific component of risk analysis.

10.6.4. Context-bound scientific knowledge

An important point about the networks of finer inference commitments, as opposed to the gross ones such as which dose–effect function to use in going from high-dose data to low-dose estimates, is that they are less clear-cut and direct in their policy implications. In these cases a commitment to one option or another cannot be directly correlated with a policy value commitment. The eventual influence upon a risk estimation is indirect and conditional upon further commitments in the network. The finer commitments are also more buried, taken for granted, and less *visible* as *choices* to the scientists concerned (and to outsiders). For example, the assumption that a given detailed method of dosing an animal produces generalizable effects, with no artifacts due to situation-specific processes, does not necessarily have an *identifiable* consequence for ensuing risk estimates and policy conclusions, which are constructed several intellectual stages later.

This point is important because it is often assumed that the inference commitments, which may have no clear scientific warrant *and* have an effect upon risk estimates, are *deliberate* value or policy choices. Consistent with Chapter 9, our analysis here suggests that they are not even necessarily always *choices*, let alone choices made with specific external policy interests in mind. This perspective on the inference-based network structure of scientific knowledge is not at all a special feature of toxicology, or chemical risk assessment. It has been advanced as a general feature of science, even in its "hardest" forms, and has been supported by a broad-based body of detailed work in sociology and the history of science.

Pinch's analysis of von Neumann's "impossibility proof" against the *hidden variables* explanation of quantum phenomena in sub-atomic interactions is a good example [64]. Pinch's historical reconstruction showed flaws in the von Neumann (1933) proof that were not identified for a long time, since it was adopted rapidly by mainstream physics anxious to justify its unease about hidden variables theory and its developing commitments to something different. Von Neumann's "proof" became the key reference point for 40 years or more to justify rejection of the "hidden variables" approach. The proof became "naturalized" as a given fact, despite resting upon chosen premises and inferences that were open to question: these disappeared from view by a process of *social* closure of the cognitive questions.

Wynne's analysis of the justifications used, again by mainstream physics, to reject a deviant theory of X-ray–matter interaction (the J phenomenon) in the 1920s offers another example [65]. Here, dubious "replications" of the crucial experiments claimed to show that the original experimental observations were spurious. The negative "replications" were embraced enthusiastically and uncritically, and they were used as authoritative, culturally rehearsed reference points right into the 1970s. The tacit social agreement to accept such "replications" as definitive and conclusive was deconstructed by a detailed historical analysis. This deconstruction did not *falsify* the mainstream's commitments to reject the J phenomenon; but it did indicate the social element in the construction of knowledge, and in the corresponding closure of questions existing in the supplementary beliefs and commitments.

Later work, for example, by Latour and Woolgar [66] and Knorr-Cetina [67], developed this kind of analysis to show how the formation of agreed commitments is integrated with the social processes of tacit persuasion or *justification* of knowledge. Thus, there is an inherent misrepresentation of the social commitments to particular inferences, standards of evidence, criteria of classification, etc., as unavoidable, given in nature, and universal. Again, however, it is important to stress that this is an incremental and diffuse process, deeper than the deliberate play of specific external interests and value commitments.

We can now bring together this interpretation of the underlying open-endedness of scientific knowledge with our previous discussion of the standardization of regulatory testing in the EP test, toxicology testing protocols, etc. The inherent open-endedness, or incompleteness of scientific knowledge, is completed by social processes operating in practical contexts, to produce *practical* scientific knowledge. The increasingly elaborate specification of scientific practices is an attempt to control social uncertainties and variations in detailed methods of observation or data production; the attempt to draw up standardized sets of inference rules is an attempt to control social uncertainties of inference commitment in the network of scientific judgments needed to convert the data into risk estimates (and hence regulations). But standardizing risk analysis in this way is tantamount to prescribing, and standardizing, in extreme detail, the real worlds of risk situations and processes. Even if it were achievable, it loses contact with reality.

The reasoning behind such standardization is to restore *authority.* The US NRC study's proposal formally to discriminate between scientific risk assessment and risk assessment policy, or policy-determined scientific inference rules, was intended to reinsulate "proper science" from an insidious invasion by implicit policy interests and concerns. Thus, a new arena was separately defined with its own norms and principles. The choice of inference rules in this arena would be guided as much as possible by current science, and *scientists* would judge whether the state of knowledge justified equal policy choice between inference options, or whether new evidence suggested transfer of such a rule out of the policy domain and into the purely scientific one, etc.

10.6.5. From private to public audiences

The argument in favor of standardization is *explication* of decision rules and assumptions. The guarantor of salvation against arbitrarily varying value commitments, and unspecified relationships between models and real risk-scenarios, is supposed to be *transparency.* Lowrance's prescriptive principles are typical and have been widely cited:

> First, [scientists] can leaven their discussions by including critical, articulate laymen in their group ... Second, they can place on record their sources of bias and potential conflicts of interest, perhaps even stating their previous public positions on the issue. Third, they can identify the components of their decisions as being either scientific facts or matters of value judgment. Fourth, they can disclose in detail the specific bases upon which their assessments and appraisals are made. Fifth, they can reveal the degree of certainty with which the various parts of the decision are known [68].

McCray has expressed a similar norm – if risk assessment truly is an inextricable mix of scientific judgment and value judgment, the best operating principle is to make sure that the assumptions made in assessing risk are routinely made explicit; in this way, they can be subjected to both scientific and political scrutiny. Risk assessments should identify each area of inference where scientific uncertainty is confronted, and should state the analytic choice(s) made in each area.

Despite offering major advances over previous approaches, McCray and Whittemore both still cleave to the assumption that all biases are from value *choices*, and transparency is the solution.

The basic problem with the "transparency" approach has already been demonstrated. As one reaches into science for open inference commitments with policy implications, these are not conscious to the expert, and they can be found extending deeper and wider, even if – as mentioned – the precise way their policy consequences fall becomes less clear-cut. The solid ground of science free from judgments and inferences that are not fully determined by logical rules and unambiguous facts recedes as it is more closely examined [69]. The *appearance* of natural factual solidity is created by legitimate, institutionalized social agreements and judgments of scientists to ignore unexplained anomalies, to resolve an ambiguity one way rather than another, etc. Scientists are comprehensively socialized into particular institutionalized sets of such judgments and inferences, and thereby become immersed in them as if they were utterly natural and logical. To refer to these as value choices is to do violence to their depth and subtlety and to suggest, quite falsely, that they can be set aside if the expert so wishes. Basing the credibility of regulation upon this premise by making transparency, open peer review, and formal statement of value choices the institutional principles may be a brittle strategy.

Seen as a cultural process, the construction of authoritative knowledge will vary from one context of use to another. The contours of defined uncertainty and fact can vary at several levels:

(1) Between different scientific traditions, with their different "natural" methodological, theoretical, and network commitments.
(2) Between different practical contexts over space and time for the *same* scientific tradition; for example, consensual views accepted as fact, such as a blood-grouping characterization based on a given technique, can be demoted to "judgment" if the technique is shown to be based on other suppositions whose warrant is not complete. Pinch has shown how in solar physics, for example, scientific definitions of the degree and location of uncertainty in the same specialty vary with the particular context of debate [70].
(3) Between regulatory systems: as indicated in Chapter 2, different regulations for the same problem occur in different systems despite the universal availability of the same research. Each system evolves different network constructions, even of "the facts", in a complex, incremental combination of inference commitments, selection of emphasis, and assigned credibility of different studies and approaches, etc.

Several conclusions to be drawn from this analysis are:

(1) The attempt to develop definitive standard observation and decision rules in science for regulation is ultimately doomed, because the process of digging out strictly open-ended inferences and commitments buried in the structure of scientific knowledge can go on potentially indefinitely unless some institutional means of closure is reestablished.
(2) There is not an objective amount of uncertainty attached to any given estimate or statement that can be stated for policy. The degree of

uncertainty perceived by an expert in a piece of knowledge depends upon its perceived *role* in a relationship or context of *use*.

(3) The overdependence upon "definitive" scientific observations or analysis for risk management generates pressure for bureaucratized, mechanistic forms of science, which have every detail of method and inference controlled to a degree that is both unrealistic to achieve and – even if unachieved – corrupting.

(4) When one takes fully into account the vast potential for the elaboration of scientific uncertainties when science is pressed to deliver authority, the costs of achieving anything approaching definitive knowledge for viable regulation are far greater than ever suspected. Thus, the transaction costs of approaches to back-end regulation dependent upon this kind of standardization through scientific risk analysis may be far greater than previously thought.

(5) The pressures to formalize risk assessment science will intensify, even in relatively informal decision cultures such as the UK [71]. However, the attempt to gain legitimacy via formalization and "transparency" should be cautioned by the realization that this may produce caricatures of scientific knowledge and method whose pretences are eventually laid bare, with further erosion of public confidence. Scientific authority and consensus rests upon socially mediated *credulity* – a readiness to suspend skepticism – as much as upon critical scepticism. Putting such scientific knowledge to work to restore eroded institutional credibility is only likely to erode the science too, by exposing its ultimately social–conventional basis.

(6) Over-elaborated "science" is being extended in a vain attempt to control institutional uncertainty and pluralism within the risk-system to be regulated. This is neither a realistic nor viable response to those structural uncertainties which entail greater costs than those normally recognized in downstream regulation.

10.7. Conclusions

In this chapter we have analyzed the extent of misfit between deep and diverse kinds of uncertainty that surround all the various dimensions of hazardous waste management, and the need to regulate *as if* the uncertainties involved were narrowly limited and thus credibly manageable within unidimensional, quantitative boundaries. Risk analysis as a technical activity is expected to reconcile this fundamental contradiction in a way that is credible and authoritative to a wide array of different actors and institutions, with their own experiences, interests, and perceptions of the issues. Unfortunately, science as a regulatory ally tends naturally toward the negation of uncertainty and diversity, and lends itself easily to bureaucratized forms, which artificially convert genuine ignorance, ambiguity, and indeterminacy into apparently controllable risks or marginal uncertainties.

A balance has to be maintained between too little structuring of the risk framework, in response to the authentic unknowns and lack of definition of the field, and overelaborate, artificial bounding and concealment of ignorance, in response to demands for definitive risk–benefit management knowledge. The former emphasis runs the risk of lack of effective control and lack of pressure to develop consistent technical and evaluative knowledge for regulation; the latter runs the converse risk of actually obstructing the search for more precise and effective regulatory knowledge. It risks concretizing the standardized frameworks that have been *artificially* cast around the unknowns and variances in the field, as if these creations were reality. Although *individual* regulators may recognize this distinction, the system may still be constructed and run as if no such distinction existed.

There are two reasons for requiring that risk analysis be formalized, i.e., that its decision rules should be transparent. The first concerns essentially *descriptive* needs; if an analysis of risks in a system or situation is to be authoritative it must be reproducible and its assumptions, data, logic, and manipulations be accessible to test by other experts. The second reason relates to the *prescriptive* role of risk analysis in regulation. The role of regulatory norms is to control behavior. However, in hazardous waste systems, unlike conventional pollution controls, this cannot be done by specifying only performance criteria that are the end-product of a regulator's risk analysis and some process of evaluation. Because of the complex *behavioral* life-cycle, which has perforce to remain partly free, boundary conditions have to be specified in terms of general criteria of hazard, materials, and actions. Many diverse and *unknown* behaviors and situations must be controlled, even if they cannot be clearly defined. The actual regulated actor, or a *local* regulatory body (or both) then has to be left to perform a "situation-specific" risk analysis of their local situation, and to do this, normative risk-analytic methods (observation rules) and criteria (decision rules) must be laid down by the central regulator for the "regulated" to follow. As we have seen, these rules can vary in precision, but what they lack in clarity must be compensated by mutual *institutional* trust and collaboration between the parties involved. Behavioral controls are then imposed on the regulated if his (in theory, centrally controlled) risk analysis shows that in a given situation certain specified risk criteria have been met.

In both aspects of its role, normative and descriptive, risk analysis has to withstand divergent interpretation, challenge, and criticism, and it has to be clear enough to be properly repeated. Hence, the observation and decision rules of science in this field are coming under increasing and more penetrative demands for full explication and precision – the transparency norm. At the same time the "coverage" of science in this area seems to be being outstripped and uncertainty is growing, thus pulling in exactly the opposite direction. As mentioned before, the scope of formal knowledge *socially required* by management is increasing faster than rigorous knowledge can expand. Therefore, the pressure continues to produce increasingly elaborate formal risk-analytic models and regulatory decision rules, yet these are based on more and more fragile foundations.

As we have argued in the more general framework of the last two chapters, the context-dependent limitations of scientific analytical formalization provide opportunities for conflicting framing, not merely observational imprecision. This deeper form of uncertainty is fueled by divergent institutional interests and perspectives, often of a subtle and unintentional kind, and by *behavioral* freedoms that influence risks in the hazardous waste cycle. We saw in Chapter 9 that uncertainty can be intractable even for relatively well structured problems. For hazardous waste life-cycles and chemical–environmental risk analyses these problems escalate. Furthermore, the very nature of science is not to embrace uncertainty and to be explicit and sceptical about inference choices, but to bury and become unself-conscious about the majority of them.

A consequence of this analysis is that science in its usual sense will inevitably fall behind in the task of reducing uncertainty in risk management and thus in buttressing the credibility of regulation. The general analysis of uncertainty and its effects upon regulatory coherence, given in the last two chapters, only sharpens the damaging effects of the heterogeneity and *behavioral* indeterminacy of the hazardous wastes risk life-cycle upon attempts at standardized or formalized risk analysis and regulation. Conversely, the perceived benefits of decentralized, flexible institutional mechanisms and related technical frameworks are multiplied. To the extent that technical certainty is a surrogate for institutional uncertainty, regulation that attempts to control the latter by indiscriminately expanding its expectations of the former places itself in severe difficulties.

In the next chapter we address more directly the implications of the widespread concern over public reactions to hazardous waste regulation – actual or anticipated – and try to offer a coherent framework for thinking about this incoherently experienced "risk perception" problem. This public dimension has so far surfaced only briefly, but it needs to be more explicitly related to the issues of scientific uncertainty, risk-analytic framing, and institutional credibility dealt with so far.

Notes

[1] Gori, G.B. (1980), The regulation of carcinogenic hazards, *Science,* **208**, 256–262. See the differences of emphasis between the US National Research Council (1983), *Risk Assessment in the Federal Government* (National

Academy Press, Washington, DC) and the London Royal Society (1983), *Risk Assessment*, (Royal Society, London); see also Lowrance, W. (1976), *Of Acceptable Risk* (Kaufman, Los Altos, CA).

[2] Within risk assessment this has led to fundamental conflicts of framework with younger disciplines, such as epidemiology, which have developed their methods and approaches within the new climate of institutional needs and aims. US Congress, Office of Technology Assessment (1981), *Assessment of Technologies for Determining Cancer Risks from the Environment*, OTA-81600811 (OTA, Washington, DC); Anderson, E.L. (1986), Quantitative approaches in use in the United States to assess cancer risk, *Risk Analysis* **6**, 139–160.

[3] Serwer, D. (1978), *The Rise of Radiation Protection* (US Brookhaven National Laboratories Report). See also the collection of papers by Taylor, L. (1977), *A History of the ICRP* (Pergamon, Oxford, UK).

[4] Nicholson, W.J. (Ed), Management of Assessed Risks for Chemical Carcinogens, *Annals of the New York Academy of Sciences*, **361** (NYAS, New York); Schmandt, J. (1984), Linking science to policy, in M. Lafolette (Ed), *Harvard Newsletter on Science, Technology and Human Values*; WHO (1982), *Evaluation and Risk Assessment of Chemicals,* Interim Document 6 (World Health Organization, Copenhagen).

[5] Lowrance, W. (Ed) (1981), *Assessment of Health Effects at Chemical Disposal Sites*, Symposium Proceedings, Life Sciences and Public Policy Program (Rockefeller University, New York, NY).

[6] The nearest parallel to the 57-year-old International Commission on Radiological Protection, for example, is the International Agency for Research on Cancer (IARC), Lyon, France, which was established with WHO support in 1970. IARC has grown more rapidly than ICRP. See IARC, *Annual Reports*; IARC (1979), *IARC Monographs on the Evaluation of the Carcinogenic Risk of Chemicals to Humans* (IARC, Lyon); and IARC (1983), *Approaches to Classifying Chemical Carcinogens According to Mechanism of Action* (IARC, Lyon).

[7] US National Academy of Sciences (1980), *The Biological Effects of Ionizing Radiations*, BEIR III Report (National Academy Press, Washington, DC).

[8] US Environmental Protection Agency, (1982), *Environmental Monitoring at Love Canal* (EPA, Washington, DC). See also, US Congress, Office of Technology Assessment (1983), *Habitability of the Love Canal Area: A Technical Memorandum* (OTA, Washington, DC).

[9] UK Granada TV, World in Action (1984), *Tried Untested*, transcript of program, London.

[10] *Ibid.*

[11] *Ibid.*

[12] Wolff, S. (1981), Cytogenetic analyses at chemical disposal sites: Problems and prospects, in Lowrance, *op. cit.* [5], pp. 62–80. Schaumberg, H. *et al.* (1981), Neurotoxic effects at chemical disposal sites, in Lowrance, *op. cit.* [5], pp. 81–104.

[13] For example, the extreme difficulties caused by the EPA-sponsored study on chromosome damage to Love Canal residents. Compare Levine's account with that of Kolata, G.B. (1980), Love Canal: False alarm caused by botched study, *Science*, **208**, 1239–1242.

[14] Conway, R.A. (Ed) (1982), *Environmental Risk Analysis for Chemicals* (van Nostrand Reinhold, New York, NY). Nichols, J. and Crawford, P. (1983), *The Management of Chemicals in the 1980s* (OECD, Paris).
[15] Gusman, S. and von Moltke, K. (1982), *Public Policy for Chemicals* (Institute for European Environmental Policy, Bonn).
[16] See [14] and [15].
[17] Ahmed, A.K. and Dominguez, G.S. (1982), The development of testing requirements under the Toxic Substances Control Act, in Conway, *op. cit.* [14], pp. 506–528.
[18] Haymore, C. (1982), Incorporating risk assessment into the Resources Conservation and Recovery Act regulatory process, in Long, F.A. and Schweizer, G.E. (Eds), *Risk Assessment and Hazardous Waste Sites*, American Chemical Society, Symposium Series 204 (ACS, Washington, DC), p. 21.
[19] *Ibid.*
[20] *Ibid.*
[21] Yakowitz, H. (1985), *Global Aspects of Hazardous Waste Management*, background paper for Expert Meeting at World Commission on Environment and Development, Geneva, May.
[22] Wynne, B. (1984), The institutional context of science, models, and policy: The IIASA energy study, *Policy Sciences* **17**, No. 3 277–320.
[23] Schaumberg, H. *et al.* (1981), *op. cit.* [12], p. 11 (quoted by Lowrance).
[24] Mackay, D. (1982), Basic properties of materials, in Conway, *op. cit.* [14], pp. 33–60.
[25] Nisbett, I.C. (1979), Ranking chemicals for testing: A priority-setting exercise under the Toxic Substances Control Act, in TSCA Interagency Testing Committee, *Scoring Chemicals for Health and Ecological Effects Testing*, Appendix B (Enviro Control, Rockville, MD).
[26] Finnecy, E. (1983), *Some Notes on Criteria and Philosophy for a Waste Nomenclature System* (UK Atomic Energy Research Establishment, Environmental Safety Group, Harwell, UK).
[27] Nisbett, *op. cit.* [25], p. B-53.
[28] Though risk–benefit trades off decreasing risk and/or increasing containment against the costs of achieving such decreases, degree-of- hazard schemes do not always proceed to the second stage of defining containment costs of situational risk reductions.
[29] See [14], [18], and [26], for example.
[30] See, e.g., references on cultural–institutional influences on the interpretation of science in policy, Chapter 2, note [19].
[31] Finnecy, *op. cit.* [26].
[32] This point is made by UK regulators, for example.
[33] Dow Chemicals (1978), Critique of US EPA promulgated hazardous wastes characteristics and listings, *Federal Register*, 40 CFR 250.
[34] Broeker, B. (Hoechst Chemicals) (1983), *Determining Acceptable Risk*, discussion paper delivered to International Conference on Chemicals and Regulation, Bellagio, Italy, August; also mimeo, Cornell University Science Technology and Society Program, Ithaca, NY.
[35] Brickman, R. Ilgen, T. and Jasanoff, S. (1985), *Controlling Chemicals: The Politics of Regulation in Four Countries* (Cornell University Press, Ithaca, NY).

[36] Finnecy, *op. cit.* [26].

[37] See UK Department of the Environment (1981), *Special Wastes: A Technical Memorandum Providing Guidance on their Definition*, Waste Management Paper No. 23 (HMSO, London).

[38] US EPA, *Federal Register*, 40 CFR 260-263, 45, 19 May 1980, pp. 33060–33137 and 40 CFR 264-265 and Hazardous waste management system: Standards for owners and operators of hazardous waste management facilities, pp. 33154–33258. Thomas, S. and Roberts, M. (1983) *RCRA: Decision Making in the EPA* (Harvard University, Graduate School of Public Health); Daniels, S. (1981), Development of realistic tests for effects and exposures of hazardous wastes, in R.A. Conway and B.C. Malloy (Eds), *Hazardous Solid Waste Testing: First Conference*, pp 345–365 (American Society for the Testing of Materials, Philadelphia).

[39] Finnecy, *op. cit.* [26].

[40] Richardson, G. and Ogus, R. (1982), *Policing Pollution* (Clarendon Press, Oxford, UK). See also Chapter 1, notes [8] and [10].

[41] See Conway, *op. cit.* [14].

[42] Lee, G.F. and Jones, R.A. (1982), in Conway, *op. cit.* [14], pp. 540, 547.

[43] See McKay, *op. cit.* [24].

[44] *Ibid.*, p. 56.

[45] Lee and Jones, *op. cit.* [42], p 542.

[46] Perket, C.L. and Webster, W.C. (1981), Literature review of batch laboratory leaching and extraction procedures, in Conway and Malloy, *op. cit.* [38], pp. 7–27; Lee and Jones, *op. cit.* [42], pp. 332–344.

[47] Fisher, S. (1981), quoted by Perket and Webster, *op. cit.* [46], p. 14.

[48] See [42] and [46].

[49] Perket and Webster, *op. cit.* [46], p. 17.

[50] *Ibid.*, p. 21.

[51] Collins, H. (1975), The seven sexes: A study in the sociology of a phenomenon, *Sociology*, **9**, 205–224. Also, Barnes, S.B. and Edge, D.O. (Eds) (1981), *Science in Context* (Open University Press, London).

[52] Data compiled by E. Finnecy, UK AERE, Harwell, from Clark and Piskin (1977). Personal communication, March 1984.

[53] Lee and Jones, in Conway, *op. cit.* [14].

[54] Perket and Webster, *op. cit.* [46].

[55] Notice the interesting parallel here with Bruno Latour's discussion of Pasteur's reorganization of the world in the image of his laboratory techniques in order to make an innovation out of penicillin: Latour, B. (1983), Give me a laboratory and I will raise the world, in K. Knorr-Cetina and M.J. Mulkay (Eds), *Science Observed*, pp. 141–170 (Sage, Beverley Hills, CA).

[56] UK Department of the Environment, *op. cit.* [37].

[57] Haymore, C. (1982), Incorporating risk assessment into the Resources Conservation and Recovery Act regulatory process, in Long and Schweizer *op. cit.* [18]. House of Lords, Debate on the Gregson Committee Report, *Hansard* (HMSO, London).

[58] Wynne, B. (1982), *Rationality and Ritual: The Windscale Inquiry and Nuclear Decisions in Britain* (British Society for the History of Science, Chalfont St. Giles, UK).

[59] ECETOC, Avenue Louise, Brussels 1983.

[60] US National Research Council (1983), *Risk Assessment in the Federal Government: Managing the Process* (National Academy Press, Washington, DC).

[61] Whittemore, A. (1983), Facts and values in risk analysis for environmental toxicants, *Risk Analysis*, **3**, 23–34.

[62] Jasanoff, S. (1984), *Recognizing Risks to Health: The Assessment of Toxic Substances in Britain and the US*, draft mss. (Cornell University, Ithaca, NY).

[63] McCray, L. (1983), Toxicological influence rules – a review; background paper for USNRC study, *op. cit.* [60].

[64] Pinch, T. (1980), What does a proof do if it does not prove? in Whitley, R.D. *et al.* (Eds), *Social Processes in the Construction of Scientific Knowledge* (Reidel, Dordrecht).

[65] Wynne, B. (1976), C.G. Barkla and the J phenomenon: A case study in the treatment of deviance in physics, *Social Studies of Science,* **6**, 87–105.

[66] Latour, B. and Woolgar, S. (1979), *Laboratory Life*, (Sage, Beverly Hills, CA).

[67] Knorr-Cetina, K. (1981), *The Manufacture of Knowledge* (Pergamon, Oxford, UK).

[68] Lowrance, W. (1976), *Of Acceptable Risk* (Kaufmann, Los Altos, CA).

[69] Hesse, M.B. (1974), *The Structure of Scientific Inference* (Cambridge University Press, London); Bloor, D.C. (1983), *Wittgenstein: A Social Theory of Knowledge* (Routledge and Kegan Paul, London); see also Barnes, S.B. and Edge, D. (Eds), (1981), *Science in Context* (Open University Press, London).

[70] Pinch, T. (1981), The sun-set: presentation of uncertainty in scientific life, *Social Studies of Science*, **1**, 23–36.

[71] The UK Hazardous Waste Inspectorate has commissioned Dames and Moore International to develop a formal, universal method of risk analysis for the large number of landfill sites in the UK, using aggregated scoring techniques, covering the full range of different risk and acceptability factors. I am grateful to Rob Holmes and David Mills for discussions on this.

CHAPTER 11

Risk Perception, Decision Analysis, and the Public Acceptance Problem

Brian Wynne

In Chapter 10, I described a growing dualism in the role of science or risk analysis in regulation. I argued that this has been caused by decreasing confidence in regulatory institutions, requiring them to engage in more elaborate *justification* of decisions. The influence of this increasing reassurance role on the structure of risk analysis may be the opposite of that required for ill-structured and heterogeneous problems. Thus, an inherently self-defeating understatement of fundamental uncertainties is intensified by this growing role.

In this chapter I analyze a related aspect of regulatory responses to the "reassurance" problem. I examine the way risk perception studies have been incorporated in decision approaches to risk management. I suggest that these approaches are fundamentally inadequate, even counterproductive, because they automatically convert concern into an extra attribute of risks in an esoteric decision technique – an abstracted, individually rooted "negative utility". This denies the objective grounding of risk definitions and perceptions in the social experiences and relationships of technologies.

I therefore show how decision-analytic techniques convert structural uncertainty and conflict that arise from authentically heterogeneous social existences into the artificial one-dimensional technical uncertainties of "expected utilities". Even recent extensions, using risk perception studies to incorporate elements of public concern, fall foul of this basic fallacy. In appearing to deal with public concern, this elaboration of managerial techniques leaves unrecognized the need to respond with institutional innovations. One such partial approach, namely formal compensation programs for siting hazardous waste faciities, is then reviewed in the context of the preceding interpretation.

11.1. Risk – From Abstraction to Context

The new public perception dimension is reflected in the now-frequent lament that decision makers are being forced to manage risks according to how many people will be frightened rather than how many will be killed. The predominant framework is that there exist objective physical risks associated with different technologies and activities; then there are subjective perceptions of those risks. Hence, the so-called "objective–perceived risk dichotomy" [1], the assumption amongst experts being that irrational emotional factors enormously multiply public judgments of the scale of some objective risks, such as nuclear power, while reducing the scale of others, such as road accidents. These "subjective biases" are taken to have extremely costly, even terminal, implications for those technologies that attract these negative associations. Measured by economic value-of-life statistics for the costs of extra control technologies per calculated life saved (extra risk avoided), different social and technological activities show enormous variations, reflecting composite "biases" from a supposedly rational overall distribution of risk reduction resources [2].

In the 1970s, such observations led to psychological research to account for these subjective biases and variations [3]. The initial assumption was that lay people were just ignorant of the risk levels of different activities, and that simple information, e.g., of comparative risks, would correct them [4]. When both resistance to such information and evidence demonstrating passable lay knowledge of quantitative risks proved this assumption false [5], the dominant "rational" view then became that people were behaving primitively, looking at risks with superstitious overtones and emotive associations [6].

The more serious work began to factor risks into multiple attributes and to identify people's differential scaling of these [7]. In addition, the systematic forms of bias in popular cognition and framing of problems, especially the treatment of statistical figures and choices, were intensively investigated [8]. The existing fields of behavioral decision theory and decision analysis developed rapidly, based upon normative models of rational individual choice modeled largely by one-off gambling decisions [9].

A *cardinal* premise of all the risk perception work was that (physical) *risks* exist separately from the context in which they are experienced. Thus, the development of the idea of risks as objective, unidimensional quantities was that they also contain intrinsic attributes. Public perceptions of these could be measured, and even quantified and compared with "objective" measures. The hope was that in this way the "public perception" dimension could be abstracted into general formulas of "acceptable risk" for different technologies. Risk attributes were regarded as subsidiary explanations for bias away from "objective" risk measures. They were not, at this stage, seen as oblique indicators of the *context* from which real risk experiences are analytically abstracted and given transformed meaning. The work of Otway's group at International Atomic Energy Agency/International Institute for Applied Systems Analysis (IAEA/IIASA) was discontinued before it could develop from the investigation of the supposed attributes of *risks* to the attributes of *technologies,* which are the source of risks [10]. However, it was then realized that the basic determinants of public reactions were not risk-specific attributes and attitudes, but more comprehensive belief-value systems within which the "risk" or technology was experienced [11]. Even so, decision analysis used psychology whilst neglecting the sociological implications of this point.

In a critical review in 1982 of "scientific" definitions of "acceptable risk" in regulation, Otway and von Winterfeldt summarized some of the more common risk attributes found in a range of psychological research on risk perceptions [12]. These included involuntariness, lack of personal control, uncertainty and disagreement, lack of personal familiarity, imaginability, delayed effects and anxiety, "societal" damage (e.g., genetic degeneration, or catastrophic potential), invisible benefits, inequitable risk–benefit distribution, and human rather than natural causes.

It is immediately evident that even some of these "risk" attributes are in reality attributes of the *social relationships* in which the risky activity takes place. That is, they are not merely attributes of the risk, nor even of the "technology", if this is viewed as context free. They are attributes of the technology *seen as a historical process of social relationships.* In consequence, the nature and boundaries of the "risk problem", even the physical risks themselves, let alone the social risks inherent in those relationships, objectively vary according to social experiences and positions. (This boundary problem was discussed in Chapter 9). This *sociological* translation of attributes makes the bridge from the two psychological approaches of psychometric factor (or attribute) analysis and the more comprehensive belief-system frameworks. The panoply of "rational" models for analyzing risk decisions are based upon the false premise that when we measure their reactions, people are (or *should* be) making decisions about physical risks reified and separated from their context in the way analysts suppose they are. This "objective risk" framework is itself a subjective artifact of analytical-policy managerial framing, which bears a questionable relationship with other, objective categories of social experience, perception, and evaluation. Indeed, the experience of "technologies", as I argue below, is rooted in empirical social experiences of decision-making *institutions* and their historically embedded relationships. The implicit conflict between normative "rational" risk decision methods and descriptive sociological accounts of the way concern is enacted in real issues focuses on whether these deeper institutional dimensions of social experience should be recognized as a legitimate part of public decision beyond discrete "utilities" or "attributes" of an imposed problem.

A concrete illustration of the distinction between a risk-based and a social relations-based approach is the experience of occupational risks in a hazardous industrial plant. Experience shows (and plain common sense would suggest anyway) that workers' responses to physical risks cannot be dissociated from their response to management attitudes and social relations in the plant [13]. Any single "event" or "decision" is located within that continual social process – abstracting a discrete decision is already to distort the basis of concern. If those social relations with management are pervaded by mistrust and hostility, the ever-present uncertainties in physical risks will be quite rationally amplified, because they are being managed by untrustworthy actors; the overall *objective* risk is correspondingly multiplied. Physical risk attributes integrate with other attributes of the technology, including its (past and present) social relations.

The basis of perception relating to the acceptability of a "risk" is, rationally, the basic social *experience,* which is not physical risks alone, not even *technology* in a context-free sense. It is the technology as embedded in an institutional web of control, seen as a social organizational process.

Thus, we reach the important watershed in the risk field, that the abstracted language of "risks" elaborated over this area of social interaction by "risk analysts" suppresses a whole dimension of social experience involved in a given technology or its risks. It thus denies legitimacy to the values and anxieties that arise from this domain. Adopting this framework in "risk management" or regulation, therefore, jeopardizes its own credibility by effectively saying to people that their own social experiences and searches for meaning do not count; or, if they do, only as separate individual subjective psychological states, often classified as neuroses or pathologies for which "therapy" may be required [14]. I argue that this self-destructive cycle occurs even where regulators attempt to take risk perception studies into account, i.e., when they are prepared to relax the failing "objective" prescriptions offered by "revealed preference" societal risk comparisons or comparative value-of-life quantifications, and systematically incorporate factors for public "concern" or "risk aversion".

11.2. The Sociological Roots of Risk Attributes

I now illustrate concretely how the dominant framework based on the psychology of risk perception has obliterated an institutional dimension. I then show how this analytic method has been incorporated into extensions of established decision-analytic approaches to regulation and risk management, with the effect of simply elaborating the basic fallacies of those frameworks whilst imagining that they are *responding* to public concern.

As indicated earlier, psychological risk perception studies have attempted to factorize the multiple elements of risks that people supposedly perceive, "weigh", and accordingly react to. One of the most sophisticated and influential research traditions is that associated with Slovic and Fischoff, and their collaborators [15]. They and others [16] have taken the many relevant attributes – up to several dozen – identified by various perception studies and, using standard data reduction techniques, eliminated duplication. They accounted for the variance in recorded attributes by reduction to just two orthogonal dimensions – "dread" and "unfamiliarity". The attributes showed high intercorrelation, and further analysis was claimed to show essentially the same two-dimensional pattern. Fischoff *et al.* called these, respectively, the "emotional" and "cognitive" dimensions of response [17] (see *Figure 11.1*). The way in which this kind of work has been interpreted is that the key dimensions – "dread" and "unfamiliarity" – are subjective *psychological* categories. Even if widely held, they are taken to be basically *individual mental states*, which bias perceptions of "real" risks. Yet there are ample indications that this is a categorical misrepresentation, and an effective suppression of objective *social* categories of experience and action. While others [18] have argued that the observations are empirically unsound, I suggest that, even assuming they are sound observations, they can and should be seen in *sociological* terms. Let us take some examples.

The "dread" dimension of nuclear power is widely associated with two particular symbols – the mushroom cloud image of annihilation by the atomic bomb and the *unseen*, all-penetrating image of radiation. Yet the supposition that this *unseen* property of radiation creates an automatic sense of dread is false. This is shown by examining *Figure 11.2*, taken from a British newspaper supplement on nuclear energy in 1956 at the opening of the first electricity-providing nuclear power station [19]. In those more trusting, paternalistic days, as the caption illustrates, the unseen nature of radiation was being socially constructed as a *positive* element of the image of nuclear power – it was part of the mystery that kept the controlling experts and elites suitably distanced, socially and intellectually, from the public, which was evidently meant to simply look on in awe. Thus, this attribute is clearly only a code for a set of social relationships, the association with one particular set or another being contextual and socially achieved. Measuring its existence by psychological research is to measure that social achievement, and not an *intrinsic* individual mental state.

It is not that the psychological observation is wrong, but that it is incomplete and, as such, may mislead. Thus, taking the fundamental category of experience and reaction to be psychological rather than social *is*

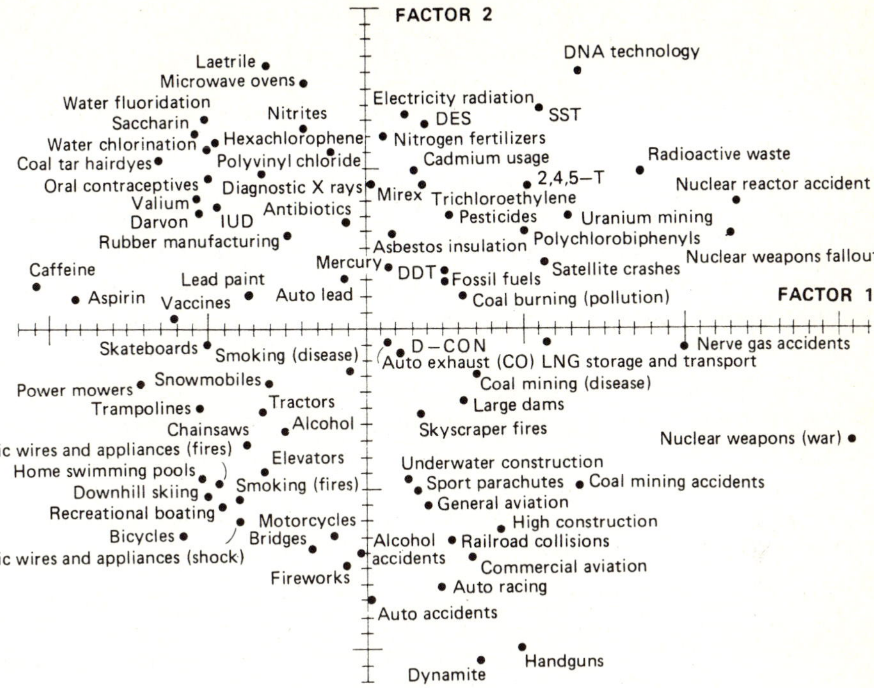

Figure 11.1. (*a*) Hazard locations on factors 1 and 2 derived from interrelationships among 16 risk characteristics. Each factor is made up of a combination of characteristics, as indicated in *Figure 11.1* (*b*) (from Slovic *et al.* [17]).

wrong, because it implies individual rather than social categories for understanding and responding to public reactions.

Taking the "atom bomb" part of the "emotive" dread now, this can be seen to express a substantial empirical logic in the institutional realities of nuclear technology. There are two equally defensible, but contradictory, rationalities in the possible connections between nuclear energy and nuclear weapons. One of these would say that the possibilities for horizontal proliferation of nuclear weapons (hence increased chances of atomic bombs annihilating us) by the spread of civil nuclear energy are nullified by the Nuclear Non-Proliferation Treaty safeguards, managed by the competent IAEA. This is a legitimate (predictive and present) behavioral judgment. An alternative judgment is that proliferation to unstable states is proceeding inexorably, with the (not exclusive, but substantial) help of global civil nuclear energy diffusion. In this equally legitimate stance, safeguards are failing and, indeed, it is unrealistic to expect that formal safeguards could ever control that proliferation. In this alternative view there is also a concrete empirical judgment about *institutions*, which allows an empirical

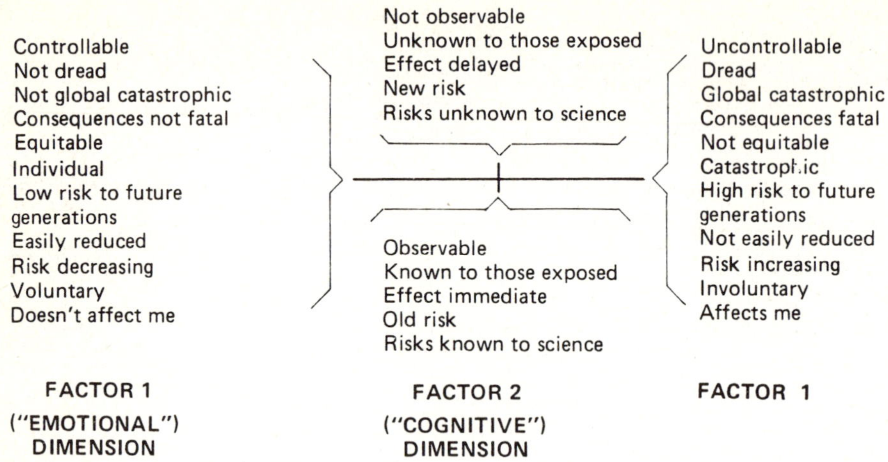

Figure 11.1. (*b*) Each factor in *Figure 11.1* (*a*) is made up of these combinations of characteristics of characteristics (from Slovic *et al.* [17]).

association between nuclear energy and nuclear weapons. Which one of these conflicting views is "correct" is beside the point in the present context. The point is that each one is equally empirically defensible and rational.

Thus, the so-called irrational emotional dread of the atomic bomb and the mushroom cloud as a risk-perception multiplier for nuclear energy can be taken as a linguistic shorthand, or symbolic language containing a cognitive dimension (namely a tacit behavioral judgment) that the institutions supposed to keep nuclear energy and nuclear weapons apart are not able to succeed. The alternative belief is also based upon questionable *social* assumptions. Its lament of the *emotiveness* of the weapons association and recognition only of direct physical risks (on which dimension there *is*, indeed, no connection) is merely an authoritarian way of dismissing those different legitimate social judgments and concerns.

We can also take an illustration from the "unfamiliarity" dimension. This may again be treated as a purely individual cognitive state, but it is more plausibly related to the social processes of decision making (or "risk management") about the technology in question. When examined more closely, conflicting "risk perceptions" in the debate about the UK's planned oxide nuclear fuels reprocessing (THORP) plant at Windscale in the 1970s, were seen to derive from different basic definitions of the *boundary* of the technology, or risk source, in question [20]. The same dislocations of basic framework can be identified in conflicts over waste disposal facility siting, reflected in the frequently expressed fear that a facility of a certain proposed size, or a research facility, will later be expanded with no control [21]. To those who were used to exclusion from decision making, the THORP issue was a temporary window on a usually alien social process. There was a

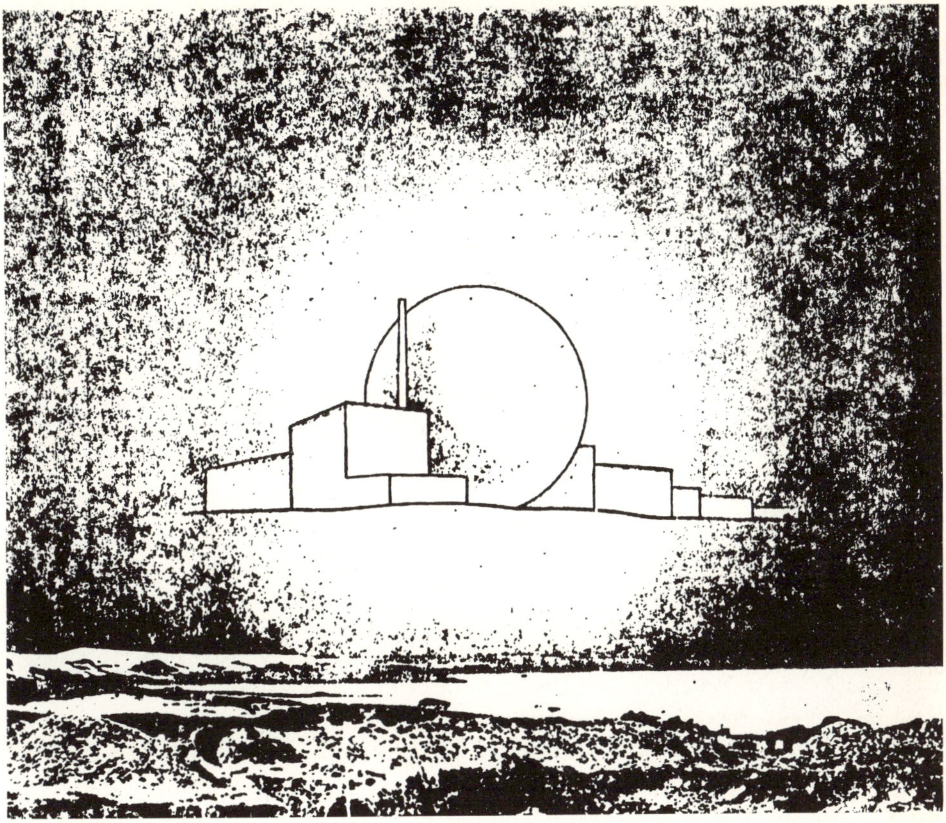

Figure 11.2. A 1956 newspaper presentation of Dounreay atomic plant, UK.

quarter-century of social exclusion to catch up on, as witnessed in the frequent opposition references to past secrecy and regulatory incompetence as an argument against the new plant. Judging the *current* and future risks rationally involved judging the risk-managing *institutions*; and the evidence for this rationally came from the concrete experience of past behavior and relationships. *The technology is a historical process, and the risks are rooted in this process.* "Unfamiliarity" is deeply misrepresented by regarding it as a risk attribute that converts into individual cognitions.

There was also a rational desire to associate all potential future fuel cycle developments (and their risks) following on from the reprocessing plant with the decision on that one plant itself, because there would perhaps (on solid experience) be no consultation with those *future* decision steps. The nuclear proponents, and the decision-making establishment generally, dismissed this as emotive and irrational; but *they* could rationally identify with and participate in any future decision steps. In decision-analytic terms they could conceive of distinct *future* decision trees, with real open-ended decisions. The public objectively could not, and thus *rationally* condensed the putative later tree and all its potential events (risks) and negative utilities onto the *present* decision. Each boundary definition for the formulation of risk problems was rational, but they were incompatible. However, rather than recognize its underlying social dimension, the authorities took the wider risk boundary framing to be the familiar emotive multiplication of the objective risks by an unfounded, irrational perceived risk multiplier. Differing *social* realities of trust or control can also underlie apparently inconsistent risk perceptions made by the same people of "the same" risks. For example, when domestic users spray far more pesticides on their own garden than they will accept to be sprayed on forests, there is a logic to this "inconsistency" that derives from their lack of control of the present and future spraying of forests. The "risk problem" in that case *objectively* extends out into an open-ended future, unless they feel they can control the responsible institutions. In their garden, they can.

Underlying all risk perceptions, the public and experts are framing *social* assumptions, which are often based in experience. These underlying social dimensions of risk are suppressed by the dominant approach, and there is no medium in which they can be reorganized, debated, and negotiated.

11.3. The Social Alienation of Effective Causes

So far I have tried to show that an important dimension of public experience and concern is omitted, and thus denied value and legitimacy, by dominant approaches to risk perception. Later I illustrate how these approaches have been embodied in extensions of conventional "rational" decision techniques in risk management, but first it is worth exploring how deep the sociological roots may go, and how they might be expressed in forms that may or may not be picked up by risk perception research, as distant signals of something more profound and complicated. I suggest that people's choices and perceptions are constrained by the systems of *explanation* they construct in an attempt to order their (disorderly) social experience. This is consistent with a sociological version of Otway and others' argument for a comprehensive belief value system approach to public reactions as opposed to attribute-utility analysis of risks and perceptions [22]. This deeper ground can be approached by unpacking one attribute reported by Slovic *et al.* as part of the "unfamiliarity" dimension [23]. This is the perceived unintelligibility and apparent arbitrariness of the *effective causes* of the risks in question – who or what, if anything, is controlling the risks that persecute us?

Anthropologists and sociologists have long recognized that we attempt to render experience consistent and orderly by creating fundamental continuity between experience of society and of nature [24]. Durkheim noted [25] that we domesticate disordered areas of experience by extending familiar explanatory categories and frameworks from more orderly areas of experience. In stable societies with little control over nature, elements of social stability and emotional security would become the categories for understanding nature. In technological societies, with apparently strong control over natural processes, we might use these as the basis for explaining otherwise troubling and disordered areas of experience, such as human nature or rapid social change. Indeed, the boundary between nature and culture is a matter of incessant argument and negotiation as we attempt to locate authority for our beliefs and prescriptions in the "inviolate and untouchable" laws of nature rather than the "mere" values or arrangements of human society [26].

When accidents happen or harm is threatened, there is wider uncertainty – and scope – for negotiation over the explanation of their ultimate cause. If fundamental, morally significant social experience is seen as atomistic and fragmented, natural processes will tend to be seen in the same framework: hierarchies are assumed to exist in nature where they are deeply etched and "natural" in social experience [27]. In modern differentiated societies our knowledge of the social causes of decisions and actions that affect us is indirect and mediated by our attachment (or denial) of credibility to different sources of explanation. Experts and expertise mediate our understanding of society and relations with each other as much as they do our relations with nature. In a complex modern society there is a need for reassurance of the ultimate order of things, a need that may be fulfilled by explanations of identifiable causes and responsibilities for events that are otherwise diffusely but intolerably threatening and capricious. When we are not provided with such credible explanations from credible sources, then for good reason we are likely to create our own.

I suggest below that the sources and dominant values behind technological development in modern society are now generally so obscure and inaccessible that the controlling causes of risks (and actual damage) are inexplicable in terms familiar to ordinary people. Social stratifications of decision making and expertise have reached such extreme degrees of alienation in many fields that the lay "consumers" of such decisions are forced to fill a void of responsibility with their own cultural constructs, symbolizing their experience of these apparently capricious, unintelligible social relations. This at least domesticates and constrains an otherwise more deeply threatening arbitrariness in experience.

11.4. Social Constructions of Responsibility – Domesticating Disorientation

In his classic account of the social and psychic devastation caused by the 1972 Buffalo Creek dam failure in the Appalachian mountains [28], Kai Erickson proposes that the reaction of the economically and politically marginal people who were victims of that "point disaster" was profoundly conditioned by their prior internalization of a state of "chronic disaster", represented in their long-term neglect by and alienation from employers and public authorities. The psychic withdrawal characteristic of extreme traumatic shock that Erickson observed was already consolidated on the community scale in the alienation and self-dependence of the community, trusting none of the agencies on whom they nevertheless depended, and thus tolerated, for economic survival. Erickson argues that what was most significant about the social aftermath of the disaster was not the personal trauma – "psychic numbing" – which everyone experienced, but the *collective* trauma, the inability of the old social networks to reestablish themselves as the framework of personal psychic convalescence and development. The people felt betrayed by the coal company that had neglected the dam whose burst caused the disaster, not because they had previously thought it a conscientious company, but because structurally, in their position, they had been forced effectively to trust it – to behave *as if* they trusted it, despite realistic appreciation of its selfish motives, past neglects, etc. This seems a suggestive general way of looking at "public acceptance", even in normal circumstances.

In Erickson's perspective the powerless always tend to defend and rationalize, thus consolidate, their own impotence and apathy because to do otherwise is to expose themselves to the greater human damage of *explicit* recognition of neglect and powerlessness. They withdraw, and justify and defend that withdrawal as consistent with cosmic principles; it becomes their culture, integrating their beliefs about cause and effect in the experiences they encounter with their established social relationships. Erickson saw the classic symptoms of trauma in the ordinary human reactions to "the age we are entering", namely "a sense of cultural disorientation, a feeling of powerlessness, a dulled apathy, and a generalized fear about the state of the universe" [29]. These are the symptoms of social experiences and roles that are highly prescribed by others, yet where the logic and structure of such prescriptions – of their own social marginality and manipulation – are obscure. They are the "ineffectuals" in Thompson's terminology [30], whose lives are prescribed by others, and who view nature and the world and their own experiences as determined, but capriciously so, by some unintelligible others. The "effective causes" of their powerlessness are socially invisible. The dam burst and flood was tantamount to the condensation onto a single and dramatic event of years of nonaffirmation, or identity stripping, by the outside world. There is a basic continuity between event-focused accidents and chronic exposures to unknown threats. This is well captured by Fowlkes and Miller:

> The most insidious feature of the unnatural disaster, then, is its capacity to become increasingly widespread at the same time that it goes unrecognized and its unidentified casualties increase proportionately. Whatever the destructive impact of chemical exposure itself, the unnatural disaster takes an equally destructive toll in the form of irreparable damage to the social fabric where exposed communities or other social groups divide and polarize in their attempts to understand and resolve the ambiguities confronting them [31].

I would only add that the diffusion of *social* uncertainty as to where such physical chemical exposures are coming from, and why, is now so extensive as to create a context of anxiety regardless of whether or not there is a real physical and/or chemical threat – and expertise is routinely unable to offer clear discrimination.

The foregoing example highlights a deep complexity in social attitudes and values that is not at all recognized in policy analysis. Single events and historical processes are a *seamless web* – discrete accidents or decision problems are interpreted consistently with already accumulated social experiences and problem agendas. In the Buffalo Creek case, the survivors seemed to have had a sense of who was responsible for the technology's havoc, but an even stronger sense of hopelessness that anything could be done about it. In this case the effective cause of their disaster was at least *seen as human agents*. But there was an ambivalence in attitudes, which seemed to reflect a sense of extra-human agency in the events – of forces beyond human control simply taking their own capricious directions; and to the people they *were* forces beyond their familiar and "controllable" realm, because, although human, they were forces belonging to the inaccessible, alien, and generally untrustworthy social world of "bosses", "decision makers", "experts", and the like.

A coal company and its management of a dam might be taken as a supreme example of a concretely visible, familiar technology, with *clear* lines of control and responsibility. If such a familiar technology can be perceived as alien, how much easier must this be for more complex technologies? Many technologies and industries typical of the modern age – nuclear power, genetic engineering, modern war technologies, toxic chemicals, and computers – lie at the opposite extreme. Their controlling human agents and relationships are far more invisible, diffuse, and socially remote. It is impossible for ordinary people to even identify, let alone identify with, the effective causes of their confusing and often troubled experience of these technologies, even if they do not produce dramatic interventions in their lives. Yet the importance of these experiences requires that people construct some working explanations so as to rationalize them one way or another.

Coal companies may not surround themselves with the elaborate language and relationships of "expertise" and the rhetoric of legitimation more associated with modern technologies. But like many other industrial and technological areas, they are pulled along in the slipstream of a general mode of legitimation that cultivates the idea of awesome, other-worldly technical power, controlled by forces beyond the bounds of ordinary nature and culture (recall *Figure 11.2*). But this disorienting relationship's corollary is an ambivalence and instability of attitudes that being alienated and indiscriminate, can suddenly flip over from *benign* (or, at least, neutral) social externality to *malign* externality.

The point is that, with the effective causes and structures of responsibility so obscured, the only possible public responses are *total* acceptance or quiescence (tinged with a lingering anxiety in the face of such supernormal and unintelligible power), or *total* rejection (tinged with a lingering fascination at the sheer technical mastery such technology may entail). There is no possibility for internal discrimination, measured criticism, or conditional, qualified responses – all possible currencies of moderation are historically obliterated, leaving behind inflexible absolutes. Furthermore, when things go wrong, the natural logic is to conclude that someone somewhere is being malevolent, rather than acting in normal ignorance [32]. In this mode, once the public recognizes the eternal fact of life that nothing enjoys zero risk, the result is a loss of public credibility; the smallest risks or problems become signals of a crucial fall from infallibility. This is tantamount to primitive thought, where the symbol is collapsed into the word, and no creative tensions exist any longer between the metaphorical skeletons of ideas and the literal versions of the metaphors. People behave *as if* the technology were *literally* an alien being from space. They have effectively been *encouraged* to think this way by past modes of promotion and legitimation of "technology". It is worth looking at some suggestive evidence.

Psychoanalysts have examined clinical cases involving condensed images of technology. These have become central surrogates to explain more complex experiences and potential responsibilities that people cannot unravel. The images, or *specters,* are not only psychic simplifiers but also analogical frameworks for social relationships. They are constructed of *technological* metaphors and images; perhaps increasingly so given the increasingly central role of such experience in daily life.

Daly defines a specter as a kind of potent, artificially created, but invisible behavioral force:

> A sense of the operation of such forces arises when men find they cannot account for emotionally significant events by ascribing them to the

> conventional sources of power and efficacy (e.g., human, natural, divine) which are believed to make things happen in the world. When such inexplicable events persist and are experienced by numbers of people, agencies are created to account for these events. These agencies are given names, made into realities, and adapted to as powerful things ...
>
> The spectral view of technology arises from a sense of domination by mysterious forces or agencies which are, or were, linked to technological enterprises but which are now apprehended as being beyond the control of any particular man or collection of men ...
>
> [People] behave as if the spirit of meeting specifications in many discreet, limited and finite human ventures had taken flight from the hands of responsible agents and become an independent reality – a reality which has come to overhang the modern world and to enter into the dynamic processes of personality – as a spectral object [33].

There is, in other words, a ritual defense mechanism – a transference of responsibility onto identifiable agents, for complex and otherwise inexplicable experiences that are too emotionally important to be ignored. Daly describes how several patients created such specters of their own biological systems, investing them with powers to decide and cut a clean swathe through otherwise overpowering ambiguities. Thus, they would obsessively refer to a simple measure, such as their pulse rate, as a guide to decision making – it was made into a source of "objective decision rules" supposedly reflecting a greater, more powerful, but impenetrable mechanism.

It is a central point of Daly's analysis that these conditions are no longer, if they ever were, restricted to clinically psychotic *individuals*. In his view, they are now *mass neuroses*, transmitted in normal processes of popular cultural dissemination. Given the kinds of symbolic action depicted in *Figure 11.2*, this is hardly surprising. Indeed, the past use of images of scientific and technical power as being from outside the realm of human interests and values has ironically cultivated an escalating search for objective social decision rules based in science. This is akin to a collective-scale version of consulting pulse rates; an example is the incessant effort to avoid the ambiguity of *negotiating* acceptability from situation to situation, by instead trying to create objective scales of "acceptable risk".

In many cases the social creation of such technological specters may be a rational reaction by people to irrational situations in which they are placed.

Consider the experience of the people of Times Beach, Missouri, who were offered $33 million for their whole town by the EPA in 1983, because it had been so pervasively poisoned by dioxin-contaminated oil and chemical wastes that it could never be cleaned up [34]. Like others before and after them, most notably the citizens of Love Canal, New York State, they suffered years of uncertainty as claim and counterclaim were swapped over their confused heads, about their health and its probable origins in the unknown contamination that had pervaded their homes and bodies. Government agencies and scientists contradicted each other repeatedly, and obscurity surrounded the effective causes of their awful predicament. If human hand had been involved, perhaps (as the defendants argued) it was still not straightforward human irresponsibility, but legitimate ignorance: nobody knew that the waste oils sprayed on dust roads at Times Beach could contain dioxins; nobody at Hooker Chemicals knew, when the old canal bed was used to dump chemicals, that it would eventually be used for residential development [35]. Even actions aimed at protection, mitigation, or compensation were experienced as exacerbating the existing alienation, fear, and mistrust. Thus, necessary tests on people naturally generated more fear as blood samples were taken, yet the results remained obscure and "private" to the experts. EPA monitoring crews entered people's homes dressed in "space invaders" protective gear while children were "just running around" in normal clothes [36]. Even some of the EPA staff involved felt uneasy about the extra alienating effect this must have had upon the people – and this was meant as a protective, remedial project!

Apart from the fragmentary cultural transmission of fear and blame, this encounter with aliens was the only *concrete* experience of the issue that many people had. Yet although it was, perhaps, an extreme version, this was not a basically abnormal experience. Most people are fragments of technological systems, which entail many connected parts whose coordination is essential, but complex and chronically problematic. However, they never experience the whole system [37]: their experience is fragmentary and bounded by their local organizational and cultural context, within which they have to make out. Finding it impossible to penetrate the boundaries of their local experience and to understand the rationalities, interests, and interactions of those whose doings structure that situation, they create shorthand images to "explain" those external agencies and their frequent unpredictability and apparent malevolence.

A graphic example of this was given by McDermott, who described a specter created by American GIs in Vietnam [38]. They were operating in the jungle, constantly sniped at or attacked by Vietcong guerillas who could not be identified and pinned down; regularly shelled and rocketed, but never even sure it was not by their own side; and received orders but never explanations from their superiors. Their experience was frightening, confusing, contradictory, and utterly obscure as to its effective causes. They could not find an enemy and they could not identify their friends. Yet they received orders and were attacked in an equally arbitrary fashion. Their (very high) risks were an inseparable combination of physical and social realities. As part of their rationalization of this predicament the GIs had condensed the potent, but diffuse and invisible, effective causes onto a single symbolic agent; they lived a relationship with a "huge fucking" gun, which lived in a hollowed-out mountain and emerged unpredictably at whim to unload death and destruction onto them, in a manner that authentically described their actual experience.

This "effective cause" of their risks was an agent beyond control, imbued with a kind of autonomous malevolent intelligence. In one major sense it was no comfort at all, but in another sense it was, because at least it offered *explanation*, cause, and responsibility. It was at least a *focus* for their fear and their blame; and it offered order to an otherwise uncontainable anxiety. It was a metaphor that represented their social relationships with those elites (and here also enemies) who remotely, invisibly, and unintelligibly controlled their fate. It was a technological specter.

Langdon Winner has discussed this general process as technological animism [39], in which "men export their own vital powers" into technologies, which then return in experience as alien and capricious. However, Winner falsely implies a lack of social stratification or cultural differentiation in this process. People are circumscribed by mystifications created not by themselves, or "men" in general, but by processes of domination whose human structure is increasingly socially complex and remote, and thus "invisible", unintelligible, and, indeed, authentically dislocated. People therefore transfer responsibility from this frustratingly intangible and impenetrable human complex onto extra-human specters. But they are symbolizing real experience, not "mere" fantasies. We are talking about transfer and condensation not so much of their own (anyway small) responsibility and power, but of the power of elites in the confusing social and technological structure around them [40].

Not only does this cognitive process artificially naturalize and consolidate the social alienation of decision makers and lay publics by placing the "effective causes" of risks and technologies apparently beyond human access, but it inevitably encourages a lack of human tolerance for ambiguity. This in turn creates a structural brittleness in the decision system, for which we are now reaping a harvest of apparently irrational "public opposition". When diffuse, but encircling responsibility is so condensed onto such technological specters whose inner social workings are inaccessible, experience has to be interpreted, and life conducted, by either total identification with or total repudiation of such simplistic specters. Thus, public "debate" and interaction becomes rigid and prone to sudden discontinuities: government itself may become less viable. As Crozier has put it, there is no authority without negotiation [41], and since such fantasies and specters preempt the possibility of negotiation by replacing and "blackboxing" more discriminating perceptions of relationships and causes, they tend to destroy even the *possibility* of legitimate, critically based, and mature authority. Social learning and more socially robust policies are preempted by this process, which is buttressed by the bureaucratic concealment of ignorance.

In this section I have advanced a much more complex picture of the problem of public concern from that portrayed by developments of the more "scientific" approaches of risk-perception psychology, decision analysis, and all related techniques. Indeed, the problem is not only more complex, but in some key respects the opposite of that implied by such approaches; this means that their way of tackling "public concern" may actually only extend it, albeit in more diffuse form.

11.5. Anxiety and Reassurance

At this point it may be helpful to differentiate analytically between two dimensions to "concern". One is more specific, and relates to the usually identifiable risks and attributes of particular decisions, accidents, or technologies. But this interacts with the other more diffuse feeling of general social uncertainty and inaccessibility to remote decision processes, which includes an alienation and lack of identity with such controlling institutions, a growing sense of unease and disorientation as to what values and principles guide their deliberations, and a growing feeling of anxiety about the coherence of familiar, cherished social institutions. The latter is the kind of experience interpreted in Section 11.4. as giving rise to a symbolic condensation of extra-human images of "effective causes" and ultimate responsibility and control. The two dimensions are continuous and my differentiation is analytical. Indeed, I suggest that the developing decision processes of handling the more specific risks and decision concerns relating to *back-end* regulation inadvertently exacerbate the deeper, more diffuse feelings of public alienation, which arise more perhaps from the incomprehensibility of *front-end* technological innovation and "control" generally. These in turn influence public perceptions of other specific decisions or events, and are only amplified when they are ignored by the specific decision framework. The ways in which regulatory bureaucracies process ignorance, structural ambiguity, and lack of control into a tenuously credible image of manageable risk, and the way in which scientific knowledge is often constructed in "reassurance", both fuel this process. They are forms of symbolic action that obscure the origins of more general social anxiety arising from the unintelligibility of front-end innovation processes. Uses of science in reassurance, and decision-analytic methods of addressing public concern, coincide in this respect. Both replace the need for institutional innovations aimed at social learning with abstract concepts, which elaborate one-dimensional managerial techniques, and thereby further entrench those alienated relationships that are arguably the basic problem in the first place.

A further implication of this analytical approach concerns the question of how social responsibility is allocated for decisions, commitments, and events that produce risks and unwanted suffering. An interesting strand of social anthropology interprets *blaming* as an important process of social cohesion and even healing [42]. Being able to identify the agents responsible for some significant misfortune or fear is, it is argued, an intrinsically powerful part of *domesticating* and coping with that misfortune. Whilst this may involve antagonism directed at a specific social target, the very same process allows a corresponding identification with society at large, not least because society at large may have helped to identify the blameworthy and assign compensation, etc. When risks and damage are attributed to "natural causes", but more especially to "unknown causes", there is no concrete focus for anxiety, anger, and retributive feelings. It is therefore difficult for them to be purged because they have no channels of expression, and the result is a diffuse but powerful sense of unease [43]. Therefore, as part of the general corrosion of credibility of decision making, it may be significant that in The Netherlands, for example, for all the thousands of past uncontrolled waste dumps, including several spectacular and expensive public health problems, very little remedial or punitive damages have been recovered from the responsible companies, who have largely gone unidentified. The same is true of the USA [44]. Even at Love Canal, where Hooker Chemicals was known to have been the dumper, responsibility was passed around endlessly between several bodies, and the whole episode almost defined as if a natural disaster [45], with no socially defined responsible party to act as a focus for purging deep anxiety and preventing the escalation of mistrust.

In the UK in 1984, the Black Report into excess childhood leukemias around the Sellafield reprocessing plant [46], two official inquiries into alleged health defects around chemical waste incineration plants [47], and an official inquiry into the causes of an explosion at a public water transfer scheme (Abbeystead), which killed 16 people and maimed several others [48], all followed a common pattern. They all claimed public reassurance by demonstrating that the *prima facie*, suspected responsible party was not the cause, and that "unknown causes" generated the damage in each case. Whilst reassuring in the localized sense that none of the operating companies was therefore officially to blame, in another equally significant way, this logic was not reassuring at all. If such tragic suffering can be visited upon people *without any identifiable effective causes*, then logically they will be averse to projects and plans that appear to expose them to even more of the same. They will to all appearances become "risk averse". Yet, arguably, had they been able to identify a cause they could have begun to reconcile themselves to their grief, and to overcome it by channeling their blame onto a specific agent. Social healing would have had, at least, a chance to begin, and the ensuing containment or reduction of the diffuse all-pervading anxiety described by Erickson, Daly, and others might have even benefited the blamed party.

There is a gaping contradiction between the *ex ante* posture, of marginal uncertainties and full control through rigorous knowledge, that is struck by decision-making bodies when they promote and regulate technologies, and the recurrent theme expressed *after* an accident. Then, the legitimate ignorance argument – "no one could possibly have known" – comes to the fore. The underlying belief seems to be that this more focused reassurance defends and legitimates the general decision-making system by showing that specific agents within it were not irresponsible or incompetent. I would argue that it does the opposite, because it further encourages the belief that unknown and uncontrollable causes are in charge – it creates greater social uncertainty. Furthermore, the lack of official acceptance of responsibility (even if nonnegligent) encourages a public sense of *malevolence* on the part of the forces in control. Thus, the process of "defensive" legitimation is eating away at its own foundations [49].

So much for a more sociological interpretation of public concern and of the interactions between its different dimensions. Let us now examine how this dimension is treated in "rational" risk management decision techniques.

11.6. Decision Analysis – Destroying Social Institutions

There are various branches of decision analysis and here I do not present a complete review. The general field combines economics and psychology to generate normative frameworks for defining rational solutions to decision problems [50]. Decisions are always defined as discrete occasions. The basic method defines a branching set of options and probable consequences, attaching subjective values to each of those expected outcomes and estimated probabilities of their occurrence. In risk analysis, the potential consequences of a technology can be factored into different dimensions – deaths (to workers or public); morbidity; environmental damage; economic damage (e.g., air pollution on crops; clean-up costs, etc.). These can be further differentiated; for example, deaths to children can be said to involve more damage than deaths to old people; collective deaths in one event may be more costly than the same total number spread over many routine events; etc. These are the "attributes" identified in risk perception work described earlier. An "index of risk" can be composed of these factors for the activity in question.

When units are attached to quantitative risks, many value choices are made, whether or not these are explicit. For exactly the same activity or "risk", deaths per employee may give a very different comparative risk picture and imply different safety norms from deaths per unit of output, or time, or resources. These value problems are familiar enough and have been tackled by several authors. A leading approach is that of Watson *et al.* [51]. They point out, correctly, that choosing an index of risk for a decision analysis is a value laden, political act; but if this can not be avoided, at least it can be made more transparent, considered, and democratic. *Decision makers* rather than *experts* can choose which individual attributes to use with what weighting in a composite definition of "risk". The weighting distribution across the range of chosen attributes would be the *utility function*, since it expresses the negative utility for that decision maker of those aspects of harm chosen to be represented. Having performed this exercise in an explicit way for competing options, and varying the operative utility function, the decision maker can then trade-off benefits, risk, and values to arrive at an optimal choice. In theory, given sufficient accountability and consensus, this could be a societal choice. Indeed, the steady shift from earlier interest in subjective expected utility analysis toward multiattribute utility analysis for public decisions has been accompanied by much work claiming to define the *objective* attributes of different technologies as a basis for more "objective" public risk decisions [52].

There is no reason, in principle, why this formal procedure should not include "public concern" or "risk aversion" as part of the index of harm, or utility function. Watson *et al.* expressly recognize "concern" in this way, and advocate its inclusion (assuming it is not based upon simple ignorance of risk magnitudes) as one attribute alongside others in a multiattribute utility analysis. "Concern", once converted into units, e.g., of money equivalence, may be weighted in alongside potential deaths and injuries of various kinds, environmental damage, etc. These negative utilities are termed "detriment" in risk–benefit optimization techniques. Even perceptions of outcome-uncertainty can be at least formally included by discounting greater perceived uncertainty against lesser uncertainty. Converting "concern" into units has been increasingly performed by eliciting people's willingness to pay to avoid a given risk increment [53].

As far as it goes, this kind of framework is an admirable attempt not to patronize public concern, but to take it seriously as a substantive, legitimate part of public risk management and policymaking. There are, of course, familiar and not at all trivial questions to do with the aggregation of qualitatively different kinds of "attribute" and experience into single scales, and with the assumptions mentioned above. However I want to identify two further problems not usually acknowledged at all. These concern the transparency or otherwise of people's problem definitions and technology definitions; they lead us to question whether this overall approach of enlarging decision-making *techniques* can ever be an adequate way to address the substantial dimension of public concern. To relate this to familiar terms, I suggest that "attributes" are distant signals of a more basic divergence in the very problem definitions that people are addressing, but which are distorted into the single framework of the decision analyst. In Kierkegaard's terms (see the Dedication, this volume), perhaps risk attributes are equivalent to the rules of aesthetics that the critic saw in the poet's "music".

The first problem is that the whole framework of Watson *et al.* takes it for granted that the technology that is the risk-source is an objective, bounded entity, definitely separable from other risk sources, and is the same to all social groups. As discussed in Chapter 9, this is never so, not even for experts. Its interpenetrations into other technical and organizational systems are manifold and its boundaries are *intrinsically* ambiguous and *socially* defined. It is therefore not only a matter of what kinds of *detriment* (including concern) to incorporate from an assumed given risk source, but of what are the appropriate boundaries and shapes of the risk-generating problem in the first place. If concern originates in the lack of recognition of divergent frameworks, it is not going to be solved by a method that incorporates it into a monopolistic framework. Because most technologies are composed of heterogeneous social networks, as well as physical entities, their defined characteristics and boundaries are different to the different social groups involved, even those "internal" to a technology, as determined by their objective social experience. Thus, they may not only be structurally conflictual, they may also not be matters of explicit definition and choice. Because their objective social experiences are different, different culturally determined perspectives will define not only what risk attributes are involved, with what weighting or utility, but also what *technologies* (and social relations) are involved. Here we can recall the examples in Chapter 9, especially the 2,4,5–T case and the Windscale reprocessing plant debate.

The usual analytic framing, of technology creating "consequences" or "expected consequences", which are "evaluated", is thus fundamentally inadequate. Multi-attribute utility analysis responds to "concern" by incorporating it in a decision technique that assumes a single, discrete, context-free problem, and thus a unitary policy-decision framework of values. It cannot recognize the social contextual properties of technology and the origins of concern in social relations, *because its fundamental structure denies context.*

Another highly influential and important approach to decision analysis, which has recently turned its attention to risks and the public concern issue, is that of Raiffa and colleagues [54]. In order to illustrate the *systematic* neglect of institutional dimensions let us examine the coal-mine rescue dilemma. Raiffa has used this to illustrate the inclusion of concern, signified by such legitimate attitudes as "regret" or "elation" over particular outcomes [55].

Three miners are trapped down a mine and may die unless a rescue is attempted. If a fourth miner attempts to rescue them, the chances of all *four* miners being killed become four to one. The decision problem is depicted in *Figure 11.3.*

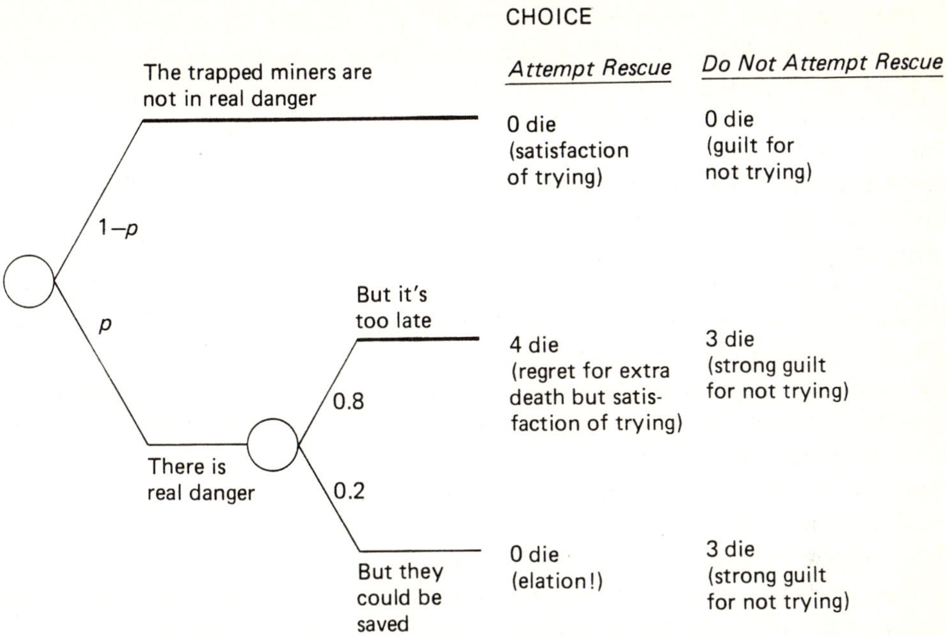

Figure 11.3. The problem of the miners (from Raiffa [54]).

The principle is that the utility for a perceived unwelcome outcome includes an *ex ante* anticipation of potential *ex post* regret. This "concern" may include guilt, for not trying to rescue the miners; it may also include an element of the decision-maker's accountability to other groups. As Raiffa puts it, "if accountability to others and regret are major concerns, then these concerns should be recognized by incorporating them into the description of consequences."

Depending upon how these new attributes are weighted,...it may or may not be "optimal" for (what is assumed to be) the decision maker to make a rescue bid. (Using the usual mathematically simple computation of the "least-cost" option.) For example, in *Figure 11.3*, let $p = 1$, and let there be a 1 in 3 chance of "total rescue" (i.e. $p = 0.75$ that all four die in a rescue bid). Then (assuming all deaths are accounted equally), if we use classical methods and neglect any other considerations, such as regret,etc., it would be equally rational to attempt or not attempt a rescue, since the expected utility of a rescue is $-0.75 \times 4 = -3$, and of a nonrescue it is also -3. If the "guilt" or "regret" for not trying were positive (i.e., negative utility) then including this would tip the scales in favor of a rescue. However, if the guilt for not trying were outweighed by the regret for an extra death at $p = 0.75$, the balance would swing back to favor a nonrescue.

Raiffa's attempt to account for factors such as guilt, regret, and accountability in such decision-analytic terms is one of the most sensitive and sophisticated attempts to develop a confessedly normative framework into one relevant to observed realities. However, there are several interconnected points that encourage similar criticisms to those already made of other decision-analytic approaches.

The normative value of such a framework is reduced more than usual by the evident fact that a range of different, conflicting decisions can be reconstructed to be consistent with the prescriptive model, according to adjustments to the now even larger and vaguer range of subjective utilities. Raiffa's response to this is that the technique's value lies in forcing decision makers to clarify their concerns, or "utilities", and the extra considerations are part of this clarification. This is fair enough. But the question here is whether the basically individualist model of "concern", "regret", etc., can be adequate for *social* decisions, where institutional values enter in a way not reducible to aggregated individual perceptions and expressible utilities. The evidence from observation is that any "decision maker" would want to "feel" a decision to be intuitively right and would then, if forced to, retrospectively set the factors to reach the "rational" solution (i.e., the one felt to be right).

Raiffa implies, for example, that a decision maker who could not quantify the utility of his or her accountability and regret considerations would be illicitly "resisting" making these things explicit, for reasons of image management. This betrays the false assumption that such considerations are, in fact, individual and *privately* clear, even if kept from explication for ulterior motives (a rationalist version of the conspiracy theory!). This approach by definition cannot appreciate that "accountability" and "regret" (like decisions generally) may be embedded diffusely, nonprecisely, but substantively in processes of social interaction, where their meaning (and thus, "utility") is being *incessantly* renegotiated, as suggested earlier. They are the currency of interaction whereby social institutions are maintained. Thus, it is crucial to note that the problem is not the requirement to make them clear, it is the requirement to make them clear in a context of *mistrust* or social uncertainty, where they may be expropriated and used inflexibly by analysts and decision makers with whom no negotiation will take place. To extract control of their meaning from such contexts of relationships and hand it over to a decision maker or decision makers with whom there is *no* real relationship may be *precisely* the problem! It is to replace the need for an enlarged social relationship of negotiation in decision making by an enlarged *technique*.

In a real coal-mining accident, with three or however many miners trapped in a mine, a real decision maker would make a rescue bid whatever the odds, because if he did otherwise and was seen, for example, to conduct a "rational" decision analysis as to whether or not to mount a rescue, the *institution* of coal mining would disappear. Men would not go down a mine unless they trusted their fellows to try to rescue them in an emergency. This is the hard evidence from social observation. In reality, any such *individual* artificially abstracted event or decision is embedded in such concrete historical–social processes. This is not "mere" context, but the *seat* of value. This kind of collective trust is a necessary condition for the very survival of the social–economic institution itself. The value of the institution (partly a recognizable economic–social activity, but *culturally* far more than can be expressed in mere words and numbers) is simply *denied* by the "rational" decision-analytic approach. There is intrinsic quality and value in the forms of social interaction that constitute coal mining as a culture, whereas decision analysis can only define the human interaction as an *instrument* in the attempt to achieve individual values.

It is hardly surprising, therefore, that in a real coal-mine rescue dilemma, the decision maker's concretely social sense of accountability would intuitively supersede decision-analytic calculation – and rightly so. The decision analyst might reply that this social relationship can be adequately represented in the individual decision maker's utilities expressing expected guilt, accountability, etc. In a case like a mining accident this might be superficially so, because in such cases these particular utilities are likely to be so strong as to make the decision to attempt rescue the clear winner. But this may be the right result for the wrong reason: the very approach still radically cuts off any *continual* relationship and regular negotiation with the social context (i.e., publics). The quality and form of such relationships, therefore, cannot be attended to as a necessary part of decision making.

Although in the coal-mining case the obliteration of continual institutionalized social relationships may result in the right decision, in many other cases the concerns and values alive in the "decision context" may not be so clear-cut or focused. There may exist the same fundamental need to recognize the authenticity and inherent value of these social bonds and their own problem definitions, but being more diffuse, it will not be recognized within the individual, event-specific utilities of decision analysis. A "successful" run of the coal-mine rescue dilemma will therefore be categorically misleading for the many more usual cases where "the" problem is less clear-cut and universally recognized. The regulatory equivalent of the coal mine case might be facility siting; yet, as I argue later, the individual utilities framework used in formal compensation methods is not adequate to deal with the realities of conflicting problem definitions, even around such apparently clear and commonly held problems.

Thus, a major problem obscured by the use of the coal mine examples is that decision analysis precludes recognition of diverse problem definitions that arise in the social context and that should be negotiated in the ongoing institutional relationships of decision making. Reducing these dimensions of social experience and interaction to individual psychological and "single-decision problem" components of concern obliterates their moral standing, and in doing so potentially undermines a basis for public trust of policy decisions and decision-making bodies.

There is an important practical conflict that underlies sociological attempts to identify the authentic categories of people's reasoning and concern. This kind of work is often taken by its critics as an uncritical defence of the *substance* of popular attitudes, beliefs, and reasons [56]. This may be true of some such work, but it is not at all a necessary part of it, and it is not part of the present exercise. To understand the *categories* of experience and thinking of other people as an essential prerequisite of being able to communicate and negotiate constructively with them, is very different from the faith that they are "correct". What must be defended is not the substance of popular attitudes, but that their own categories of meaning authentically and legitimately *exist* and these may extend to deeply uneasy attitudes toward decision-making institutions. Of course, to see the consistency and resilience of belief systems once set in their proper social context is more respectful than to assume that they are flawed through unnecessary ignorance, mental failings, emotional maladjustment, or whatever; and this perception is often accompanied by a greater modesty about the limits of our own rational frameworks and the institutional relationships they reflect and justify. As analysts and managers, our frameworks of rationality appear detached and purely instrumental (as in decision-analytical models), because that is the social meaning of *our* public knowledge – detachment from and manipulation of others [57].

The frameworks of more localized units of social familiarity see risks in trusting the schemes of the social managers simply because the latter effectively ask for a blanket erosion of familiar social bonds, to be replaced by a blind trust in themselves; and then we complain of "irrationality" and "emotionalism" when we do not receive such unqualified trust and credulity! The relationship of the publics of industrial society to the decision makers is, as I have said elsewhere, similar to that of the peasant farmer in the Third World faced with the self-appointed "rational" economic plans of the Western economic planner for improvement of that peasant farmer's lot [58]. There is plenty of evidence to show that, while they too have socially bounded knowledge, in their own terms, the peasantry know relevant things that the experts do not recognize. Included in that local experience may be knowledge that makes the expert's framing of the issue inadequate or irrelevant. The conclusion is not that we should justify and encourage further social alienation by celebrating a "superior" rationality of the "technological peasantry", any more than that of the agricultural peasantry. But the denial of the basic integrity of those categories of experience by self-appointed "rational" approaches is inadvertently the most likely cause of further alienation and concern.

One reason for this kind of basic category mistake in the decision-analytic approach is that its origins lie in the psychology of choice, which is founded on isolated decision models taken almost exclusively from gambling. Not only does this enshrine the individual decision-maker fallacy; it also enshrines the assumption that decisions and attitudes do actually exist in isolation from a social context – that they do not signify or interact with historical and cross-cutting relationships, decision-streams, and problems. Lopes has admirably summed up this limitation that pervades the whole field:

> The simple static lottery or gamble is as indispensable to research on risk as is the fruitfly to genetics[...]we psychologists are a bit trapped by our own proficiency at being good experimentalists. We realize the importance of control and so we are drawn to those tasks in which we can exercise control. Hence our preoccupation with simple, static lotteries [59].

Such highly simplified "decision simulations" systematically exclude such factors as past experience and relationships, or effects on future options, or any kind of different problem definition. When they are then found not to correspond with real-world choices and behavior, the rational decision axioms that underly the simple experiments are not questioned – reality itself is, as people are dubbed irrational, and therapy or "information campaigns" are called for. Yet, as Lopes suggests, "People who violate [rational decision] axioms may simply and with good cause be trying to consider facts about the world that the axioms ignore" [60].

When we turn to the real world, we find that not only isolated individuals violate such tidy rational behavioral axioms, but that whole institutions, organizations, and societies do so.

11.7. Rescuing Social Values

Many people are found to be hesitant about even giving credence to the framework of values embedded in such decision simulations. These may include private consultant–client (decision-maker) sessions to elicit utilities, and questionnaires. The latter may elicit utilities from a larger population, ask respondents to participate in a formal rational choice game (like the coal-mine rescue or an equivalent gamble); or they may ask people to express their utilities via "willingness to pay" questions (to avoid proposed risk increments). As already indicated in the earlier discussion of Raiffa's work, the reaction of decision analysts and "experts" is normally that people who resist such exercises are deviant, irrational or otherwise, and need corrective therapy [61]. Yet this often intuitive unease with such frameworks indicates a deeper problem, which these "rational" approaches ignore. The skepticism may signify a deeper feeling that the individual utilities framework already requires its participants to accept *prior* values, ones that are *not* offered in the agenda of choice. To even enter the game is, therefore, automatically to give this whole moral framework a social authority, without the necessary prior explication and negotiation. As it stands, the whole framework may beg the most important value questions.

What are these more fundamental questions? They are signaled by, but extend further than, questions about arbitrarily imposed problem definitions. They concern the ultimate motivations and meanings of social interaction and human nature. The sociological orientation employed here, and the rational choice framework of risk decision making developed from economics and psychology, are fundamentally at odds over this deeper issue (which, incidentally, combines empirical–normative or descriptive–prescriptive dimensions). Regardless of various perceived differences between psychological and economic approaches [62] the decision-analytic approaches all assume – or require – that human beings are essentially complete as isolated individuals – they have their existing values and utilities, and interact with others only as a process of exchange (a market) for maximizing, optimizing, or satisfying individual gratification. This is the "essentialist" model of human nature [63].

Sociological work that has actually examined real situations shows that a fundamental "utility" for people is the maintenance of established and trusted social ties, loyalties, and groups, as the basic fabric of personal identity. This is the "interactionist" or "relational" model of human nature. In this view social interaction is not merely a means of optimizing prior individual values, but is of inherent value in itself; it is also the *ends*, that is the process of *completion* of individual values, which would otherwise remain open-ended and incomplete [64]. Intrinsic value lies in certain forms of social interaction as such, and not only in the ends achieved by the interaction; it is thus *inherently* incapable of precise, "scientific", and static expression as required by "rational" decision-analytical methods and by the conventional psychology of risk perception. This mode of value expression is a constantly recurring process because completion can never finally occur; meanings and values have to be constantly tested, repaired, renegotiated, and developed through autonomous social interaction. Values and perceptions are therefore *inherently* social and open-ended to some degree; the requirements of precise, once-and-for-all "objective" specifications of values and utilities are themselves a reflection of an alien and fundamentally incompatible view of human relationships and public life. To the people on the receiving end it is a major social risk to give assent to such a moral–social framework.

It is important to note that from the sociological perspective there are three *separate* flaws in the rational individual, or essentialist, approach. The first is the familiar one that economic terms alone are inadequate to grasp the categories of value and meaning in which people think, act, and interact. Even when noneconomic realities are recognized, they are converted into economic terms. The second is the artificial abstraction of discrete decision problems from their institutional setting, where negotiation of different problem definitions is the essence. These two have been raised before. The third flaw is that, even if the definitions of utility and risk are expanded to try to incorporate factors such as "concern" and "regret", these are still defined as *individual* utilities. Social entities are still "valued" only derivatively, as *instruments*; they are seen only as markets of exchange between such "complete" individuals, optimizing this now-expanded range of utilities. In denying that the process of social interaction itself has value (because it is regarded only as instrumental means), the essentialist approach embodied in all the "rational" methods denies the very basis of meaning that people still have as partly their own. It therefore permits inadvertent social identity stripping and is intrinsically unlikely to be able to foster public identification and acceptance.

Even when handled with due sensitivity and liberal intent, as exhibited by Fischoff, Watson, Keeney, Raiffa and others, this basic approach cannot be directly extended to address the institutional elements of public concern that I am trying to explore. The sociological framework holds that the most basic value or meaning, motivating public and personal life, is the continual maintenance of the social and cultural units with which people identify, and to which they give personal loyalty [65]. This precedes economic motivations, because, as the field of institutional economics implies, the latter first need such a collective context in which to exist, just as do any other ends. This "social imperative" could *in theory* be explicated as a personal utility; but, even apart from the nonindependence of utilities problem, this would involve such a drastic redescription into individualist terms, that it would destroy what it was meant to represent. Furthermore, as I have argued elsewhere following Kekes [66], there is a normative rationality of social behavior that includes a general objective. This cannot be parceled up into discrete, disconnected decision events. This is the defence of one's social and personal autonomy from potential outside control. It is manifested in a "passive" rationality, which eschews the "active" rationality of framing values, goals, and choices, in others' terms, for them to expropriate and manipulate. Unwelcome control encroaches via the explication of formal, *static* definitions of values, utilities, etc., to be taken off as inputs to alien decision techniques and processes. Contrary to the central faith of decision analysis, freedom may be tantamount to maintaining *imprecision* in values, which allows flexibility to negotiate them amongst one's own complex social networks.

In addition to historicity, decisions and risks for real people have *interlinkage across the issues* as defined by analysts or decision makers. Some of these social interlinkages and incompatible problem frameworks are distantly indicated by the *attributes* analytically associated with given risks. Thus, even the progressive versions of rational policy and decision analysis that try to recognize public "concern", do so within a potentially self-defeating framework in which "concern" is defined as an individual trait that can be dealt with by enlarging a managerial decision-analytic technique, even though it may be aspects of the managerial decision-analytic relationship itself that are causing the concern. It is a form of identity stripping by policymakers from the very people it is supposed to serve. Elaborating the decision technique may only elaborate the problem, not the solution, by even more comprehensively expropriating the lay person's meanings into an alien framework represented by the social relationships of control and management themselves.

11.8. Empirical Explorations: Compensation as Risk Perception Decision Analysis

In this chapter I have suggested that elaborating decision-analytic techniques by factoring in "concern" is worse than no use if that concern is, at least indirectly, related to the use of such techniques in the first place.

One response is to explore ways of adapting the institutional structures and relationships instead; for example, by more openly recognizing scientific ignorance, structural conflict, and lack of control, and to adopt social structures that have a broader "control" of technology. Another institutional response is to place more responsibility to define the significance of the risks in local (risk–receiver) hands, as was attempted in the Massachusetts hazardous wastes facility siting initiative and others that followed it. This approach attempted to replace formal decision analysis by direct *empirical* solutions to the economic regulatory question of "How much are they willing to pay to avoid a certain risk increment" (or its inverse, "How much would they have to be paid to accept it?"). It is worth briefly considering this important approach here, to see how its general premises about public perceptions and decisions relate to those already discussed.

The idea of compensating people on whom extra risks fall – as in communities asked to play host to hazardous facilities – is not especially new. Whatever it has been called, past developments have often been accompanied by mitigating factors, ranging from design changes, through local sports facilities or redeveloped wildlife areas, to direct and indirect financial transfers. Providing local jobs and spin-off business is a form of "compensation". However, since the late 1970s, there has been an attempt to systematize and refine the principles of this approach, at least for the important facility-siting aspect of regulation. The most definitive work in this direction so far has been in the USA, especially that of O'Hare *et al.* [67]. It offers a major advantage, in principle, over the approaches criticized before, because of its strongly empirical core – indeed, the approach was embodied by O'Hare in the Massachusetts Hazardous Waste Facility Siting Act of 1980, and has been repeated in several similar Acts in other states.

The starting point for the "compensation" approach is the baneful role of the statutory Environmental Impact Analysis, which pretends to be a definitive environmental and socioeconomic risk analysis of development, but which in reality is satisfactory and credible to none of the various interested parties. Intransigence is built in to the process, whereby definitive claims are (as is always possible) picked apart by whichever parties wish to do so in the analysis of the impact statement. If, on the other hand, parties can be shown that they have something to trade in negotiation, and something to gain by it, and if they are helped to develop their own information on impacts and risks pertinent to *them*, then they can *concretely* define a well-informed "willingness to pay" or utility function, in the process of real negotiations.

The approach recognizes a range of different currencies of negotiation more subtle than money, and attempts to clarify under what circumstances different currencies may be valid. It also recognizes that some issues simply may not be amenable to compensation processes. According to O'Hare *et al.* [68], the characteristics of a "compensable" siting issue include lack of existing mistrust between parties and impacts that are clearly traceable. This already hints at problems. Before discussing these, however, one further aspect is worth highlighting for its connection with already established questions.

O'Hare *et al.* recognized that decisions normally involve relationships that have a history and a future. Indeed, in real issues, decisions as such are difficult to identify. In calculating strategies, therefore, parties to a decision take into account expectations created by previous experiences of the other party. If it acted in an authoritarian fashion last time, they are unlikely to expect a viable compensation, so may not even bother to enter negotiation, instead adopting an intransigent stance from the outset. Likewise, present actions have a future effect, what are called "demoralization costs". Therefore, carefully organized compensation may be a crucial element, not only for current decisions. Contextual relationships, e.g., in the future, must be considered. This is a valuable insight and already a major advance from formal risk-decision analysis, which (except via categorically inadequate terms like "regret") recognizes no such context. In tacitly recognizing the continuing social relations of such decisions, in which such factors as "demoralization" are carried, the compensation approach is a break from the elaboration of managerial techniques criticized before; and it offers potential for engagement in the alternative, which those approaches conceal, of considering the options for "institutional repackaging" of projects, regulations, procedures, and so on.

However, this is where the underlying problems and questions reemerge, because although the compensation approach broadens the currency of negotiation from money alone, it is still wholly framed within a conception of narrow market relationships, where "we expect the public's participation in the political and legal process to be motivated by concern for individual utility [69]." Thus, the range of "compensation" options is restricted by this framework. If a more "interactive", sociological framework were used, the conceivable options could be extended to include structural changes in the forms of social control of the development in question and any like it (as opposed to one-off design changes, for example). Also, irreversible improvements in the institutional process of debate, analysis, and decision could themselves be considered as "compensation" (whatever the *specific* outcome) in a different political culture (and perhaps even the USA one) that did not solely recognize individual utility. These would be entirely consistent with the need to build *future* trust or to avoid "demoralization".

A further elaboration of this "social relations" dimension would be the radical alteration of these relationships and expectations by the promulgation of the same development proposal, or regulations, but with an explicit public recognition that developers (perhaps *the* developer) and decision makers had acted corruptly, incompetently, or insensitively in the past, but had now recognized these errors and had attempted to learn from them. It might then ask for social conciliation, a new, more modest start, and commit itself to (what would presently be) an utterly deviant course of acknowledging ignorance of risk-implications where that existed. This would be attempting to share control, and it would also be asking for shared responsibility in making decisions in the face of such realities. Whilst this looks absurdly infeasible seen as a sudden step, viewed as a goal and a gradually emergent alternative to present trends it might be more realistic and attractive, and consistent with principles of social learning as opposed to political deskilling.

In any case, the point is that despite valuable breaks with more formal analytic approaches, the existing individualist framework of the compensation approach has a restricted capacity for conceptualizing wider institutional options as part of risk-related regulations or developments to which publics have to react. Nor does this framework allow the possibility that in negotiation, issues, perceptions, and evaluations cross-connect and interact, but in different textures for different parties. Thus, the recognized problem of how to identify which range of impacts to negotiate over is actually a rephrasing of the question of how to bound and structure *the problem,* or the definition of the risk source, in the first place – a central theme of earlier critical discussion.

The logical need to recognize equally legitimate, different basic problem definitions can be seen by examining the question of information. The assumptions about knowledge in the O'Hare *et al.* approach are interesting to explore. They recognize that, as direct *users,* interested parties are more likely to develop the precise kind of risk–benefit information they need than is an analysis that claims to be definitive for all and sundry. (This has long been recognized as the rationale for adversary procedures.) Greater efficiency and the crucial ingredient of market intelligence appears to result. This is an advance on conventional one-dimensional assumptions about information and "facts".

However, there is no reason why this approach should stop at different parties' information development within a common problem definition. A central reason why parties differ as to the kind of impacts deemed important, and as to the kind of *information* they best need for a decision, is that they are actually addressing a different decision problem, because it is set by them in a different network of connecting problems, experiences, and constraints. A decision problem may superficially appear the same at a common intersection of two (or more) problem networks (e.g., of parties to a siting issue), but the network connections of the problem to associated problems are different for each party. That is, the meaning and definition of "the" problem at the intersect is very different, even though it may appear "the same". This is the underlying social generation of irresolvable ambiguity, with which conventional rational decision analysis also cannot cope. Such frameworks simply cannot carry the requisite information to represent the authentic problems. Genuine information richness requires that these deeper problem networks be brought out. The deeper roots of public risk perceptions would then be discerned.

Allowing for pluralism of information, as do O'Hare *et al.*, so as to carry more overall information in the "decision system", should therefore logically be allowed to lead to the explication and definition of the different and implicit problem networks of the different parties. This is the most information-rich, market-efficient solution. One can therefore derive a sociological framework from the (undeveloped) principles of O'Hare *et al.* on information.

The "lack of existing mistrust" requirement for compensation to be a feasible risk-decision process also falls foul of a more substantially *sociological* basis than the "individual utilities" framework can recognize. This is illustrated by the troubling experience of one aspect of the Massachusetts Siting Act. Under the law, a prospective developer must file a Notice of Intent (NOI) with the State's Hazardous Waste Facility Site Safety Council, which was brought into being by the Act. In an attempt to have formal licensing *after* a negotiated agreement on principles (so as to avoid the early and often fatal polarization that licensing hearings engender), the Council then makes a very rapid evaluation of whether the plan is "feasible and deserving of state assistance". This was designed as no more than a very crude sieve, to filter out frivolous projects registered only so as to qualify for land-acquisition development funding or whatever, and any other "obvious" means, e.g., unnecessary capacity or undesirable agents. It was never designed as a detailed analysis (the Council had only 15 days to decide), nor as indicating a presumption in favor of development, and thus emphatically was not a focus of conflict – the intent was to *avoid* such early polarization. Yet the NOI has constantly provoked vigorous and detailed opposition, effectively defining it as a licensing step.

This contradiction in official definitions of the situation evidently surprised O'Hare *et al.*, who saw it as evidence of public misunderstanding of the merely preliminary filter role of the NOI "feasible and deserving" review, and of the need to better communicate its proper role. But this interpretation indicates a lack of appreciation of an authentic social rationality, that *every* step in a decision process, whatever the official definition thereof, builds presumptions and momentum for later steps, and so must be energetically fought. The step signifies far more *informally* than it does *formally* – the gap is the social relations dimension. There is also here a dimension of mistrust, which underlies the previous point, that problems and perceptions rationally interpenetrate one another in networks. Thus, a more diffuse preexisting sense of anxiety and suspicion of "authority" has already created a climate of mistrust when a developer makes a move, even if the developer is unknown to that community. This is consistent with the interpretation earlier of the perceived arbitrary, nonhuman nature of "the efficient causes" in people's lives and their risks. Even known, "positive image" developers may thus have a background threshold of mistrust to overcome.

By partly acknowledging the context and historicity of decisions, but then reducing this to individual utility functions, the compensation approach as so far developed shares many key limitations of the formal risk-decision analytic approaches. When the individual decision model is scaled up to collective decisions, even when different attributes and different utilities are recognized and elicited (analytically or empirically), the automatic assumption is made that every individual (and individual group) in the collective public policy setting is focusing on the same problem, with the same intensity and the same relationship to "other problems" (e.g., future compensation) as every other actor. This obscures the crucial *social* fact, that people are plying different routes through their own networks of problems, interconnecting differently with other ones, and intersecting only at certain points (e.g., when a so-called "risk issue" emerges) with the different problem networks of other groups. The network affects the shape of the group's problem at the intersect. Allowing groups to express their utilities in different currencies of compensation is not an answer to this issue, because the "utilities" are still *individual,* and related to a single, imposed problem definition. An adequate answer would need to facilitate the expression of different prior problem definitions and concerns. In its turn, since social control (too little, too much, or the wrong sort) is part of the problem for some groups at least, this may require changes in the social relations – i.e., institutional control – of the developments being sited or the technologies being regulated.

11.9. Conclusions

The main theme of this chapter has been that the dominant approaches to risk management and regulation have begun to take seriously the "public perceptions" issue, but that they share a central flaw. Whether formal analytical or formal empirical (compensation), they assume that individuals are the only seat of value, and that social interaction and its patterns of loyalty, identity, and trust are merely secondary instruments for the optimization of these primary values – individual utilities. Concern or "risk aversion" is then assumed to be solely an individual cognitive–emotional entity, which can be reduced to and factored into such individual utility functions without qualitative transformation of its meaning. The social rooting of concern in the threat to cherished social relationships and identification is inconceivable to this approach, because social relationships do not have *intrinsic* value in themselves. This analytical framework does not merely omit part of reality, but systematically, by *definition*, denies value and meaning to a part of reality that may be highly meaningful to people; it is, therefore, not just incomplete, but antagonistic.

I have argued that this approach has been buttressed by an unnecessarily psychologistic interpretation of good psychological work which indicates, even if it does not directly show, the *social* grounding of risk perception. In various forms of decision analysis relating to risk decisions, even in progressive attempts to address public concern, these social relationships are not only not recognized, but they are effectively said to be worthless, when the very experience of "worthlessness" via the expropriation of meanings into alien, inaccessible systems of social control (such as decision analysis) is arguably already the generator of that diffuse concern in the first place. This social alienation of the "effective causes" of risks, the condensation of this experience of unintelligible social control onto extra-human symbols [70] and, the consequent brittleness and polarization of policy debate, was then illustrated from empirical and theoretical work in sociology and psychiatry. The associated processes of social embrittlement or political deskilling are consistent with the growing use of science for public reassurance (as analyzed at length in Chapter 10), which, being based upon false premises, may actually contribute to rising background anxiety. These reassurance roles of science amplify the social inaccessibility of the institutional decision processes that they are defending, because they conceal rather than explicate fundamental conflicts, uncertainties, and inherent ignorance.

Overall, the analysis in this chapter indicates that there is a much deeper, more pervasive, and intractable problem of public concern than normal approaches recognize, even remotely. This deeper anxiety is always an integral part of the concern engendered in specific issues.

It could be argued that since this "concern" and the associated credibility problems *always* plague us, whether in public or interpersonal situations, we should not become too neurotic about it. Deriving from this is a practical argument against the kind of analysis I have given, that there is always bound to be such social stratification, and exposing the basic sociological roots of concern over risks is only to encourage a kind of fundamentalism that would destroy all public institutional authority, and thus social life as we know it. This is a legitimate point of view and deserves to be taken seriously. I therefore return to it in the final chapter. The essence of my reply is that I have criticized certain deeply embedded myths of rationality and authority in public decision making, not because they are myths, but because they are *inadequate* myths, degenerative rather than enabling. All it is necessary to say here is that the abundant existence of increasingly active and, on its own terms, successful public opposition to regulation of hazardous wastes does not prove that public acceptance or credibility is decreasing, from whatever cause. An equally plausible suggestion is that public "acceptance" has never amounted to much more than passive toleration, and that the significant change is not in that (always rather low) level of public credulity, but in what we are now demanding of it. The rapid changes and increasingly inaccessible social control of the technological innovations that "drive" those changes may simply be exposing an uncomfortable fact about public trust in decision making institutions that has hitherto been more easily concealed.

The last three chapters have given a rather theoretical analysis of science, institutional credibility, and public risk perception in regulation. This chapter may seem to have digressed wildly from the main theme of the book, but I have tried to show a fundamental identity between the way science is being caught up in institutional processes of credibility management and the way "rationality" is employed to structure institutional responses to public concern. These problems go far wider than hazardous wastes to the more general question of how institutions handle uncertainty and conflict in the modern climate. The "risk-averse" approach, of suppressing ignorance and denying conflicting rationalities, predominates at present, but badly structured risk problems especially call for an alternative strategic stance in which these are *accomodated*, not suppressed with increasing strain and cost to credibility anyway. In the final chapter I knit this argument together with a discussion of the more empirical comparative case study work on hazardous wastes, so as to develop certain suggestions about the practical development of approaches to hazardous waste management and, indeed, to regulation generally.

Notes

[1] Wynne, B. (1983), *Public Perceptions of Risk – Interpreting the Objective-Perceived Risk Dichotomy*, Working Paper WP-83-117 (International Institute for Applied Systems Analysis, Laxenburg, Austria), also published as (1984) Public perceptions of risk, in J. Surrey (Ed), *The Urban Transportation of Irradiated Fuel* (Macmillan, London); Slovic, P., Fischoff, B., and Lichtenstein, S. (1980) Facts and fears: Understanding perceived risk, in R. Schwing and W.A. Albers (Eds), *Societal Risk Assessment: How Safe is Safe Enough?* (Plenum, New York, NY).

[2] For some of the classic expressions of the "objective risk" stance, see Lord Rothschild (1979), Risk, *Atom,* **268**, 30–35; Starr, C. and Whipple, C. (1980) Risks of risk decisions, *Science,* **208**, 1114–1119; Starr, C. (1969) Social benefit means technological risk, *Science,* **165**, 1232–1238; Morgan, M.G. (1981), Probing the question of technology-induced risk, *IEEE Spectrum,* November, 58–64, and Choosing and managing technology-induced risk, *IEEE Spectrum,* December, 53–60. The economic value-of-life comparisons of risk aversion have been developed by Jones-Lee and others; see Jones-Lee, M. (1985), The value of life and safety: A survey of recent developments, *Geneva Papers on Risk and Insurance,* **10**, 141–173; Jones-Lee, M. (Ed) (1982), *The Value of Life and Safety* (North-Holland, Amsterdam).

[3] Such as the joint IAEA–IIASA risk perception project begun in 1974.

[4] See, e.g., Starr (1969) and Rothschild, *op. cit.*, [2].

[5] Lichtenstein, S. *et al.* (1978), Judged frequency of lethal events, *Journal of Experimental Psychology: Human Learning and Memory,* **4** 551–578. Fischoff *et al.* (1981), Lay foibles and expert fables in judgments about risks, in T. O'Riordan and R.K. Turner (Eds), *Progress in Resource Management and Environmental Planning*, Vol. 3 (Wiley, Chichester, UK). It is interesting that the same psychological data on people's accuracy of estimation of risk levels could be interpreted by an analyst (Renn) as pretty good, and another (Slovic) as poor. See Renn, O. and Swaton, E. (1984), Psychological and sociological approaches to study risk perception, *Environment International,* **10**, 557–575. Slovic, P. (1986), Informing and educating the public about risk, in H. Kunreuther and P. Kleindorfer (Eds) (1987), *Insuring and Managing Hazardous Risks: From Seveso to Bhopal and Beyond,* pp. 307–348 (Springer, Berlin).

[6] Often in the heated context of the nuclear debate. See, e.g., Hoyle, F. (1977), *Energy or Extinction* (Heinemann, London). The tradition lives on, as illustrated by J.H. Fremlin, (1985), *The Risks of Power Production*, (MacMillan, London).

[7] See, e.g., Otway, H.J. and Fishbein, M. (1976), *The Determinants of Attitude Formation: An Application to Nuclear Power*, Research Memorandum RM-76-80 (International Institute for Applied Systems Analysis, Laxenburg, Austria); Otway, H.J. and Fishbein, M. (1977), *Public Attitudes and Decision Making*, Research Memorandum RM-77-54 (Laxenburg, Austria); Otway, H.J., Maurer, D., and Thomas, K. (1978), Nuclear power: The question of public acceptance, *Futures,* **10**, 109–118; Otway, H.J. and Pahner, P.H. (1976), Risk assessment, *Futures,* **8**, 87–98.

[8] Tversky, A. and Kahnemann, D. (1974), Judgment under uncertainty: Heuristics and biases, *Science,* **195**, 1124–1131; Causal schemas in judgments under uncertainty, in M. Fishbein (Ed) (1980), *Progress in Social Psychology* (Erlbaum, Hillsdale, NJ) Kahnemann, D., Slovic, P., and Tversky, A. (Eds) (1982), *Judgment Under Uncertainty: Heuristics and Biases* (Cambridge University Press, New York, NY).

[9] Wright, G. (1984), *Behavioral Decision Theory* (Penguin, Harmondsworth, UK), Lopes, L. (1983), Some thoughts on the psychological concept of risk, *Journal of Experimental Psychology: Human Perception and Performance,* **9**, 137–144.

[10] Otway, H.J. *et al., op. cit.*, [7]; Vlek, C. and Stallen, P. (1981), Judging risks and benefits in the small and in the large, *Organizational Behavior and Human Performance,* **28**, 235–271; Renn, O. (1983), Technology, risk and public perception, *Angewandte Systemanalyse,* **4**, 50–65.

[11] Otway, H.J. *et al, op. cit.*, [7]; Otway, H.J. and Thomas, K. (1982), Reflections on risk analysis and policy, *Risk Analysis,* **2**, 69–82.

[12] Otway, H.J. and von Winterfeldt, D. (1982), Beyond acceptable risk: On the social acceptability of technologies, *Policy Sciences,* **14**, 247–256.

[13] Nelkin, D. and Brown, M. (1984), *Workers at Risk* (Sage, Beverley Hills, CA).

[14] See [52] and [59].

[15] Slovic, P. *et al.* (1982), *Acceptable Risk* (Cambridge University Press, New York, NY).

[16] *Op.cit.* note [10]

[17] Slovic, P. *et al.* (1984), Behavioral decision theory perspectives on risk and safety, *Acta Psychologica,* **56**, 183–203.

[18] Otway and Thomas, *op. cit.*, [11].

[19] *Financial Times* [London] (1956), Special Supplement on Atomic Power, 11 October.

[20] Wynne, B. (1982), *Rationality and Ritual: The Windscale Inquiry and Nuclear Decisions in Britain* (British Society for the History of Science, Chalfont St. Giles, UK).

[21] An observation based on participation at several public meetings and on press reports of public inquiries into nuclear waste disposal research plans in the UK.

[22] Otway and Thomas, *op. cit.*, [11].

[23] Slovic *et al.*, *op. cit.*, [17]; See also *Figure 10.1.*

[24] Douglas, M. (1966), *Natural Symbols* (Penguin, Harmondsworth, UK); Douglas, M. (1975), *Implicit Meanings* (Routledge and Kegan Paul, London); Horton, R. and Finnegan, R. (Eds) (1975), *Modes of Thought* (Routledge and Kegan Paul, London).

[25] Durkheim, E. (1954), *The Elementary Forms of Religious Life* (University of Chicago Press, Chicago, IL).

[26] See, e.g., Ezrahi, Y. (1974), The authority of science in politics, in E. Mendelsohn and A. Thackeray (Eds), *Science and Values*, pp. 74–97 (Humanities Press, New York, NY); Barnes, S.B. and Edge, D.O. (Eds) (1982), *Science in Context*, Part V (Open University Press, London).

[27] Barnes, S.B. and Shapin, S.A. (Eds) (1979), *Natural Order* (Sage, Beverley Hills, CA).

[28] Erickson, K.T. (1976), *Everything in its Path: The Destruction of a Community in the Buffalo Creek Dam Disaster* (Simon and Schuster, New York); Edelstein, M. (1982), *The Social and Psychological Impacts of Groundwater Contamination in the Legler Section of Jackson, New Jersey* (Ramapo College, NJ).

[29] Erickson, *ibid.*, p. 271.

[30] Thompson, M. (1983), Postscript: A cultural basis for comparison, in H. Kunreuther and J. Linnerooth, (1983), *Risk Analysis and Decision Processes: The Siting of LEG Facilities in Four Countries* (Springer, Berlin).

[31] Fowlkes, M. and Miller, P. (1983), *Towards a Sociology of Unnatural Disaster: The Case of Love Canal*, Paper presented to the Annual Meeting of the American Sociological Association, August 1983, Washington, DC.

[32] For valuable discussions of this, see Baier, A. (1986), Poisoning the wells, and Hacking, I. (1986), Culpable ignorance and interference effects, both in D. MacLean (Ed), *Values at Risk* (Rowman and Allanheld, Towota, NJ).

[33] Daly, K. (1970), The specters of technicism, *Psychiatry,* **33**, 417–431.

[34] Garmon, L. (1983) Dioxin in Missouri: Troubled times, *Science News,* January 22, pp. 60–65.

[35] Levine, A. (1982), *Love Canal: Science, Politics and People* (Lexington, Boston, MA).

[36] Garmon, *op. cit.*, [34].

[37] See Perrow, C. (1984), *Normal Accidents* (Basic Books, New York) for an organizational analysis relating to this point. See also Wynne, B. (1983), *Technology as culture*, Working Paper WP-83-118 (International Institute for Applied Systems Analysis, Laxenburg, Austria).

[38] McDermott, J. (1974), *Technology and Man's Future* (St. Martin's Press, New York, NY).

[39] Winner, L. (1977), *Autonomous Technology*, pp. 33–35 (MIT Press, Cambridge, MA).

[40] This analysis does not entail criticism of individual decision makers and experts. Indeed, the processes of condensation I have outlined imply more control and responsibility to their symbolic counterparts than any such elites actually have, or would claim to have. I am describing the implications of institutional structures and processes, not of individual actions.

[41] Crozier, M. (1980), Les Dévéloppements futurs de la Bureaucratie, *Courier du Personnel* [Commission of the European Communities], **416,** 29 July, 13–20.

[42] For the relation between blame, accountability, and social cohesion, see discussions in Baier, *op. cit.*, [32]; Douglas, M. and Wildavsky, A. (1982), *Risk and Culture* (University of California Press, Berkeley, CA); Douglas, M. (1983), *Risk Acceptability According to the Social Sciences*, (Russell Sage, New York).

[43] Notice that this is somewhat contrary to a common view that human-made risks are evaluated more negatively than natural ones. However, I do not believe it is as simple as this. Risks may be evaluated more negatively, but still be more of a vehicle for social cohesion when they can be attributed to human causes as opposed to natural or "unknown" causes.

[44] US General Accounting Office (1985), *Illegal Disposal of Hazardous Waste: Difficult to Detect or Deter,* GAO/RCED-85-2, Ch. 2 (GAO, Washington, DC). Only 16% of clean-up costs have been retrieved from past dumpers.

[45] Levine, *op. cit.*, [35].

[46] Black, D. (Chairman) (1984), *Investigation of the Possible Increased Incidence of Cancer in West Cumbria* (HMSO, London).

[47] Lenihan, J.M. (Chairman) (1984), *Inquiry into Allegations of Increased Incidence of Health Defects in the Vicinity of the ReChem Waste Incinerator Plant, Bonnybridge* (Scottish Office, Edinburgh); UK Health and Safety Executive, Industrial Air Pollution Inspectorate (1984), *The Operation of the ReChem Waste Incinerator Plant, Pontypool* (HMSO, London).

[48] UK Health and Safety Executive (1985), *The Abbeystead Explosion* (HMSO, London).

[49] The encouragement of public feelings that malevolent forces are in charge is clearly indicated in interviews carried out by a colleague, Lyn Hague, with victims of the Abbeystead disaster. See also Erickson *op. cit.*, [28]; Baier, *op. cit.*, [32]; and Levine, *op. cit.* [35].

[50] For broader reviews, see Wright, G. (1984), *Behavioral Decision Theory* (Penguin, Harmondsworth, UK); Sjoberg, L. *et al.* (Eds) (1985), *Human Decision Making* (Bodafors, Doxa); Bell, D., Keeney, R., and Raiffa, H. (Eds) (1977), *Conflicting Objectives in Decisions,* (Wiley, Chichester, UK).

[51] Watson, S., Fischoff, B., and Hope, C. (1984), Defining risk, *Policy Sciences,* **17** 193–210; Stallen and Tomas have shown that, even from a "technical" stance, "concern" is an uncontrolled attribute if measured only by "weight" or intensity; it has concretely different structural qualities. see Stallen, P.J. and Tomas, A. (1985), *Public Concern about Industrial Hazards*, paper to the Annual Meeting of the Society for Risk Analysis, Washington, DC, October.

[52] See, e.g., Hohenemser, C., Kates, R., and Slovic, P. (1983), The nature of technological hazard, *Science,* **220**, 378–384; Ricci, P., Sagan, L., and Whipple, C. (1983), *Technological Risk Assessment* (Alphen, Netherlands).

[53] See, e.g., Hammerton, M., Jones-Lee, M., and Abbott, V. (1982), The consistency and coherence of attitudes to physical risk, *Journal of Transport Economics and Policy,* **16**, 181–199; Hammerton, M., Jones-Lee, M., and Abbott, V. (1982), Equity and public risk: Some empirical results, *Operations Research,* **30**, 203–207; Brown, R.A. and Green, C.H. (1981), Threats to health and safety: Perceived risk and willingness to pay, *Social Sciences and Medicine,* **15**, 67–75; Jones-Lee, M., Hammerton, M., and Philips, P. (1985), The value of safety: Results of a national sample survey, *Economic Journal,* **95**, 49–72.

[54] Raiffa, H. (1985), Back from prospect theory to utility theory, in Thompson, M., Wierzbicki, A., and Grauer, M. (Eds), *Plural Rationalities and Interactive Decision Processes* (Springer, Berlin).

[55] The "regret theory" extension of classical (subjective expected utility) decision analysis was developed in response to Kahnemann and Tversky's demonstration of the descriptive limitations of that theory. Kahneman, D. and Tversky, A. (1979), Prospect theory: An analysis of decisions under risk,

Econometrica, **47**, 263–292.

[56] This view is more often expressed verbally than argued in writing. I have had it expressed to me several times in conversations at risk seminars and conferences.

[57] Bernstein, B. (1971), *Class, Codes and Control,* Vol. 1 (Routledge and Kegan Paul, London).

[58] Wynne, B. (1982), Institutional mythologies and dual societies in the management of risk, in H. Kunreuther and E. Ley (Eds), *The Risk Assessment Controversy: An Institutional Perspective* (Springer, Berlin).

[59] Lopes, *op. cit.*, [9].

[60] *Ibid.*, p. 142.

[61] Remarks by a senior staff member of the UK Health and Safety Executive at a seminar, University College, London, October 1985: "There may be some to whom it is usually repugnant to ask such [willingness to pay] questions ... what kind of people are these?" – implying a certain minority maladjustment from reasonable behavior.

[62] Though a broad split is increasingly apparent in the psychological approaches between those who align with economic and quasi-economic frameworks, and those who align with more sociological ones.

[63] See Unger, R. (1976), *Knowledge and Politics* (Free Press, New York NY); Unger, R. (1976), *Law and Modern Society* (Free Press, New York, NY). As I see it, there is a strong connection between the notions of "relational" values or rationality of Unger, "passive" rationality of Kekes, and the idea of Elster, of binding oneself to certain precommitments as a response to recognized weakness. If social bonding and commitment is a fundamental need brought about by the *vulnerability* of individuals, then human interaction is a fundamental moral category and the quality of such interactions is itself a category of value; see Elster, J. (1984), *Ulysses and the Sirens: Studies in Rationality and Irrationality* (Cambridge University Press, London); Hollis, M. and Neil, E. (1975), *Rational Economic Man* (Cambridge University Press, London); Chein, I. (1972), *The Science of Behaviour and the Image of Man* (Tavistock, London).

[64] Technically, one would say that utilities cannot be independent, because specific values and problems are being defined and weighed in relation to wider values. A basic axiom of rational decision analysis is false.

[65] See Douglas, *op. cit.*, [42]; also MacLean, *op. cit.*, [32]; Short, J. (1984), The social fabric at risk: Towards the social transformation of Risk Analysis, *American Sociological Review,* **49**, 711–725.

[66] See Wynne, B. (1982), *op. cit.*, Ch. 9, Kekes, J. (1976), *A Justification of Rationality* (Albany Press, New York).

[67] O'Hare, M. *et al.* (1983), *Facility Siting and Public Opposition* (van Nostrand Reinhold, New York, NY).

[68] *Ibid*, p. 94.

[69] The methodological analogy of this point is the soft systems methodology developed by Peter Checkland and colleagues, whereby the main task is to elicit the implicit world views and problem networks of the different people or social clusters involved, and then help make *accommodations* among them; see Checkland, P. (1981), *Systems Thinking, Systems Practice* (Wiley, Chichester, UK); Checkland, P. (1985), From optimizing to learning: A development of systems thinking for the 1990s, *Journal of the Operational Research Society,* **36**, 757–767.

[70] See Turnbull, C. (1985), *Wayward Servants* (Eyre and Spotiswood, London); Douglas, M. (1975), *Implicit Meanings* (Routledge and Kegan Paul, London); Duncan, H.D. (1968), *Symbols in Society* (Oxford University Press, London); Barnes, S.B. and Shapin, S.A. (Eds) (1979), *Natural Order* (Sage, Beverly Hills, CA).

CHAPTER 12

Summary and Conclusions

Brian Wynne

The main aim of this book has been to examine risk management in the new context of rising public concern about risks and the bodies supposed to be controlling them. We have attempted to pursue this objective by comparative empirical analysis of a relatively ill-structured risk management issue, where the problems of implementation and the basic definitions themselves are especially prominent. The shift of focus in this book to include comparative analysis of *forms of legitimation* as well as of specific regulations or procedures is justified by the evident fact that the fastest-growing problem for regulators is to make any regulation decision *work*, which includes making it implementable and *credible*. Since images of science and rationality are embedded in structures and processes of authority, it should be hardly surprising to find that not only technical regulations, but even basic normative definitions of what constitutes "science" or "rational method", are shaped by national political cultures. I have introduced the key idea that science, or risk analysis, is playing two roles in regulation. It is involved simultaneously in a descriptive (empirical) discourse – measuring risks, control mechanisms, etc. – and in a (symbolic) discourse of *persuasion* or *justification* – negotiating the credibility of risk analyses, regulations, etc. These discourses are not separate, but interpenetrate. Furthermore, although the arenas may change, this dualism also exists in the practice of science not involved in public issues.

Our overall analytical ambitions have inevitably left a somewhat bifurcated book, with Chapters 4–8 focusing on case studies, and Chapters 9–11 on the more abstract analysis of interactions between regulatory uses of rationality, public risk perceptions, and dominant "rational" modes of response to public concern. The latter chapters are crucial to understand how the regulatory practices of risk management are to be seen in relation to public perceptions. The interpretation provided here indicates that an adequate response to that concern would need to go beyond conventional notions of "back end" regulation and challenge dominant cultural beliefs about the sanctity of technological innovation.

Here I first draw together the main points that arise from each chapter, and then attempt to integrate the more theoretical questions of later chapters with the earlier more empirical materials. The later chapters provide a basis for thinking about the significance of the relatively new, and fundamentally different practical context of public justification which regulation is now encountering. Even in the USA, where this dimension has been more pronounced, it is only really the last decade that has seen its rise to predominance in the ethos of government, and even then largely around the two youngest regulatory bodies that are striving for credibility, OSHA and EPA.

Rather than repeat what are now fairly well-worn analytical tracks in the comparison of regulation, therefore, I attempt to develop a different perspective by asking questions about the role of science and rational decision methods as a discourse of public authority. I try to show that rational knowledge is organically rooted in social relationships, as part of a broader discourse of authority. Although differences in institutional relationships, cultural "styles", or administrative tradition can be seen systematically to influence specific approaches to technical choices in regulation, they also influence the forms of political discourse about those choices – their social meaning and their validity or legitimation.

The notion of effectiveness, which often motivates comparative policy analysis, is not easy to define, let alone fill out with realistic information on its degree of actual fulfillment. All that can be said, at least in the case of hazardous wastes, is that every country has good reason to ask itself whether it has a right to expect public assent. Effectiveness is an inherently nebulous yardstick, but it necessarily includes questions about whether regulation can command public credibility (and for more reason than the real point that practical compliance is likely to be better if regulations and related arrangements are widely regarded as the best possible). It is certainly the case that credibility of regulation depends upon more than just the material effectiveness of waste management.

Risk assessment as a technical–analytical activity has been developed in this dualistic context of analysis pervaded by a concern for persuasion way beyond the boundaries of a private technical sub-culture. The legitimacy of this combination of roles, or interests, should not obscure the extreme complication that it creates, especially when (as has been true) there is little awareness of or reflection upon it. This is especially problematic when we take into account the interactions between the forms of analysis–persuasion and public experience and response. The general line of argument in this chapter is that, granted the significant differences of institutional "style" between regulatory systems, even in modes of attempted legitimation, there are some common processes embedded within all these different institutional frameworks. Thus, in completing this book a discussion of the comparative analysis of technical institutional interactions is followed by a discussion of some general principles for appropriate institutional mechanisms for regulation; but this is then enlarged into a more strategic discussion of appropriate response to the growing "legitimation crisis" within which risk analysis and science more generally are being (largely destructively) entangled.

I regard the last strand as critical for improved practical effects, however we define them in detail. In case it is seen as academic, daily experience indicates that the chemical waste issue may follow the paralyzing and destructive route of the nuclear waste problem, unless different approaches can be developed. Yet if this does happen, even in part, there will be far more pervasive effects on whole economies, communities, and political systems. Despite all the fandango, nuclear energy is a relatively insignificant part of the industrial scene compared with chemicals in general.

Before returning to the regulation–legitimation discussion, however, we pick out the main points from the previous chapters.

12.1. The Main Points So Far

Risk analysis as a formal scientific attempt to quantify risks and evaluate controlling mechanisms is being increasingly used – or at least, looked to for use – in regulating hazardous wastes. At first sight this appears to conflict with the inherent structural properties of this particular issue, which are incompatible with the precision and standardization that formal analysis involves. These issue properties were described in Chapter 3. This therefore raises the question of why there is such an interest in more formal, scientific risk analysis. Of course, part of the answer is that even inherently ill-defined risk-generating processes can still benefit in parts from a more uniform, quantified definition of the risks and their controlling factors. Nevertheless, we noted that formal risk analysis, often using elaborate models, is virtually always developed, used, or demanded when *credibility* is at stake, and conflicting policy positions are being advocated. More formal precision, it is assumed, will close the argument. More cynically, formal models are seen as a form of white magic to "prove" whatever stance policy elites wish to establish in public minds.

The symbolic role of science in projecting public definitions of managed risks, technical controls, reliable methods, etc., involves the propagation of *universalistic* models of risk processes and situations, analytical methods, and controlling responses. Standardized and precise risk models, test protocols, etc., are the result. In order to arrive at something resembling an analytical model, sheer guesswork of breathtaking proportions has frequently been undertaken, and the variability and uncertainties of real risk situations are heavily understated. Thus, the significance of Chapter 3 is to point out how hazardous waste is an issue *particularly* incongenial to risk management approaches that reflect these universalistic tendencies. Some political cultures emphasize these more than others, but the point about the public credibility problem is that normal policy responses everywhere are accelerating the impetus toward analytic standardization, uniformity, and precision (as, e.g., in formal models), with potentially damaging results to credibility and realism. Decision-rule transparency is often equated with standardization. Chapter 3 introduced the important distinction between "intrinsic" and "situation-specific" risks, and the importance here of the complex and indeterminate technical–behavioral life-cycle of wastes. It also raised a point developed in Chapters 9 and 10, that a risk analysis is always a *simulation* of a postulated risk-generating process. Whether the analytic model corresponds to reality, and how flexible it can be in reflecting variability in reality, are very different questions from its "internal" consistency and robustness [1]. Large-scale ignorance of the basic structural framing of analytical problems is falsely represented in conventional scientific approaches as a marginal, probabilistic uncertainty in analytical parameters and data.

Thus, the complex *behavioral* life-cycle, heterogeneity, and lack of "structure" in the hazardous waste issue make it a different order of problem from conventional pollution issues, especially in the light of the "public acceptance" dimension. The fact that "hazardous waste" leaves a factory on the back of a wagon, to be exchanged between social–economic actors, rather than in the air or water is of the utmost significance to implementation, public reaction, and the role of risk analysis. Other important structural features analyzed in Chapter 3 were:

(1) Paradoxically, *regulation* of hazardous waste requires a new industry [of treatment and disposal (T&D)] to be developed, yet also closely regulated. Fostering a new industry under such close regulation and extreme public concern is a major policy dilemma, and allows the existence of institutionally consolidated but fundamentally conflicting perceptions and issue definitions, especially between policy center and periphery.

(2) Hazardous waste is very far from a unitary regulatory issue. The diversity of waste producers, types, conditions, movements, disposal methods, and economic relationships suggests extreme heterogeneity of problem definitions, risk types, and therefore, possibly, also of appropriate regulatory policies.

(3) The implications of this heterogeneity are further amplified by the complex *behavioral* life-cycle of hazardous waste, in which it may change its (socially conferred) nature. This more direct behavioral responsibility and the "packaged" or "concentrated" property of hazardous waste means that more intense, localized health or environmental effects are possible (with consequent effects on public reaction), and more direct, possibly dramatic social intervention is feasible (e.g., political refusal of cross-border transfer of wastes). Apart from anything else, these factors render the issue more liable to social instability and surprise in its management, and intensify the policy "need" to project reassurance and consistency.

(4) The complex, variable physical–behavioral life-cycle of waste means that there is no such thing as *the* "intrinsic" hazard of a given waste. The status of "hazard" or "waste" is variable according to the material's life-cycle *situation*, including the socioeconomic definition used by its handlers. This is a combination of social, cultural, and physical–technical determinants.

Thus, risk management has to find a practical way of resolving a tension between system-wide risk management considerations (like public reassurance, a centralized role) and situation-specific risk definitions and controls.

Hazardous waste is a "last frontier" issue, in that it was regulated later than "dispersive" waste emission to air and water. The attempt to internalize properly in production costs the "externalities" of environmental and health damage has been undermined by the transfer of waste, which was previously dispersed, into (unregulated or loosely regulated) *concentrated* hazardous wastes, in the form of filter sludges, slags, dusts, etc. Enforcing proper T&D here therefore has no alternative but to try to force the costs back "upstream", to internalize them in production costs, because there is no other medium left to which the externalities can be diverted. Hence, there is even more pressure on the regulation sphere to dilute prevailing scientific ignorance and lack of control into a false sense of acceptably managed risks. There are also potentially more severe conflicts between industrial and environmental interests, with regulators trying to manage a correspondingly more unstable middle ground. The considerations of Chapters 9 and 10 thus grow in significance for this issue.

These and other inherent structural issue properties of hazardous wastes are also significant for their direct interpenetration with cultural factors and the social dimensions of risk perception and institutional credibility. These connections are developed through the case study chapters. In the Dutch case in Chapter 4, we highlight the point that environmental issues, like public policy decision-making generally, are *socially* constructed. They have no naturally given boundaries, shape, or *logic*. Even after "definition" via formal legislation and enactment of precise regulations, different actors in the regulatory arena are still able to define the issue in diverse ways, according to their cultural–organizational situation. This extends even to subtle redefinition of technical regulatory definitions and meanings into diverse *practical* frameworks, and this negotiability is extremely resilient to the apparent constraints of increasing uniformity and precision of the official "standard" regulatory criteria. Rules never contain an *intrinsic* logic of application to specific cases. The idea that one can obtain more effective rules by focusing effort on strengthening the *rules* rather than on the means of their practical interpretation is false [2].

This general point is important, especially for this book's emphasis upon *implementation.* The Dutch case shows a large component of regulation to be that of central bodies trying to control local regulatory bodies, as well as regulators controlling waste handlers. In reality, whatever values may have been "officially" negotiated into predominance in the formal regulatory policy, they have almost to be renegotiated all over again at the point of implementation and enforcement. The rationality of implementation and enforcement is not one of defective perception or diligence at local levels, but of the local management of a set of conflicting constraints and values that create a different technical–institutional problematique from that envisaged in policy*making* at more central levels [3].

This perspective consolidates the point made throughout the present book, and consistent with a major theme in recent years in policy analysis, that it is important to integrate analysis of *implementation* processes with that of policy- or rule-making *per se* [4]. This has become especially important in the light of growing political recognition of implementation "failures" in many areas, and the consequent loss of authority on the part of regulatory institutions generally. For hazardous wastes indeed, their heterogeneity and extensive *behavioral* life-cycles make it difficult logically to conceive of the traditional sharp split between the main phase of centralized regulatory policymaking, followed by the relatively trivial "technical" phase of policy implementation. Policy- or rule-*making* in the hazardous waste case effectively has to be dispersed into the more decentralized and incremental arenas of "implementation", because this intrinsic situational diversity is so dominant [5]. As already noted, even control over what is to count as waste has to be partly relinquished to autonomous actors.

After showing how the inherent heterogeneity of the issue encourages and accentuates the conflicting rationalities of problem definition, in Chapter 5 we then applied the same analytical orientation to the central technical pillar of hazardous waste regulation; namely, the definition of "hazardous waste" for legal regulatory purposes. In the Dutch case, centrally defined quantitative hazard criteria were shown to be interpreted flexibly and divergently at local levels, according to local social factors. Hazard classification systems are generally the first and key regulatory step, since they define the population of wastes (and, thus, the waste producers, handlers, and disposers) that must be subjected to controls. Especially in view of the growing *international* transport of waste, these frameworks need to be clear, precise, and consistent. Yet, an analysis of the process of establishment of hazard classification systems in the FRG, USA, Austria, UK, and the Netherlands shows that each "local" technical hazard classification is different, because different *technical* methods and combinations have been employed, embodying different national (and in the FRG case, state) administrative norms, interests, and social assumptions. This is a very important insight for practical policy, because both within and between sovereign regulatory systems the usual response to implementation loopholes and public disillusionment has been to call for more technical precision and uniformity in defining "hazardous waste". This could be regarded as a symbolic projection, or reassurance ritual, but the fact is that regulators also act as if they believe in it, which thus affects material regulatory practice, and is not merely "external" rhetoric [6].

As the Dutch case showed within a single regulatory system, so Chapter 5 shows comparatively, between systems, that technical norms and criteria embody corresponding implicit behavioral–institutional prescriptions or assumptions. Technical knowledge is not context free. What was *informally* exercised diversity of practical interpretation within the Dutch system is more structurally embedded and systematic at the international comparative level. At both levels technical standards are socially constructed, as social languages reflecting their institutional setting. At neither level, however, is it yet recognized that *institutional diversity* is tacitly determining what is taken to be only *technical* lack of rigor or coordination.

A practical conclusion from this analysis is that attempts to plug growing international "leakages" of hazardous wastes, just like local implementation lapses, will not succeed by continuing to entertain the fallacy that a more rigorous standardization of technical hazard classifications is the solution. Regulatory dislocation is due to more than the lack of technical rigor or standardization alone. This, it seems, can be carried so far, but then runs up against the implicit need to standardize the different national *institutional* assumptions, cultural traditions of decision making, etc., embedded in different national technical frameworks.

The misleading drift of current policy thinking is allowed to survive because of the belief that there is an intrinsic natural meaning and definition of "hazardous waste", out there in nature, which should be revealed uniformly by more rigorous scientific scrutiny, whatever the local starting point. Our analysis shows in a very concrete, empirical way that this concept of scientific knowledge and method as prior to and separate from social factors – facts clearly separated from values – is fallacious and damaging [7]. This corresponds with insights gained from social and historical analyses of science, that scientific knowledge is always context bound and unable to provide a universal basis of values or norms. The "rules" represented by scientific theories and concepts do not contain the meaning of their practical application to situations [8], and elaborating those theories, concepts, or rules does *not* overcome this inherent difficulty. It is not a viable alternative to making the means of practical application more resilient.

The belief that science can reveal "the" precise, universal definition of hazardous waste leads to intensified attempts to standardize international hazardous waste classifications, with the attendant growth of international regulatory bureaucracy and bureaucratized scientific analysis to support the definitions (as critically analyzed in Chapter 10). An alternative is to reduce the amounts of hazardous waste that the *international* system has to carry. In other words, there may be more benefit than previously recognized by resisting the pressure for a free international commercial movement of wastes and instead seeking the institutional mechanisms for better local couplings of waste arisings with T&D facilities in the same region.

In addition, and addressing part of the same overall problem, the insight that there is no scientific, context-free definition of "hazardous" or "waste" means that there may be greater benefits than usually perceived through *institutionally* reducing the indeterminacy, complexity, and multiparty nature of a typical waste life-cycle, which cannot be adequately controlled by *technical* definitions.

In Chapter 6, we examined two states of the FRG – Bavaria and Hessen – that have attempted to do this by statutory public ownership of wastes (and the attendant risks), consigning them to publicly managed T&D companies. In this institutional option, the regulatory responsibility effectively merges with the T&D management operation, and the whole process of risk discrimination and matching with appropriate T&D options becomes more organizationally unified, internal to a public company. It is therefore less behaviorally uncertain and, at the same time, more available for regulatory oversight because it is physically (and organizationally) less dispersed and heterogeneous. There are, of course, other problems with this institutional arrangement; and in the European context it has become controversial because the waste export bans imposed on local waste producers to ensure adequate input to the state's public facilities are said to contravene the principles of free trade given in the Treaty of Rome. Nevertheless, without prejudice to these other aspects of the overall argument, our analysis tends to support something like this kind of institutional structure, because of its regional coherence and the freshly perceived extra need (following Chapter 5) to reduce the pressure on international movements of waste due to the *intrinsic* limitations of technical harmonization.

In Chapter 7 we examined the regulatory system of the UK, which is highly decentralized in both this and its T&D industry. In further contrast with the FRG states analyzed in Chapter 6, UK regulation and T&D are decoupled, T&D being nearly totally in private industrial hands. Implementation of regulatory standards and definitions at the local level was the focus of this chapter, with special attention to the combined effects of:

(1) Extreme local responsibility, even for the interpretation of highly technical but not *practically* precise hazard definitions.
(2) A burden of proof in enforcement upon the regulator, not the waste producer or handler.
(3) A regulatory policy focus on T&D site licensing (the back end of the waste life-cycle) rather than on the clear classification and control of waste *arisings*.

It was found that, in principle, the discretionary, flexible character of the UK use of technical standards, and the institutional dispersion of regulatory responsibility, correspond with the structural issue properties and risk factors described in Chapter 3. However, in the UK this institutional dispersion is structured so as to leave small local authorities with responsibility, yet limited institutional control. The responsible local authorities simply do not have the concentration of resources and legal–administrative instruments that are necessary to give effective practical interpretation of complex legal–technical definitions, nor to perform all the associated informal interactions with industry that such a discretionary system demands. Their institutional uncertainties are large. The options are either to make the *institutional* basis of regulation stronger (e.g., by consolidating into regional groups with stronger professional training, etc.) or to alter the *technical* instruments of regulation to be more universal and simple, but less flexible standards, such as specific concentration thresholds for chemical constituents (as in the Dutch system).

The UK case shows particularly clearly an important general point, which we emphasize, about the *interaction* of technical and institutional factors in regulation. Technical standards are a code for prescribed behavioral relationships and rules. To the extent that institutional traditions and relationships are predictable and trustworthy to those involved, elaborate and inflexible codes are not necessary in order to regulate behavior (which is the real focus of regulation in managing risks). It can be "regulated" acceptably as behavior unfolds, via *ad hoc*, informal negotiation [9]. Imprecise technical norms, or incomplete decision rules, could therefore be seen as *necessary* to uphold long-established institutional–cultural traditions in the UK system. However, this UK culture of mutual institutional confidence and informality decreasingly incorporates the local authorities (who are the key regulatory actors). The lack of transparency of risk-assessment decision rules in the UK is also leading to concern about inconsistency and lack of public acceptability. The UK style of regulation appears, therefore, to be entering a period of flux, signified by a chronic dispute over the proper nature of technical standards, but more fundamentally about the proper institutional distribution of regulatory resources and authority.

Although Europe is often taken to be uniform in comparative research, we argue that the Netherlands has, in important respects, an opposite institutional climate to that of the UK, especially between different actors *within* regulation; hence the Dutch use precise, inflexible technical standards. The US setting is even more comprehensively unpredictable and lacking in interinstitutional trust. Hence, the most elaborately precise and formal, all-embracing, uniform, and inflexible technical regulations were established, even if they were arguably unenforceable in practice, as witnessed by the need for Congressional reauthorization [10]. The UK's informal, discretionary system is only viable to the extent that it can successfully incorporate the many local waste disposal authorities, and its apparent failure to do this could push it toward the Dutch or US use of (more elaborate) technical surrogates for the growing *institutional* uncertainty.

An important example of the general institutional determination of practical technical policies is the UK commitment to the landfill codisposal of toxic wastes with domestic wastes, which is regarded as a deviant and backward policy by many other countries. The UK view that this system carries an acceptably low risk seems to be based upon an undeclared assumption of an ideal *social world of implementation* of site selection, licensing, and inspection and monitoring of nearly 5 000 landfills by nearly 200 local authorities, with a free-enterprise handling and disposal infrastructure. What may be an acceptable, locally variable but highly optimized risk system in theory, may be worse overall than one that assumes a dubious implementation world, and simply restricts landfill *a priori*. The apparently lesser *technical* optimality of this approach (e.g., in the USA, and FRG) may be justified by the different *institutional* premises that structure it. Whatever the relative merits and realities of different *technical* policies, their existence and effectiveness can only be understood by reference to their institutional and cultural context.

A further important insight from the UK case study was the fundamentally different perceptions of their mutual relationships by the major institutional actors in regulation. This was reflected in the conflict of views between industry and local authorities about the universality ("objectively ascertainable fact") or private control ("exclusive knowledge") of specialist information about wastes. Industry argued that the burden of proof in defining special wastes should not rest with the waste producer, because in legal formality the regulator has access to the necessary regulatory knowledge. Central government supported the industry view. In practice, however, this *formal* definition of industry–local authority relationships, and the corresponding social distribution of key knowledge, borders on fantasy. As local authority representatives argued, the idea that a typical local authority is on an equal information footing with all its waste producers for all waste streams is preposterous. The constraints and realities that define the *local* rationality of implementation are inconsistent with the rationality embodied in official definitions and justifications of regulations.

Another symptom of deeper institutional conflicts in perceptions of the very *meaning* of regulation and its corresponding social relationships involved the conflicting interpretations of the role of the UK consignment note system, regardless of its *formally* defined role.

These and other examples indicate the important point – developed in Chapter 9 – that institutional conflict or "structural uncertainty" among social actors is more fundamental than the technical imprecision or uncertainty that is the currency of conventional risk assessment. Scientists and regulatory bureaucracies tend automatically to process the former into an appearance of only the latter. As argued in Chapters 9–11, this eventually *undermines* their public credibility and also underexpresses the back-end ignorance and costs of current waste management. It is an inadequate channel of feed-back from back end to front end, and its inadequacy is systematically (though not conspiratorially) re-created in the prevailing language about uncertainty.

The Hungarian case study described in Chapter 8 illustrates further institutional dimensions of the management of hazardous wastes. It emphasizes the basic point that arose in the Dutch case, that all the paraphernalia of formal risk analysis and associated regulations are quite irrelevant without the necessary industrial infrastructure for actually *practicing* controlled T&D; and these are, in turn, impossible without the necessary investments. This also underscores the point made earlier, that central governments define the hazardous waste issue first and foremost as a problem of industrial innovation and the defense of a fledgling (T&D) industry, while local authorities and other groups see it as a regulatory problem. From the latter vantage point, the T&D industry is seen more as part of the problem than as part of the solution. Granted their common priority to nurture an inherently fragile industry, however, governments have adopted very different practical approaches; those that have grasped the public investment nettle (usually in partnership with T&D customers, that is, waste producers) have, on the whole, performed better at providing a stable T&D capacity than have those, like the USA and the UK (and, before 1985, the Netherlands) that have left it to private enterprise. The evidence is that public investment in one form or another will become increasingly necessary to ensure a viable and progressive T&D industry. Furthermore, this political–economic initiative is a more fundamental step than the secondary question of what specific regulations are to be established.

Despite a lack of T&D infrastructure, and relatively slight research resources, Hungary also has an elaborate hazardous waste classification scheme, akin to formal risk analysis. However, it was borrowed from the Austrian and FRG systems rather than developed within the Hungarian context. Whether this is relevant to the Hungarian spectrum of industrial production or to its institutional processes of regulation generally is a moot point. There is, for example, an interesting institutional dimension specific to a centrally planned economy that potentially alters the whole focus of regulation. Such systems have an established practice of collecting production process data, which could help provide better upstream waste regulation, thus reducing waste arisings. This access to production decisions is largely impossible in market or mixed economies because competition (absent in Hungarian industry) causes extreme anxiety about keeping such "internal" information private. Even accurate waste-arisings information is difficult to obtain, because of the same alleged fears of industrial espionage. Thus, the appropriate and feasible *point* of regulation in the overall life-cycle also varies with the general political–economic system. In general, the Hungarian institutional climate encourages classification of *process* types according to some kind of toxic waste output index, whereas Western economies focus on waste classification or disposal-site classification.

Each point of intervention in the overall behavioral system involves regulation of a different set of specific practices, actors, and relationships – it therefore requires different information, different forms of risk analysis, and a different set of technical and institutional instruments. Development of appropriate technical tools has to be sensitive to such differences, which are not always clear-cut, as significant national variances within the EC's toxic and dangerous waste framework demonstrate.

It is interesting to note that Hungary has seen public opposition to hazardous waste facility siting plans, and this opposition has been successful. It may be worth considering this as a reflection of a different balance between the cultural awareness of environmental risks and the available levels of economic investment from the typical situation in market economies. The standard notion that "environmentalism" only comes with "post-industrial" affluence and values might be undermined by a close examination of this phenomenon. Unfortunately, we did not have the opportunity to follow this up.

12.2. Uncertainties – Institutionalizing Ignorance

In Chapters 9–11, I attempt to abstract some more general, theoretical insights and to analyze more fully the public credibility dimension. The prime focus is upon the nature of uncertainties and the interaction between technical and institutional uncertainties in regulation. Following especially Chapter 5, Chapter 9 shows that the conventional policy approach generally attributes implementation weaknesses to uncertainty (e.g., in the volumes of waste arisings) and peripheral lapses in performance. The uncertainty is assumed to be a lack of observational precision or analytical rigor, including the lack of clarity as to legal–technical definitions. This corresponds with observations on the Dutch and US systems, that institutional *behavioral* uncertainty between the relevant *parties* leads to the elaboration of inflexible *technical* norms to try to reduce the uncertainty. This generates a "need" for more precise and universal risk analysis; and the generation of *risk knowledge* via methods of analysis and testing also becomes more inflexibly prescribed by regulatory bureaucracies, whether or not these formal models and methods correspond with real risk situations.

We show that, in reality, even apparently concrete unambiguously discoverable facts, such as volumes of waste arisings, are unavoidably subject to social construction through the ways that institutional actors and interests define terms (such as "hazardous" or "waste") and make necessary supplementary judgments. Many of these are behavioral judgments that are indeterminate and represent implicit conflicts of basic perspective, or *structural* uncertainty in the system, rather than the lack of technical resolution.

In Chapter 9 this fundamental distinction between technical and structural uncertainty was clarified by considering technologies or risk-generating systems as organizational–technical *networks*. In theory such multipolar networks are fully coordinated, but this is impossible in practice, and cross-cutting commitments and rationalities mean that conflict and lack of coordination is built into the system, even if inadvertently. With this perspective on technology it is also easier to see that expert disagreement is not merely a function of lack of data and precision or "transcience" (which is what the traditional focus on zero–infinity risks naturally highlights); it is *endemic*, because it involves different definitions of the risk-generating system or technology. As an automatic property of scientific method, technical experts tend to strip their risk definitions of the implicit social–organizational assumptions and parameters that underpin and frame the technical model that is the basis of their formal risk analysis. Although it may have less serious implications, even for *highly structured* technologies, I argue that expert disagreement can be seen as a function of different expert definitions of the relevant technology. In more heterogeneous, dispersed systems, such as pesticides use and hazardous waste management, these system definitions or problem structures can become far more seriously dislocated. Participants' experiences of a dispersed technological system are *objectively* different, in terms of both technical parameters and social relationships. Indeed, the very notion of expertise becomes more problematic because no single discipline can claim to comprehend all the dispersed nodes of the system and its interconnections, and some highly relevant forms of knowledge of the system's operation or effects may hardly be organized at all [11]. This point was most graphically illustrated in Chapter 9 with the example of the conflict between scientific toxicologists and farm workers (plus their clinical doctors) over the safety of 2,4,5–T.

The point of the more theoretical analysis offered in Chapter 9 and outlined above is to prepare the way for a very different interpretation, given in Chapter 11, of the public perception (and, thus, public rationality) problem from that which dominates current policy thinking and "scientific" risk-perception studies. The approach in Chapter 9 is consistent with the argument that modern technologies do not have "impacts" on an essentially *passive*, static physical and social environment – they create "interference effects" by *active*, but unforeseen, interference with other proliferating technological networks [12]. The degree of ignorance and *inevitable* unforeseeability of such effects is of a different, less controllable quality than the former sort of impact; yet it is not reflected in any risk-analytic methods. The standard approaches:

(1) Gloss over and misrepresent ignorance and *structural* uncertainty as probabilistic uncertainty in "known" relationships.
(2) Attempt subjective probability analysis of expert judgments using Bayesian statistical methods.
(3) Bracket off ignorance and structural uncertainty as not a part of risk analysis, but as a policy responsibility.

A typical example of the first kind is where plant discharges of toxic materials lead to exposures. Models of environmental movement, transformation, accumulation, etc., are developed and human doses estimated. Dose–effect models are then used to estimate health damage. Analyses of uncertainties then center, often elaborately, upon "data gaps and modeling assumptions" [13], once the pollutants, exposure pathways, and population of concern have been identified. The larger uncertainties (more authentically called *ignorance*, as outlined before) as to whether the model whose uncertainties are being analyzed is even modeling the relevant pollutants, pathways, or populations are not within the frame of attention. This was exemplified in the official 1984 inquiry into the incidence of excess childhood leukemias around the controversial Sellafield (Windscale) UK nuclear reprocessing plant [14]. The inquiry (and all its critics) focused *entirely* upon the plant's (large) marine discharges of radioactivity, and concluded, from known discharge rates and environmental and dose–effect models, that the observed excess cancer rates were 40 times higher than estimated from the models. Uncertainty in the models was analyzed and regarded as unlikely to bridge such a factor of 40. While this particular uncertainty analysis is itself questionable, the more significant point here is that the whole of the analysis and ensuing debate accepted this particular scientific frame of reference. At least one other potential pathway – totally different – was entirely ignored; namely, the "dirty worker effect" whereby employees are contaminated at the plant, bring home the contamination, and their families suffer exposure.

Another example was the explosion of an underground valve house in the Abbeystead UK water transfer scheme after civil engineers had vented the water tunnel. A methane explosion killed 16 and maimed the rest of a party of visitors being shown the plant (ironically, for reassurance about flooding). The engineers, operating within a physical engineering cognitive framework, had totally ignored the possibility of methane generation in the tunnel, even though in other social and professional networks this appears to have been standard knowledge. Interestingly, also, the elaborate European "Seveso Directive", designed to control risks at hazardous chemicals plants, contains no provisions that would have prevented this accident, because the installation was not to anyone's knowledge using or storing hazardous chemicals (there were not even any "no-smoking" signs).

This overall approach, therefore, focuses upon *known* uncertainties only – that is, uncertainties that can be reduced to probabilistic methods. It then acts as though it has thereby comprehended *all* uncertainties.

The second approach to uncertainty, which fences it off as not the responsibility of risk analysis, is still only conscious of *definable* uncertainty – of what is knowably unknowable – such as which dose–effect model should be used to estimate low-dose human effects? It sometimes recognizes that analysis may be intrinsically evaluative, but then refers to the choice of inferences across *known* uncertainties, as if this represents the only type of ignorance. Ricci and Cirillo [15] do, at least, acknowledge a distinction between "differences among assumptions about the technology under study" and "scientific" analysis of causes and effects. But the former are put down to "the realities of life" and dismissed, then the latter are sorted out for the usual kinds of uncertainty, using examples that again already take a given model of environmental pathway and exposure for granted. The final conclusion is that the "inability to know" (as distinct from "uncertainty") is restricted to known unknowables (such as the dose–effect function question). Only these, with no hint of deeper cognitive framework dislocations, are considered as the source of the "uneasy communion between science and policy". Indeed, it is significant that this particular treatment of uncertainty is determined by an explicit model of the "natural" social relationship between risk analyst and policymaker (or risk analysis and policy), where being explicit about too many uncertainties is seen as the main problem. Who identifies the ignorance behind the analyst's *initial* premises and risk problem definitions, and what different institutional forms of analysis–policy–public relationships might improve consideration of this prior question are not conceivable concerns.

The conventional definition of uncertainties (and of rationality) corresponds with particular taken-for-granted institutional relationships of decision making. Yet it is in this institutional arena, where knowledge and social relationships interact, that we need to analyze options. The dimensions of interaction between technical and institutional factors (including behavioral uncertainties and social–cultural determinants of rationalities), as analyzed in this book, are a necessary framework for the fundamental reconsideration and redevelopment of this dominant approach to risk management.

12.3. Processing Uncertainty – Dialectics of Credibility

In Chapter 10 we explored the implications of the conventional approach to uncertainty and rationality, and we critically evaluated the assumption that more science will reduce uncertainty and thus will provide more authoritative regulation. This was linked to the distinction drawn in Chapter 3, between "intrinsic" risks and "situational" risks for the same waste. Even the attempt to specify intrinsic risk parameters or to specify precise testing methods and laboratory protocols for "definitive" risk classification involves an implicit universal definition of a standard risk situation or "scenario". No matter how precise and repeatable such parameters may be (and there are always problems of defining proper replication [16]), the inevitable lack of correlation between that implicit scenario and actual diverse risk situations remains a major problem for realistic risk management, and this gap undermines the credibility of regulation. Indeed, the more the precision and standardization, the greater the gap between the model and (diverse) situational realities. The pressure to produce standardized, precise, and inflexible scientific frameworks of risk assessment to "cover" such heterogeneous and *structurally* uncertain systems leads, it was argued, to forms of regulatory science that *undermine* regulatory credibility rather than sustain it. The inconsistency between the pretensions of such "caricatured" scientific frameworks and the extent of expert guesswork and, even, in some instances the outright fabrication that is needed to create these frameworks, is becoming too large for them to be a credible symbolic projection in the face of empirical failure to give effective regulation.

In some regulatory cultures, such as that of the UK, there is less institutional pressure than elsewhere to employ universal, precise tests and risk criteria, and therefore more discretionary freedom locally to tailor practical regulation to variable risk situations. However, although in the UK it may be easier to recognize the situational variability of risks, the conventional view of science still causes structural conflict to be misperceived as technical uncertainty. Thus, even here, the standard reaction to loss of public credibility is to try to reduce the appearance of technical uncertainty by intensifying statements about expert consensus on the lack of evidence of harm, and by holding official inquiries that drip with such reassuring perspectives [17]. Unfortunately for the credibility of such exercises, their limitations are increasingly being exposed by critical scrutiny, often quasi-public "peer review". Thus, the limiting assumptions of the UK PAC's 2,4,5–T risk analysis were eventually exposed, to the discredit of official expertise. Analysis (again from outside normal peer review) of Sir Douglas Black's report into excess childhood cancers near the Sellafield nuclear reprocessing plant has exposed that committee's undue reliance for its reassuring conclusions upon inputs (again exempted from peer review) concerning radiation effects from the National Radiological Protection Board [18]. Yet the NRPB relies upon several assumptions whose aggregate uncertainties are greater than the theoretical discrepancy, used to dismiss the connection, between Sellafield discharges and excess leukemias. Some of the points embodied in the NRPB assumptions take the form of structural uncertainties and downright ignorance. Both were socially processed and artificially reduced by the Black inquiry into marginal *technical* uncertainty, and attempted symbolic reassurance. The fact that such processes of reduction are usually by no means deliberate is all the more witness to the depth of institutionalization of the false view of uncertainty and the role of science that I have described. To reiterate, even in the UK political culture, which is able to use scientific knowledge in a less adversarial, less elaborated, and more informal way, there is a dominant assumption or symbolic suggestion that more scientific investigation leads to less uncertainty, more rational consensus, and more public acceptability. This is now open to doubt, as the public image of secure scientific method and definitive problem definition is contradicted by increasingly competent critical reviews in public. Local authorities, traditionally subservient to national technical expertise, have been increasingly involved in developments of independent critical expertise.

Overall, therefore, although the institutional styles of science differ between the UK and the USA (see also Section 12.4), there is a common underlying view that uncertainty in regulatory authority is technical, and can be resolved by intensification (or purification) of technical effort and greater expert authority. Expert ignorance, *structural* uncertainty, or objective conflict of perspectives and rationalities are not recognized and their implications are therefore left to wreak uncontrolled damage on the viability of regulation.

In Chapter 11 we extended the critical perspective to analyze the dominant "rational" response to the public concern issue. This response, developed through "scientific" risk perception studies and their incorporation into various extensions of decision analysis, was shown to suffer from similar basic limitations as does the general regulatory use of science. This approach neglects the potential for development of negotiation, broader social learning, and the more constructive and consistent expressions of public concern. It was argued that the dominant approach represents a form of political deskilling – ultimately provocative and self-defeating. The dominant interpretations and uses of psychological risk-perception work do not recognize the fundamentally social basis of individual values. Therefore the intrinsic values of people's familiar social and cultural environments are automatically denied. In cutting out this dimension, this approach suppresses appreciation of structural conflict and ignorance, and hence reduces the *resilience* of regulation.

The three more abstract chapters (9–11), relating to rationality, uncertainty, and public concern, were developed without preconceptions as to the practical implications of their analytical orientation. The overall conclusion drawn here is that the problems raised by the complexities of public concern and credibility and the failings of *orthodox* uses of science and rationality (including the symbolic uses), considerably strengthen a tentative conclusion in favor of strong regional institutions, with public management of T&D in collaboration with regional industries, and restrictions upon unnecessary interregional movements of waste. A major reason for this is the need, which emerges from our analysis, to develop greater public identification with, and a sense of responsibility for, regulatory practices that are part of a more comprehensive industrial strategy altogether – that is, including more concern for upstream production as well. The analysis of the bureaucratic processing of uncertainty suggests that the costs of back-end uncertainties are higher than is normally recognized, which in turn means greater attention must be paid to, e.g., waste-reducing production decisions. Only in this way, we conclude, can the mounting difficulties of public reactions be resolved; and this implies more integrated regional responsibilities. However, before we return to practical suggestions, and bearing in mind the need to respect local differences, we need first to discuss what the *comparative* institutional analysis has been able to offer.

12.4. Comparative Political Cultures – What Can We Learn?

The comparative empirical content of this book naturally raises the question of what has been learned specifically from a *comparative* framework. One of the main points of the discussion in Chapter 2 was that the comparative analysis of institutions or political cultures has shown why apparently technical decisions about the risks of particular chemicals, drugs, practices, or technologies have resulted in very different regulations in different systems. The reasons for such differences, of course, are that the decisions are not purely technical decisions. Social and cultural factors pervade them. Even if described in purely technical terms, they arise from different institutional settings, and their role is to regulate or influence *behavior*, which means influencing social and economic relationships. Not only do they emerge from, but they also enter established local social structures of interaction and authority, adding one more dimension to several that already exist. In order to *change* and control behavior in the intended ways, they must also *coincide* with established practices and relationships in other senses. They must, in other words, find the best balance between such a perfect fit with existing practices that no one notices them – they have no effect – and such a complete misfit that no one recognizes them as feasible or legitimate – also no effect.

The first value of a comparative *institutional* analysis is to demonstrate more clearly the origins of divergent regulatory decisions and practices. "Technical" decisions in different institutional contexts are shaped by, and need to satisfy, different modes of organizational interaction, administrative procedures, cultural traditions, etc. They are, in other words, addressing a different cluster of specific problem definitions in each context. There are good reasons why the Netherlands, for example, uses precise, inflexible concentration thresholds to define hazardous wastes, while the UK uses only imprecise criteria, and the FRG comprehensively specifies waste types, but without concentration thresholds. What is "the best" approach is relative to the context of *surrounding* norms, practices, and constraints, including cultural attitudes, economic behavior, and general administrative traditions.

Because large volumes of waste are transported between regulatory systems, the practical importance of this insight is greater in the case of hazardous wastes than for many other issues, such as road safety, pesticides, or contraceptive pills. In addition to indirect questions of equality of trading conditions, the direct risks of losing wastes, and the consequent public health hazards in the process of transfer, are significant in themselves. They are exacerbated by the technical discrepancies in standards, modes of regulation, etc. However, the political cultural framework shows that *technical* knowledge and practice reflect institutional–cultural factors that cannot be homogenized across nations. This underlines the point of Chapter 5, that practical policy emphasis should be devoted to reducing the pressure on the international system by restricting international waste movements. This is preferable to the illusory goal of eradicating loopholes by developing more precise universal technical definitions and standards. Of course, it would imply other conditions to make it work – we return to these later.

In addition, the political cultural framework can generate important insights about other "discrepancies" among national regulatory practices. Some of these are difficult to explain without employing a cultural framework. Why, for example, is the same ignorance of the mysteries of landfill chemistry (which may naturally transform wastes into better, or worse, products) treated as benign in the UK, yet threatening in the USA, especially in California? Different physical conditions cannot fully explain the apparently large discrepancies of policy attitude. Why does the USA adopt a comprehensive, "all-or-nothing" strategy of regulation, while the UK proceeds *ad hoc* and incrementally; and why does the USA adopt elaborate, precise hazardous waste lists and criteria and other regulations, while the

UK uses imprecise, negotiable criteria? Why have only the USA and the Netherlands reacted to the discovery of past uncontrolled waste dumps by passing emergency clean-up legislation, when they are by no means the only countries to have suffered the problem? Why have the Dutch established an elaborate formal system of inflexible numerical standards for defining hazardous waste, when they do not even have significant industrial treatment and disposal infrastructure in place to deal with the wastes?

The first question above can be rephrased as a question about social or cultural responses to uncertainty in nature. Anthropological insight would argue that the dominant ethos of UK regulatory culture is one of confidence in the rather informal, elite social milieu of decision making; this is buttressed by extreme forms of integrative social ritual to support its authority in mystique and social exclusion, for example, by concealing conflicts within the elite from public view [19]. Constitutional and legal tradition support this paternalistic cultural system, for example, by collective cabinet responsibility, the Official Secrets Act, etc. The fundamental ethos has been that social uncertainties between parties can be negotiated in confidence and privacy as events transpire – there is no need for grand, inflexible projections of norms or sanctions to constrain others, in a legalistic fashion, well into the future. Uncertainty in nature dissolves into this social milieu, and becomes an element of flexible, confident *ad hoc* management, as does any other uncertainty. Thus, nature is assumed to be an essentially benign sink, not merely diluting, but *actively cleaning up* many of our otherwise risky waste practices. The UK culture requires a relatively quiescent public, prepared to engage in the rituals of integration, such as public inquiries, or at least to spectate in silence. It also requires an integrated government system. In the hazardous wastes case, at least, both of these conditions may be eroding, with potentially far-reaching cultural and regulatory implications.

The US response to uncertainty in nature tends to be the opposite of the dominant UK response, and can be explained sociologically by reference to the opposite tendencies in US political culture. Here, there is an almost institutionalized mistrust between actors in the policy sphere (as reflected in the adversarial legal culture), and a suspicion of all authority. There is, therefore, greater *social* uncertainty between parties given partial and competing powers in the system – the executive, the legislature, and the judiciary (as well as powerful public interest and lobbying groups of various kinds). Uncertainty in nature is a metaphor for uncertainty in social or policy affairs – it must be assumed to be full of adverse intent or potential, so it must be immediately and uncompromisingly controlled. One cannot trust how uncertainty might be used by other policy actors. Uncertainty is a hostage to fortune in the USA. Thompson has likewise characterized the US political culture of regulation as a fragmentary *sectist* culture, dominated by competing, well-bounded groups, anti-authority, suspicious, zealous, and with a strong tendency to impute bad faith on the part of competitors – in short, a set of witchcraft-invoking sects, each anxious to maintain its boundaries and identity against a "malevolent" outside world [20]. The institutions of government are themselves more fragmented than most European ones and, in a historically populist culture, much more responsive to the sectist social environment. While some environmentalist groups become established and more hierarchical as they become more closely identified with "the establishment", conversely, some government agencies and industries behave more like sects themselves as they become more embattled and defensive.

This anthropological analysis of institutional styles and political cultures can also explain the underlying "all-or-nothing" syndrome of US regulation, where the adversarial sectist culture forces not only universalistic, numerical, inflexibility upon regulators – "equal treatment for all". – but also an "all at the same time" political requirement, however infeasible in practice. Precise numerical norms are, of course, consistent with a lack of trust: no room for evasion must be given (or appear to be given). An associated element of the same cultural style may be a tendency observable in the USA, to go hard for idealized regulatory solutions, such as "zero landfill" and even "zero waste production". This tends to produce extreme outcomes of localized progress toward such strict ideals, but interspersed with a widespread lack of control. The same cultural elements are manifest in public opposition to T&D facility siting, which has approached a public siting veto. The result appears to be more industrial innovation to produce less wastes, combined with more illegal waste dumping.

A further value of the political cultural framework is to suggest how deeply ingrained may be elements of decision-making style. There is, for example, a current fashion to praise the efficiency of the UK "consensus" mode as an alternative to the extreme elaboration of the US system [21]. However, it is at least possible that the US regulatory culture is consistent with a competitive, dynamic economy. Probably few of those who bemoan the "inefficiency" of US regulatory culture would be happy with the less competitive, paternalistic *economic* systems of Europe.

The uses and styles of science are also shaped by such institutional cultural patterns [22]. With more integrative policy rituals, more confident and integrated policy institutions and power elites, and in the absence of strong sectist penetration, UK policy can maintain authority without the need for elaborate, precise, and specific scientific justifications. This leaves scientists and policy institutions free to use discretion, to formulate policies in general terms, and to be unafraid of ignorance, because it is less likely to be used as a resource by critics. In the UK, the American (and less so, the Dutch) use of precise universal regulatory numbers and tests would not only be curious, but also a potential *institutional threat* to the social position of scientists, because it would imply a reduction in their discretionary power and a radical confinement of their institutional role in policy. This is the cultural undercurrent of a British scientist's derogatory remark that numerical concentration limits for specifying hazardous wastes – as used in the Netherlands – is "a system which can be run by monkeys" [23].

The FRG appears to have a similarly authoritative central institutional framework for scientific expertise in policy, with hierarchical federal scientific bodies free from formal review and criticism, as in the UK [24]. However, whereas the whole network of policy institutions in the UK is hierarchical and *informally* close-knit, the FRG's postwar federal constitution breaks down this nexus, placing real autonomy and power (over other policy, administrative, and legal levers, such as export bans on wastes) in the hands of strong regional governments and courts. Even so, strong collectivist cultural norms, as reflected in the accepted authority of the LAGA (the joint committee of state waste experts), and significant industrial voluntary compliance, underlie and defend the effectiveness of formal regulatory arrangements. It is also interesting that the experience of public conflict over the siting of new facilities has led to requests from state authorities for greater federal intervention and standardisation.

The Netherlands is an interesting and complex case for comparative political cultural analysis. Although it has typical European parliamentary–executive unity, the Netherlands also has long traditions of coalition government, which (unlike single-party government) requires issue-by-issue negotiation of parliamentary and cabinet votes. This tends to encourage fundamentalist, single-issue oriented politics, which is also a dominant feature of US political culture. This in turn allows interest groups to specialize and become more influential in that area. Furthermore, the Netherlands government administration, unlike that of the UK, is fragmented at the central level (the "tubed" policy system), with more radical interest-group penetration into individual ministries, and more routine strategic leaking of "confidential" policy information. There is more open conflict among ministries than is allowed to see light of day in the UK. Public interest environmentalist groups in the Netherlands are articulate and influential, with even formal recognition in government advisory committees. Scientific expertise is, as in the USA, but not in the UK and FRG, relatively more evenly distributed amongst contending interests. This pluralist political culture has structural affinities with that of the USA (although informal Dutch government–industry collaboration is more close-knit and coherent), and produces similar regulatory results. For example, there seems to be a similar "all-or-nothing" syndrome in the Dutch inability thus far to establish a T&D infrastructure, despite progressive legislation in 1979; and there is a reflection of the *institutional* uncertainty in the Dutch use of precise concentration limits and numerical soil cleaning standards as regulatory instruments in this field.

It may also be a reflection of political cultural factors that only the Dutch and US systems have reacted to the past waste-dumps problem by giving it official legislative recognition. Both countries saw extreme public reaction, amounting to political crisis, where no others did. The implementation difficulties of these "clean-up" programs, and of the other US and Dutch modes of regulation, suggest a stronger role of public symbolic action in policy than in those countries that are less influenced by institutional uncertainty and "sectist" styles of organization. Edelman's notion of politics as symbolic action [25] is useful to understand why some political cultures tend more than others to articulate policy objectives and promises that are less realistic, in response to more extreme needs of symbolic legitimation to contending policy actors.

One can also use the anthropological framework to analyze certain tensions and dynamics within institutional frameworks. For example, in Chapter 7 we explained the increasingly adversarial position of the UK local waste-disposal authorities with respect to central government and industry as they have found it impossible to exercise the burdens of regulation placed upon them. Their predominant exclusion from the informal elite networks of negotiation and lobbying over policy, and from access to the higher reaches of technical expertise, has seen them move away from the hierarchy of integrated government and toward the style of outsider, sect-like groups, demanding more clearcut, less negotiable forms of regulation, because they are increasingly placed by circumstances in that social position. In dynamics such as these we can see the interactions between organizational patterns and technical discourse in both its empirical and symbolic (legitimizing) aspects.

12.5. Political Cultures, Regulatory Styles, and Legitimation

The above anthropological analysis of political cultures is consistent with all that has gone before in this book. It provides a framework for understanding the relationship between technical discourse and tacit institutional prescriptions or concerns.

It is also consistent with our analysis of the tacit concern for credibility, legitimation, and the maintenance of cherished forms of social interaction that thread explicit discourse about risks and regulations. However, to understand its relationship with the legitimation issue we need first to overcome a problem with the political cultures framework; that is, a tendency to reduce all questions to its own terms. In other words, for example, the "adversarial" US cultural model or the "consensual" UK one are often thought to define styles of decision-making interaction *monolithically* throughout their domain, without consideration of the cross-cutting influence of other factors, some of which may be contingent, others straddling *different* cultural domains, and of comparable significance [26].

In Chapter 3 we emphasized the importance of issue characteristics that may be regarded as universal, wherever the issue falls. We also analyzed at length the *general* legitimation problem, and the misleading views of scientific rationality and technology, which *universally* pervade regulation. There are also variable styles and tendencies *within* each supposedly monolithic cultural system, such as between different regulatory agencies, according to factors that include their age, institutional "confidence", the kind of activity and actors they regulate, etc. [27]. As noted earlier, specific regulatory initiatives are additions to a pre-existing milieu of interaction, which is already "regulating" behavior in a variety of ways – market interactions, legal and administrative traditions, cultural norms, pressure group lobbying, other legislation and formal regulations, etc. These are not *replaced* by new regulations. Clearly, the specific pattern of regulation will be particular to an evolved milieu of this kind, including the technical characteristics of the activity under regulation. Some elements of regulation, to be credible or feasible, will have to be consistent with established wider patterns of institutional interaction, custom, etc. This is part of their negotiation of legitimacy. These wider patterns tend to be the very parameters that are picked up in comparison of political cultures – the use of numerical or imprecise standards, adversarial or consensual decision processes, etc.

This generates a question that has been raised before, though in a rather different way. If comparative political cultural analysis shows that different regulatory policy *decisions* are often reached on the same issue by different cultures, does this mean that actual regulatory *effects* differ much in the end? In other words, how far do the elements of political culture picked up in a comparative policy analysis reflect the domain of legitimation, or symbolic action, more than the real-world outcomes, even including specific policy decisions (which may be reshaped by the real world of their implementation). Of course, legitimation of a given set of institutional relations and processes *is* a regulatory outcome; it is not "merely" symbolic action, supposedly divorced from the material world behind it. But unlike specific outcomes it is indirect, and continuous.

To a significant extent, therefore, political cultural analysis offers analysis of *modes of legitimation* in different regimes. It indicates the significance that people in society place, not only upon decision outcomes, but also upon *the quality of the social interactions* leading to – and beyond – decisions. It also indicates the tacit processes of social identity-maintenance through sequential specific issues and decisions. But these processes are more differentiated, complex, and dynamic than rigid use of the monolithic political cultural comparative framework would allow.

The complexity of the legitimation dimension is indicated by examining one comparative aspect, namely, the different images of science that predominate in regulatory discourse in the USA and UK. McCray has pointed out [28] that different implicit models of what science is, and where its boundaries with value questions lie, pervade US debates on the analysis of risk. The dominant model in the USA is of a formal science, where methods are fully specifiable, usually quantifiable, and where a heavy emphasis is placed upon the explicability and external accountability of detailed reasoning and observation. This model, or ideology, of science is consistent with the dominant political culture of mutual mistrust and public accountability of expertise and all authority. The dominant UK model used in public debate is its opposite – science is largely a craft activity involving accumulated experience and refined intuition, which cannot be formally specified, codified, and externally checked. Only other mature experts can judge, and their judgments are legitimately inaccessible except to a privileged (and socially trustworthy) few. This model of science supports all the key institutional properties of UK regulation – expert discretion; attenuated peer review; only indirect and limited public accountability; the mysterious fusion of policy and scientific considerations in regulation; and informal negotiation of standards and practices, sheltering behind these ideological defenses.

These alternative stereotypes of scientific authority have a long history of interaction [29], and they can be seen as modes of legitimation, each appropriate to its own wider political culture. As a symbolic language of legitimation, each is partially true, and partially false; partially descriptive, and partially prescriptive. It would appear from this that the influence of political cultures goes even deeper than the processes of using science in negotiating institutional credibility described earlier. It seems as if different political cultures employ fundamentally different forms of science. However I argue that these different forms are common to the more fundamental ways in which uncertainty and scientific rationality are interpreted in industrial societies. The images of science outlined above, although they do reflect institutional realities in each culture, also play a *rhetorical* role. The predominant style and content in using science in different local cultures naturally reflects elements of that culture, but the *process* is universal. The tendency for US regulation, for example, to use science in a more standardizing way, being less sensitive to the needs for situational variation, both affects real regulatory outcomes and is a function of the political culture and its modes of legitimation or symbolic action. The empirical and symbolic worlds are interconnected.

The political cultural framework, used carefully, makes several important contributions. But in addition to cross-cutting factors that differentiate its analytic categories, we have shown that there are further questions which go deeper than that framework. These are to do with the general purchase of images of rationality that encompass even different specific images of science in different institutional settings and political cultures. These more general processes have to do with the most basic ways in which rationality, uncertainty, and technology are defined in industrial society, definitions that constrain our vision of options and responsibilities in relation to technology.

12.6. Science and the Dialectics of Credibility

A central issue of this book has been the social processing of uncertainty. I started with the term in its broadest sense, encompassing social uncertainty, relating to incompatible world views and problem definitions, and technical uncertainty, or imprecision within a given problem definition or value framework. The main argument has been that regulatory bodies tend to respond to situations typified by the former, as if they were only the latter. Chapters 9–11 offered three connected arguments:

(1) Scientific approaches in risk analysis assume a given risk system or technology – a taken-for-granted problem definition.
(2) Scientific approaches in regulation also assume that scientific models and methods are of universal applicability, and that uncertainties are marginal within those frameworks.
(3) Rational approaches to risk perceptions and decisions again assume a common problem setting and see divergent perceptions as marginal adjustments to the attributes of a taken-for-granted risk source or technology.

These elements reinforce each other to create a pervasive mythology of risk management, in which more fundamental conflict, diversity, and ignorance are either inconceivable or, if encountered, illegitimate. The central pillar of this mythology is that rationality is ultimately unitary, and if it could be properly established would lead to the common acceptance of a procedural framework within which public decisions on regulation could be made and enacted. This is supported by the established idea of science, that its essential *modus operandi* is uncertainty seeking – that science automatically roots out and explores the uncertainties in existing knowledge. In this view, if science is harnessed to regulation, then the fundamental uncertainties are at least within our conception, and can be contained within a structured decision methodology. This belief, inculcated by the influential Mertonian and Popperian traditions in sociology and philosophy of science [30], implies that science is always driven by the aim of critically testing and revising its own basic frameworks and premises. However, this reassuring view has been fundamentally revised in the past decade or more [31]. In the newer view, scientific knowledge is developed within traditions that allow the framing of coherent and precise scientific questions. The productive questions are defined *within* a taken-for-granted theoretical and methodological framework, whose authority is not normally questioned. The tradition contains not only it's own rules of "scientific method", but also specific explanatory terms and technical practices. In other words, *recognized* uncertainties are limited to puzzles and anomalies within a framework of certainty as to the fundamentals of that scientific tradition.

To take an example, the use of *in vitro* experiments on cell damage to define radiation or chemical carcinogenic risks has, perforce, to assume that there are no significant intercellular or system-wide mechanisms that affect processes of damage or repair. The normal processes of science define the uncertainties *within* that premise; they do not critically explore the premise itself for the uncertainties lying beneath and beyond it. To take another example from Chapter 10, the EP test for defining toxic waste takes for granted that analytical uncertainties relate to details of laboratory method that have not been exactly replicated, rather than to the more fundamental question of whether the conditions of laboratory testing accurately simulate real-world waste conditions.

Within science, knowledge develops by the creation of an artificial conviction that the basic framework is correct. This certainty as to the fundamentals allows attention to be focused on narrower problems and uncertainties. Anomalous signals are often interpreted by *ad hoc* additions to explanatory idioms, so as to defend existing theoretical commitments, even to the extent of rationalizing difficult data away as spurious results. Rather than being the subject of critical scrutiny, existing knowledge becomes part of the normative framework for shaping acceptable problems, and for evaluating *new* evidence and ideas. This "functional dogma", to use Kuhn's terms [32], is not a guilty secret or a weakness of science – it is a *productive* property. However, when science is used in public policy this fundamental tendency to take its existing frameworks for granted leads to an under-recognition of more *basic* ignorance, or conflict of frameworks. Whereas this might have no wider implications in science outside public policy, in public domains it has direct implications for the social acceptance and credibility of expertise.

Thus, whatever *specific* image of scientific rationality predominates – the US "public accountability" or the UK "priesthood" model – styling regulation on conventional "rational" grounds leads the process badly astray. It encourages the bureaucratic processing of ignorance, of latent conflict of perspectives and world views, and of other more basic uncertainties, into an emasculated version of marginal uncertainties, supposedly manageable within a single, rational decision framework.

In this way conventional scientific rationality is implicated in a pervasive legitimation crisis [33]. Following Habermas, one can see that, in postwar times especially, political legitimation has been achieved (via images of rationality) through active ideological cultivation of *alienation* on the part of the mass public. Technology has been portrayed as coming from science, in a realm beyond normal access and experience. However, this ideological process has destroyed any *plenum* of political discourse in which to try to establish common values and meanings in the face of rapid technological change and public disorientation. Communicative interaction, in the context of common but taken-for-granted values, is breaking down, and has to be replaced by communicative *discourse,* where new conjunctions of meaning and value are re-negotiated. Instead, in the face of credibility and legitimation problems, we have an intensification of the rationalistic ideology that created the mess in the first place. A practical example of this is the intensified scientific effort to reduce technical uncertainties (e.g., to define "hazardous wastes" and risk criteria more precisely), within a basic framework that remains unquestioned. This is closely associated with intensifying attempts to gain public credibility and reassurance by reference to the greater control offered by more science and risk analysis. In order to gain credibility there is a drive to intensify scientific approaches; yet this intensification creates less credible scientific knowledge (as evidenced in Chapter 10) *and* imposes alien categories of problem definition and social values. Existing modes of legitimation, intensified by the perceived need for more public justification, thus create more alienation – that is, more legitimation crises. This is a self-fueling downward spiral into social brittleness and non-viability.

12.7. Implementation and Credibility

An important part of the dynamics of legitimation and credibility is the relationship between center and periphery in regulatory processes. This can also be seen as the relationship between policymaking and implementation. The institutional gap between central policymaking and peripheral implementation is all the more serious because of the underlying processes in the legitimation domain, outlined above. The typical response to the difficulties of *local* implementation – facility siting problems, enforcement deficits, intersystem loopholes, etc. – is to respond from the center. The aggregate effect of local difficulties is greater central policy concern and control, which leads to the intensification of scientific effort – more precise definitions, standardized analyses, etc. – and to a corresponding escalation of the scientific rhetoric of legitimation. This only tends to disconnect science even more from diverse, concrete local experiences and situations, making implementation more difficult.

We can now see why I have stressed implementation processes in this book. Both the institutional distance and the linear rational view of the relationship between central policymaking and peripheral enforcement are seen to be all the more damaging as the public credibility or legitimation issue grows. Central government bodies respond to legitimation problems with the false idea that the implementation lapses that are to blame, are only due to deficits (of technical precision or competence, or of other resources) at the periphery. This is contradicted by the wealth of empirical research, including that described in this book, which shows that local enforcement and implementation have a different set of *social* constraints and conflicts to negotiate, which are different from those envisaged by the center and implied in the linear model [34].

Diver provides a discussion of the complex rationalities of regulation in its different aspects, and the nondetermined character of policy rules for implementation [35]. He points out that even attempts in recent years to define increasingly precise and less discretionary decision rules for regulation in US environmental, health, and safety areas, including the specification of allowed scientific inferences and models, still leave certain ambiguities as to where the line should be drawn between compliance costs, regulatory (and other transaction) costs, and estimated regulatory benefits. There is always a residual value indeterminacy, even if it has to be bought by ingenious reclassification [36].

In a rational hierarchical model of regulation, implementation should take place in a context where legislation and statute making have already resolved the significant value conflicts and goal ambiguities, leaving only technical rule following (the rules may specify how to balance conflicting objectives, as in cost–benefit discharge optimization). In reality, the implementation context is never so clear-cut, and value conflicts remain after legislation, to be renegotiated through the ambiguous phrases of the law. This is especially the case in European contexts, but also even in the USA. When regulators face multiple and conflicting goals represented by different parties, they quite reasonably respond in a "partially committed" way. Rather than attempt once and for all to resolve the conflict, they break it down into sequential negotiations, incremental (and reversible) adjustments, conflict-*absorbing* procedures (e.g., major public inquiries in the UK), and reactive strategies. This is not irrational. Indeed, it may well be a source of resilience in the system because it keeps genuinely conflicting values in continual negotiation with one another, adapting to new circumstances and balances of interest as they arise.

Local rationalities of compromise and ambivalence, far from being a mere deficit of competence, therefore, may actually be a source of regulatory resilience, avoiding the unviable implications of a literal translation of central policy languages, which are embroiled in symbolic reassurance processes. The question is whether escalation of the central legitimizing language of scientific rationality strengthens the position of local institutions or not. The conclusion of our analysis is that it only tends to make their situational problems worse. This is a further strong argument in favor of more decentralized, regional regulation, on the crucial condition that regulatory decentralization is supported by adequate resources, powers and other factors described later.

12.8. Toward the Reestablishment of Context

So far, we have argued that, although they may have made for progress in the past, conventional models of science embedded deep in the existing approaches to regulation are becoming counterproductive in the new context of public justification. This argument was amplified in Chapter 11 by showing that a "rational" approach to the interpretation of public concern also violates the ultimate grounding of that concern. Despite its liberal intention of incorporating public concern into decision analyses, the approach converts what are fundamentally different frameworks of experience and problems into extra "attributes" within the assumed and imposed single problem definition of the policymaker. This is an identical form of response to the scientific one outlined earlier. A framework (which implies certain value commitments) is taken for granted, and uncertainty – here attributes, probabilities, and weightings – is recognized and analyzed only within that framework. Different frameworks of meaning and experience remain unexplored, only distantly – and mutely – signaled by reference to multiple "attributes" of the taken-for-granted technology and its "natural" social relationships of management. The end result is that, in response to a problem fundamentally to do with the quality of social relationships, the technical "knowledge" of experts is elaborated yet further, while the institutional relationships themselves remain unexamined and unreformed.

Rational approaches to policy management, especially the decision-analytic developments of economics and psychology, are fundamentally authoritarian, and ultimately unviable in any significant policy issues, because they inadvertently relegate all social experience that is different from theirs to an analytical *context*, which they assume can be comprehended by marginal adjustments to the existing framework (by adding attributes). It is taken to be a natural equilibrium about which only marginal adjustments are conceivable. That different equilibria altogether might be feasible is simply not a consideration.

In their present form, therefore, the scientific foundations of risk analysis and the economic and psychological foundations of risk perception–decision analysis, tend to add frills and adornments endlessly, like Baroque art, almost as an escape from reality rather than in appreciation of it.

Our interpretation of technology, risk analysis, regulation, and the social roots of public risk perceptions indicates that on pragmatic grounds alone (quite apart from moral or political judgments), the context cannot be adequately comprehended by such marginal elaborations in analytic models and their fixed underlying frameworks. The elaboration only entrenches the existing social relationships of analysis and decision-making, which are already part of the problem. In human systems, context *is* substance. And with modern forms of technology, "context" is increasing in scale and significance, while being *decreasingly* comprehended by conventional analytical approaches.

We can restate and perhaps clarify the implications of this point by using the language of information processing, as we did in the discussion of compensation for siting risks in Chapter 11. The usual situation is one in which existing analytical methods attempt to channel fundamentally diverse strands of social information into a single framework for decision making. In terms of *Figure 12.1*, regulatory bodies channel their evaluation of the human and environmental costs, risks, and benefits of regulation into formal feedbacks in standards, advice, mandatory rules, fiscal incentives, etc. (feedback A). These are negotiated and justified in a continual interaction with expertise, public reactions, industrial lobbying, etc. The bureaucratic processing and reduction of structural ignorance and conflict into the language of "manageable risks" cannot be fully concealed from the public, because they eventually see the consequences of ignorance and the inadequacy of the scientific language of justification. The full implications of real back-end ignorance and conflict are not transmitted upstream in formal regulatory signals to the point of production and innovation, because of the false description of these as manageable by conventional means. Public realization of the inadequacy of this process is obstructed from mature expression: it is strangled by the impoverished languages of scientific risk analysis and risk perception–decision analysis, and by general cultural norms of parochialism and self-interest. Thus obstructed (and localized), it seeks other outlets, which results in sporadic eruptions of apparently *extreme* reaction, e.g., to facility siting, via feedback B in *Figure 12.1.* This analysis is supported by the observation of many chemical company executives [37], that they are far more influenced in production decisions by their judgment of the future of direct public reactions than they are by formal regulations. Thus, we have a situation in which public experience is being thwarted of constructive expression via the language and process of regulation, which effectively, but inadequately, shelters the front end of innovation from social access. The result is the worst of all worlds: public quiescence artificially cultivated by the language of regulation; suppressed concern; and unpredictable, increasingly intransigent expressions of that concern when it does find occasional outlets. This is not a viable strategy.

The conventional regulatory channel, as reflected in the dominant language (and relationships) of rationality, is simply incapable of transmitting the necessary richness of the original information. The main substance of this information which the conventional scientific approach neglects, is the

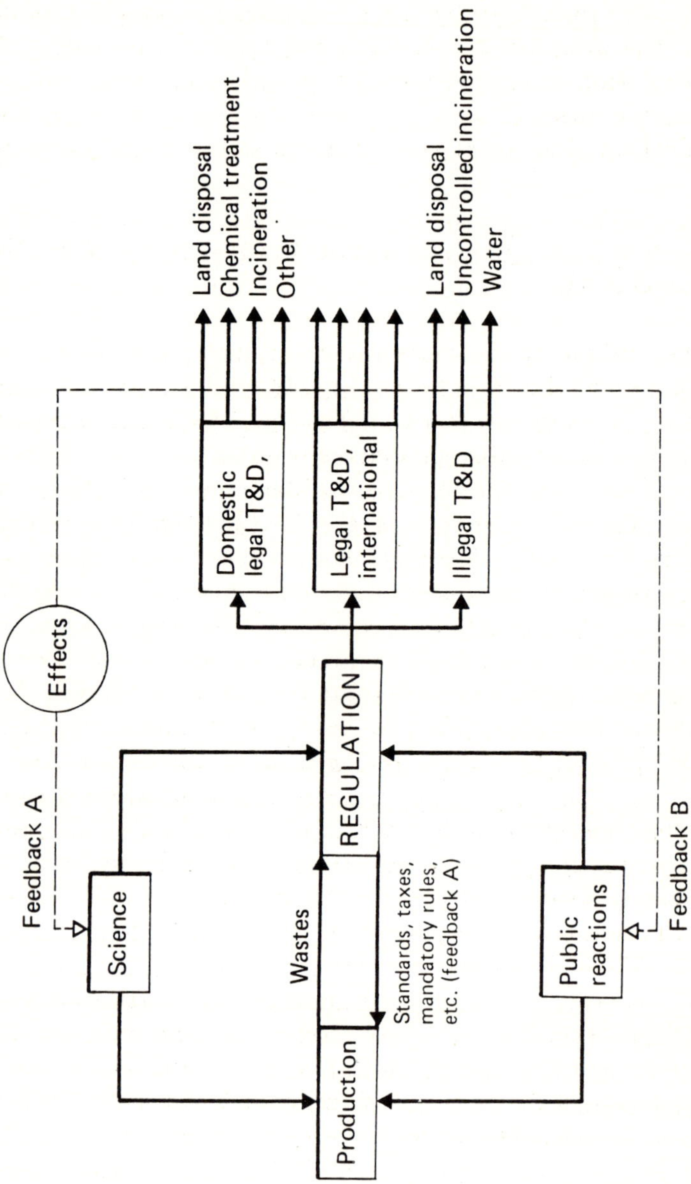

Figure 12.1. Hazardous wastes: A schematic system of major flows of materials, information, and influence.

structural uncertainty in the system, which back-end regulation cannot manage. Therefore, the mechanisms do not produce decisions that command credibility. Even if any given decision were in some abstract sense "correct", the processes in which it is embedded are not. In an alternative approach, the overloaded and distorting channel would be restructured to allow articulation and more direct negotiation among different frameworks and, thus, the recognition of the need to absorb some of the structural indeterminacy and uncertainty in more *up-stream* regulatory attention.

12.9. Uncertainty, Pragmatic Rationality, and Social Learning

The above critique of the conventional abstract, "Utopian" rationality [38] includes the point that this rationality obliterates questions of the quality of processes or *interactions* ("context") by focusing, instead, upon the rationality of decisions as separate products or events.

All of our arguments have shifted the orientation toward open-endedness. The starting point for this is the recognition, which was offered long ago by Tribe [39], that technology itself is more appropriately viewed as *process* rather than as hardware, and that its most significant consequences are nearly always the least direct and least foreseeable. This is a lethal criticism of rationalistic technology-assessment programs, especially when combined with the point that technologies engender unforeseeable social changes that alter our patterns of social interaction and our values. Therefore, deciding about technologies is as much a matter of choosing which social values we wish to live with, as of pretending to evaluate and identify its consequences in the framework of taken-for-granted values. There ought rationally to be a dialectical interaction between technological progress and social learning, whereby each adapts, adapts to, and encourages the (nonpredetermined) development of the other.

Institutional processes that help us to recognize intrinsic ignorance about consequences; conflicting frameworks of experience and value; and different fundamental definitions of technology and problem structures generally, are a necessary element of that social learning. This implies recognition that the overall direction of technology is not locked into an inevitable historical logic, untouchable by human aspiration or concern. That is, regulation is, or can be, about more than efficient "cleaning up" after technological innovation. However, the more critical rationality opened up by our analysis of the problems of conventional rationality needs to be given at least some shape, to point to alternative ways of progress. For example, by definition, ignorance cannot be carried in knowledge, so how *are* we to handle and respect it? What, if anything, is to give the open-endedness a sense of order and direction, if we are to release ourselves from the admittedly baneful ideological grip of conventional rationality and technological determinism?

There are no answers in the simple mold of existing givens. The essential difference is akin to that between liberal systems and critical theory – between "adaptive management" and "political learning". What we have criticized above is "adaptive management", which stretches conventional rationality within existing methods and institutions to incorporate public concern. The social control of front-end technological innovation is essentially concealed and protected by the mythology of conventional rationality which pervades regulation. This understates the ignorance and lack of understanding of the consequences and interactions of technology, in the ways described above; in so doing, it understates the overall costs and social uncertainties that should be accounted against the upstream innovation decisions. By thus undermining the constructive transmission of social intervention from regulation to innovation, it allows technological change to continue as impenetrably and as apparently (to the vast majority, authentically) uncontrollably as ever. If it is not dismissed as irrationality, public concern arising from this social inaccessibility of technical change is interpreted by this regulatory rationality as a marginal extra weighting to specific technological risk attributes.

This is "adaptive management" within conventional rationality. It not only arbitrarily propagates and defends a myth of technological determinism; in the very same process it *undermines* and suppresses political learning. As Forrester has put it [40], the *critical* rationality opened up as an alternative here "leads us beyond the consideration of transforming the value, whatever that is, of each outcome and utility – to practical political considerations of political identity and direction."

We can substantiate this more reflexive and interactive model of rationality by reference to modern understandings of science. From the outside, scientific knowledge looks hard – it is *revealed* by disciplined method. Yet detailed internal social analysis, including long-term participant observation of laboratory life [41], has shown that key terms do not have natural revealed meanings – they are negotiated between practitioners (who may be very few in number in any given specialty). The application of existing established rules of theory or method to new cases is never totally clear-cut, and assignment of observations and behaviors to explanatory classifications can involve choices among ambiguous alternatives. Whether or not a given experiment and its observations have *really* been replicated usually relies upon social agreement as to which elements of practice are important and which are not. Practice is so complex that it is impossible for a competent expert to be aware of every detail of experiment and inference.

None of this account implies that scientific knowledge is *reducible* to sociological processes – nature has a say; but technical or natural constraints do not alone *determine* knowledge outcomes. Social processes are of necessity involved in the creation, maintenance, and cultural reproduction of coherent bodies of scientific knowledge. The point for the present discussion is that the internal social and technical achievement of scientific consensus has been pervasively misunderstood and misrepresented as *revealed* knowledge. The rich world of social negotiation charted by the sociology of science has not been visible to the public. Indeed, misrepresentation by defenders of crude rationalism scares the uncommitted back into the rationalist fold by counterposing monolithic scientific rationalism against the (1930s Marxist) view that scientific knowledge is only a function of dominant social–ideological interests [42]. This false Hobson's choice utterly obliterates the far more complex and pertinent implications that can be derived from the sociology of science.

However, the increased involvement of science in public decisionmaking since World War II and the decreasing conclusiveness of scientific claims in such arenas have led to a closer public scrutiny and visibility of the internal process of science. This has, in turn, led to wider recognition that these processes are not as tidily rule-bound and methodical as the "hard" image implied, and that the precise use of inference rules, observation terms, etc., is open to negotiation. Values may therefore enter the process of scientific knowledge construction and development, without necessarily leading to its degeneration. Nor are they necessarily linked to particular social interests [43]. Scientific definitions of hazard and waste in hazardous waste are evident examples; so, too, is the negotiability of the framework used to define the risk system in analyzing, say, the risks of 2,4,5–T usage. Exclusive use of controlled laboratory toxicology-test data imposes one framework, but one that is incomplete and impoverished. The scientific negotiation of terms is equivalent to the negotiation of the reach and fit of such frames of meaning, and when these are used in policy they imply value commitments, even if the scientists themselves are scrupulously impartial. Some have assumed that this kind of negotiation e.g., about inferences, only occurs where policy-relevant values are in play (see the "risk assessment policy" choices discussed in Chapter 10). However, the social negotiation of such commitments appears to stretch much deeper than this, into the heart of scientific knowledge, which has no apparent (or, at least, no clear-cut) policy implications. Thus, the aim must be to establish institutional relationships and the associated images of scientific rationality, which do not pretend that science can be reduced to social values choices, and which equally do not imagine that science always provides "natural" truths if kept pure of ideological contamination. The notions of inference policies (e.g., over which models to use in human low dose–effect extrapolation) and of "transcience" are useful, so long as the institutional processes allow unprejudiced exploration of the scope and boundaries of such policy–scientific arenas, if necessary into hitherto purely "scientific" areas. The ideas of robust and pragmatic rationality developed by Ezrahi, Ravetz, and Rip [44] appear to be promising in this respect.

If science used in the public arena tries to maintain the conventional image of "revealed" knowledge and natural consensus, this will encourage erosion of credibility, as our discussion of legitimation crises has explained. We need new institutional processes that open up *internal* scientific processes to progressive dialectical interaction with external expressions of social values. This is emphatically not to argue that science should be subjected to democratic referenda. It is more that a new style of regulatory scientific knowledge needs to be developed, which is inherently more provisional and *expressly* incomplete, leaving room for external input negotiated in from other frames of social experience. This emphasizes the necessary correspondence between a new, more open-textured style of *knowledge* and new, more open-textured *social relationships of decision making.* More shared responsibility for regulation, as indeed the current fashion for "deregulation" implies, requires more shared influence in the cultivation of associated knowledge, which inevitably means a *manifestly* less "closed" system of knowledge.

Instead of this, under present circumstances, as parts of science are drawn in as an authority in the management of complex heterogeneous risk systems, so the conventional images only further help to exclude relevant pieces of empirical experience and information, as well as legitimate values, from the decision process. The 2,4,5–T case described in Chapter 9 is a good example. So, too, is the whooping cough vaccine issue, where pressure for a comprehensive uptake led to reassuring assertions about the risks. But these reassurances used mean figures ignoring the much higher risk to individuals with contraindications, such as allergic sensitivity. This suppression of social learning concealed the need for a debate of the moral dilemma between the *social* good of full eradication and the higher *private* risks of brain damage for a significant sensitive subpopulation that would occur in achieving the social goal. The eventual public diffusion of the sensitivity point led, in any case, to erosion in the credibility of the official scientific point of view, and of the program based upon it: better to have encouraged the debate in the first place. The same kind of basic options exist in most areas between socially closed, baroque bootstrapping (degeneration) or an institutional opening toward broader negotiation, even of "scientific" regulatory frameworks. In the hazardous waste field these options are reflected in the ubiquitous tension between unrealistic, universalistic risk models and criteria (encouraged by existing modes of legitimation) and situational authenticity.

The problem of finding an overall normative orientation to guide interaction results from recognition of the legitimate indeterminacy in the more open-ended framework of public rationality advocated here. Critical theory argues that scientific rationality has become an ideology of social control in industrial societies, foreclosing open-endedness in the myths of technological determinism that conceal and protect the power structures involved in technological commitments. One critical view of such symbolic action [45] tends to see the use of "objective" scientific images to justify the exercise of power as an elite conspiracy. In this view, the symbolic domain of political language and procedures (the social equivalent of "body language") and the empirical real world are decoupled, the former being manipulated to mystify and protect the latter. Critical theory sees the processes of communicative action – the use of language and symbolic action generally – as more integrated into empirical reality. It, too, is part of the empirical domain; this is reflected in the term communicative *action* [46]. This distinction is significant, not because critical theory denies mystification and the systematic distortion of public communication, but because it gives a clue to a normative framework that can offer guidance to improvement. In other words, the idea of communication *action* suggests that we can replace the now sterile and self-defeating norms of conventional rationality in regulation with the norms of *improving the forms of public communication, and thus all-round political learning.* What we have at present is its opposite – political deskilling, which results in sporadic, uncontrolled, and damaging outbursts of public hostility, further polarization, and brittleness in the overall system.

Alternative *process* models are available, even if – of necessity – they are not so clearly defined. Indeed, the decision-analytic methods of Raiffa and colleagues criticized in Chapter 11 could, with careful adaptation, be consistent with such a new "social heuristic". Von Winterfeldt's value-tree analysis [47] has been used to elicit value structures in a more public domain, among competing groups in policy issues. The soft systems methodology of Checkland and colleagues [48] is another proved approach to the constructive elicitation of unconscious problem-value frameworks held by different parties to decision issues, which encourages a social learning *process* framework for interaction, rather than the conventional "rational" utilitarian one. An underlying inspiration of the "soft systems" method is the theory of "appreciative systems" developed by Vickers [49].

None of these (and other similar) frameworks guarantees a *progressive* articulation of the collective values in the context of concrete decisions – indeed, one suspects that in order to be successful the *formal* elements of such techniques are crucially buttressed by considerable craft skill and sensitivity on the part of their practitioners. This is only another statement of the truth that there will always be a gap between empirical social practice and the public language describing it. Indeed, it is necessary and desirable that such a tension should exist. The moral heuristic contained in the myth of rationality is a combination of "description" and "prescription". It may "describe" a consensual state that is achievable, but in so doing it avoids another potentially realized state, of the bare confrontation of starkly incompatible ultimate values. To the extent that the myth *describes* one state plausibly and mentally annihilates the other, it is, of course, *prescribing* it. The alternative approach of critical rationality does not claim to destroy myth, nor the tension between description and prescription that myth embodies. Part of the repair job on public rationality is to reestablish a *progressive* tension, encouraging the development of a collective social insight. The present version is more of a degenerative shell that is losing any positive moral purchase.

Another kind of model for the critical heuristic of public rationality is one suggested, for example, by Forester [50] and Majone [51]. This is the analogy with legal discourse. Here, progress is made by constructive tension between accumulated tradition and new interpretation. The body of legal knowledge is never so rule bound that it cannot be creatively adapted and developed to reach out and appreciate new circumstances. It is *open textured,* and thus open to the social negotiation of new expressions of value and meaning. Being open, it *encourages* mature social articulation of such values. There is dialectical interaction and development between authoritative expert knowledge based in legal tradition, and critical social values. It retains authority just because it does not become too dogmatic, rule bound, socially exclusive, and *authoritarian.* At its best, it is an appreciative system [52].

But there is a pertinent difference between such processes at their best and the more closed ways in which policy *scientific* expertise relates to its context. Penetration does occur, of course, but the very fact that it tends to be "penetration", not interaction, reflects an inherent institutional closure, an incipient polarization, and the opposite of social learning. The very fact that the role of science is tied to *risk*-based problem definitions (as opposed to a social, network-based recognition of diverse problem definitions) means that science is already *institutionally* harnessed to prejudiced values and authoritarian styles of decision making, however neutral scientists try to be.

An argument against the kind of analysis given above is that there is bound to be social stratification and exclusion in even the most democratic decision-making processes. Arguing for the opening of existing institutional processes of expertise toward broader negotiation of the underlying frameworks, and exposing the basic sociological roots of the tensions over risks, as in Chapter 11, is therefore only to encourage a kind of fundamentalism that would destroy all public institutional authority. This is a legitimate point of view, deserving a proper reply.

My response is to suggest that we look as dispassionately as possible at the related problems of uncertainty and public reactions, which are arguably two of the most pervasive modern concerns of policymakers. Despite the intensified attentions of experts, uncertainties are felt to be increasing relentlessly. Despite apparent overall increases in public safety, people still often resist the "improvements" offered by promoters of this or that chemical product or technology. The answer to these conundrums, and to the above counter-argument, is that uncertainties are only felt to be more pervasive or significant because of the (growing) commitments, and the ambitions for control, that they threaten. Uncertainty as such has not increased, as if it were some objectively existing entity. The scale, pace, and *vulnerability* of the interlocked technological, financial, organizational, and emotional commitments represented in modern industrial societies are what have increased, and it is these that are multiplying the *sense* (and cost) of uncertainty – that is, threat. The problem is not uncertainty, but *vulnerability*, brought about by the rate and scale of those commitments.

Likewise, I would argue, following Chapter 11, that the public concern problem is not one of fundamentally *waning* public credulity and trust. There is every reason to believe that this trust in decision making has never been anything better than a reluctant or guarded tolerance. The present hue and cry about public perceptions may only be uncovering uncomfortable facts about the *typical* state of public trust in experts, elites, and the like – that it was *always* pretty low. I would argue that what has changed is not levels of public credulity or trust, *but the demands we make* of that credulity by the rate and inscrutability of reasons for the technological and social changes that are going on.

Thus, the arguments about opening up regulatory discourse to acknowledge more fundamental uncertainties, conflicts, and divergent underlying problem structures, have nothing to do with destroying authority *per se*. They have to do with keeping a growing authoritarianism, vulnerability, and alienation in check, and with defending "scientific rationality" as now institutionalized from its own self-defeating tendencies. They are concerned with reestablishing *legitimate* and *viable* authority in the technological sphere.

I have shown that the currently dominant approach is not merely based on paternalistic *individual* attitudes among decision-making elites, but is far more deeply rooted in *institutional* practices, languages, and analytic categories. Indeed, the worry is that many individuals in policymaking, regulation, and industry recognize, at a personal level, the processes I have described in which huge ignorance and the lack of real control is concealed by the forms of science and public language of justification and reassurance. Many policymakers sense the problems, but feel trapped and unable to see any alternative way of doing things. If the view of public attitudes given above is reasonable, then *they* do not believe the myths either. So, we have the baroque situation where *no one* believes them. They have, in other words, outlived their usefulness.

A shift toward the more open-ended dialectical mode of regulation will not remove the tension between our public language and the reality it is claimed to represent. It will not, in other words, do away with myths in public affairs – that is not the point. The essential aim, as McNeill has aptly put it [53], is the care and repair of our public myths. The current mythology of rationality in risk management and regulation is on a degenerative path, and is in need of radical repair toward a less orderly, but more enriched – and enriching – successor; one with social purchase.

12.10. Some Practical Conclusions

It may seem odd to leap from highly general, abstract issues back to the problems of hazardous waste. However, the issue of public acceptance is central, and is bound to grow in importance. It affects all other practical aspects in one way or another, and not only facility siting. The uses of science are also central, and directly affect technical instruments and norms. Some implications are more direct than others, so that the following vary in their specificity. Overall, they are offered to provoke thought rather than as inescapable conclusions from our analysis. They are based upon defensible arguments, but they do not claim to be indisputable. The universal validity of any observations ought also to be qualified by the political and cultural considerations discussed before.

One general practicality seems to be that public concern over and involvement in chemical waste management are going to grow, and cannot be ignored. Therefore, the questions of how to develop social learning constructively rather than exacerbate polarization through this concern seem to be crucial in the medium term. Some countries, such as the FRG, Austria, and Denmark, have partly grasped the nettle by encouraging domestic waste disposers to take more interest in and responsibility for *their own* household waste. This should have at least three beneficial effects:

(1) It increases knowledge.
(2) It puts industrial waste hazards in more measured perspective.
(3) It should actually improve waste regulation.

All of these also tend to encourage a sense of shared responsibility, which must be better than the dominant notion elsewhere, that "social learning" is fostered by more effort in public relations to show how well things are being managed by "the experts".

Other countries seem to fear that encouraging people to take responsibility and to develop interest is tantamount to inviting them to interfere in expert management of a complex system that they do not understand. Better to encourage them to stay uninterested. This seems an ultimately futile approach.

Other means of developing more resilient social learning systems of regulation are discussed in the previous section. However, we would argue that the dimension of public credibility generally strengthens the argument, already taken from our institutional analysis, in favor of mainly decentralized, but more integrated systems of regulatory responsibility. The argument is that public opposition can only be mitigated by public identification with, and a sense of responsibility for, what is going on. Social learning situations that involve critical debate are only feasible in such contexts. At the same time the structural features of hazardous waste life-cycles also militate in favor of strong local management. Precisely how local is not amenable to general prescription, but depends upon existing structures of government, cultural attitudes of identification, industrial concentration, etc. However, the *Länder* in FRG, the Provinces in the Netherlands, the states in the USA, and something like the disbanded metropolitan councils in the UK (akin to regional amalgamations of most existing UK local waste disposal authorities) would seem to be appropriate. Again, this also depends on whether they are envisaged as T&D managers or as regulators of (private sector) T&D managers.

The problem that is then exacerbated, of course, is the inconsistency that allows large loopholes *between* regional authorities, and economic–regulatory strains as the better facilities are undermined by all the business going to cheaper options elsewhere. Legislated *minimum* standards are a familiar, limited means of dealing with this difficulty, but there seems to be a strong parallel argument in favor of reducing movements of waste between regions and nations as much as is possible. This not only encourages a certain local sense of responsibility for wastes, which might help place public support behind enforcement and new facility siting, but also it accords with the inherent limits to technical harmonization of the regulatory frameworks of different systems, analyzed in Chapter 5. Also, given that hazardous waste can be moved around far more easily than can production (the waste source) itself, its management is more vulnerable to shifts in public opinion (e.g., against "foreign" waste). It would seem a prudent insurance against likely developments in public opinion to plan now for reduced intersystem movements of waste and for correspondingly more thorough localized arrangements, than to have them suddenly enforced in the future.

Restriction of intersystem waste movements and regional management raise the questions of finance and control. Our analysis suggests that those systems that rely on private free-market investment for their T&D capacity have problems in obtaining adequate, stable investment in more capital-intensive, "high-technology" T&D facilities. Again, countries and regions will differ according to the size and nature of their domestic waste market, but a general current in favor of joint public–private facilities investment emerges from our studies. Involving local waste producers in the investment encourages them to use the facilities, especially if rebates are offered to those who do enter into partnership.

The question of who *manages* such facilities (and the associated pick-up and transport systems) is separate. Again, our analysis favors single public-company management, because this reduces the complexity and behavioral uncertainties that otherwise plague the hazardous waste life-cycle. Regulation (e.g., inspection and registration) is also thereby made far more feasible. Of course, countervailing inefficiencies may be generated in waste recovery and recycling, and a close comparison of the Hessian, Bavarian and UK recycling industries would be valuable in this regard.

We claim that the significance of these qualitatively different uncertainties for hazardous wastes has not been clearly identified before. All strategies that reduce them therefore increase in value, that is, in the perceived cost it is worth incurring to achieve this reduction. Simpler institutional life-cycles is one strategy, so too is upstream waste reduction.

So far these suggestions deal mainly with T&D industrial infrastructures, and not with regulation as conventionally seen. The nature of hazardous waste generation and movement makes it logical to manage waste T&D itself as the major aspect in its *regulation*. However, it could well be that the strengthening of local regulatory bodies (e.g., at provincial level in the Netherlands, or local authority regional amalgamation in the UK) could create bodies with sufficient resources and expertise to deal with private waste T&D companies adequately, without having to enter into T&D management themselves. At the moment this does not seem to be true (there are very few exceptions).

Whatever the merits of public or private T&D management, another strong point from our analysis has been that adequate regulatory knowledge to define the fantastic variety of risk situations is never going to be held by even a well-endowed local body. They therefore need to be aided by a well-informed, interested but measured public involvement. This will be all the more true if T&D is in private hands, because the waste life-cycles will be more complex and variable. A major point arising from our (and other) analysis of implementation is that enforcement powers are not automatically established by the creation of *formal* rules and powers in legislation. The shape and outcomes of local enforcement bargaining is open to considerable influence by articulate, organized public opinion and lobbying at the local level; indeed, it is undermined by lack of such public interest. Yet apparent lack of interest should not be read as lack of concern, which may suddenly crystallize and erupt in unpredictable and unconstructive ways. Developing more sustained and measured local public interest – as may be conceptualized in the foregoing ideas of critical rationality and social learning, and encouraged by more integrated regional responsibilities – could therefore be in the interests of all parties, including industry.

Self-regulation by waste producers and handlers is also inevitable in part. Necessity should, as far as possible, be converted into virtue by more open-textured relationships between (enhanced) local regulatory bodies, local industries, and local populations and/or workforces. Of course, conflict will be never far away, if not endemic, but at least if constructively channeled via the heuristics outlined earlier, it would avoid the extremes of alienation that are festering away in the increasingly brittle and ultimately unviable systems that predominate today.

These "practical" ideas are deliberately sketchy, because as broad principles they would need to be worked into a more concrete, detailed shape in different local circumstances by experienced local practitioners. As a final practical point I return to an earlier observation, that a new perspective on familiar problems – to see them in a new way, maybe as new problems – is also a practical contribution. It seems appropriate to finish by referring to a theme that has bridged the two parts of this book, namely the complex and fundamental correspondence between knowledge and social relationships. I hope the reader will find it as valuable as I have done to turn to the thoughts of someone who was taxed with practical problems very far indeed from those of regulation, but who was exploring the same question – what is "telling the truth"? Dietrich Bonhoeffer's conclusion, following a long tradition in moral philosophy, was that "what is true" is, in the end, relative to the evaluation of the kind of human relationship that the knowledge supports [54]. Conventional scientific rationality, as it is used in risk management and regulation, is germane or "true" to a particular kind of social relationship in policy regulation and technology. Social authority is assumed to be naturally and legitimately based upon narrow bodies of expertise, which are assumed to be able to *reveal* truths and "best" decisions. The horizon of moral or political competence and choice is absolutely restricted to how best to clean up at the back end of technologies. Where these technologies come from, why, from what earlier assumptions, values, commitments, and foregone alternatives at the stage of innovation, are taken to be entirely other questions, normally answered – or rather, preempted – by the myths of technological determinism that are supported by conventional scientific rationality in regulation. This only consolidates a kind of gross public schizophrenia in which we are supposed to confine our ultimate horizons to a detailed examination of the print, while others invisibly manage the story.

The conventional "truth" used in policy corresponds with social relationships that are determined by the protected social interests that control technology and that are restricted to individual private utility. The *social* experience of alienation, of being increasingly controlled by arbitrary, even capricious forces of unintelligible technological and related social change, is reduced to the framework of those individualistic relationships – and is thereby morally invalidated. Yet this kind of social experience – and concrete versions of it in many specific issues – lies at the heart of "extreme" public reactions to rationalist policy management. So we need a new, more open-textured kind of knowledge, or discourse, corresponding to a recognition of the neglected dimensions of social relationships. This more open-textured discourse, and its institutional embodiments, will naturally recognize that the exploration of human meaning and value is not exhausted in the control and distribution of risks (even if this is somehow "equitable"), but may also include engagement in the direction and development of technology.

Notes

[1] For a discussion of some of these problems in analytical models for policy, and the institutional pressures behind them, see the special issue (1984) of *Policy Sciences,* **17** (4).

[2] See, e.g., Douglas, M. (Ed) (1973), *Rules and Meanings* (Penguin, Harmondsworth, UK); Twining, W. and Miers, D. (1976), *How to Do Things with Rules* (Weidenfeld and Nicholson, London); Goffman, E. (1974), *Frame Analysis* (Harper and Row, New York, NY); Giddens, A. (1976) *New Rules of Sociological Method* (Hutchinson, London). While, of course, it does not prove its validity, the growth of the Wittgenstein industry in the past decade testifies to the immense influence of "interpretive" approaches in philosophy and social science.

[3] See especially Diver, C. (1980), A theory of regulatory enforcement, *Public Policy,* **28**, 257–301; Hawkins, K. (1985), *Environment and Enforcement* (Macmillan, London). See also the notes and references for Chapter 2.

[4] Jenkins, W. (1978), *Policy Analysis: An Organizational Perspective* (Martin and Robertson, London); Majone, G. (1985), *The Uses of Policy Analysis* (Yale University Press, New Haven, CT).

[5] As has also been noted, the same result is arrived at, but for different reasons, with larger, more centralized technology areas, like nuclear power, because each "implementation" step is so huge, financially and politically, that it *makes* policy rather than enacts it.

[6] For the notions of symblic action and public ritual used here, see Wynne, B. (1982), *Rationality and Ritual: The Windscale Inquiry and Nuclear Decisions in Britain* (British Society for the History of Science, Chalfont St. Giles, UK).

[7] Recent expressions of this general stance are given separately by Weinberg and Huber. Weinberg's attempt to fence off a realm of uncertainty and dispute into the special subarea of *transcience* still presupposes that the realm of facts – or fuzzy facts – is in principle prior to social values; Huber seems to believe that those risks which are "creatures of science and engineering" are the least problematic ("predictable and inherently self-limiting"), because of the more intense scientific risk research surrounding them – nuclear power and commercial drugs, for example. See Weinberg, A.M. (1985), Science and its limits: The regulator's dilemma, *Issues in Science and Technology*, **II**(1): 59–72; and Huber, P. (1985), *Ibid.*, pp. 73–82.

[8] See, e.g., Barnes, S.B. (1982), *T.S. Kuhn and Social Science* (Macmillan, London); Barnes, S.B. and Edge, D.O. (Eds) (1982), *Science in Context* (Open University Press, London); Law, J. and Lodge, P. (1984), *Science for Social Scientists* (Macmillan, London); Collins, H.M. (1985), *Changing Order* (Sage, London).

[9] For an original – and still pertinent – exploration of the correspondence between the degree of elaboration of knowledge and language, and the degree of social (un)familiarity or trust among its users, see Bernstein, B. (1971), *Class, Codes and Control*, Vol. I, *Theoretical Studies Towards a Sociology of Language* (Routledge and Kegan Paul, London). The evaluative question in which Bernstein became embroiled, of which kind of code is "richer", is irrelevant here.

[10] *Hazardous Waste Control and Enforcement Act,* 1983 (Bill, HR 2867). See the *Report of the Committee on Energy and Commerce*, US House of Representatives, 98th Congress, Report 98-198, May 17, Washington, DC. Reports from the Congressional Office of Technology Assessment, the US Government's General Accounting office, and various policy researchers have all emphasized the failure of the original 1976 RCRA to tackle the hazardous waste problem in a realistic fashion. For various views, see the Congressional Hearings, HR 2867, 22/24 March 1983, before the Subcommittee on Commerce, Transportation and Tourism of the Committee on Energy and Commerce, 98th Congress, 1st Session, Serial No. 98–32, Washington, DC, 1983.

[11] A clear example of this has been revealed in research by Lyn Hague (Lancaster) on the Abbeystead (UK) water-transfer scheme explosion in 1983. The scheme was designed and operated in apparent ignorance of the likelihood of methane generation in a tunnel, and a subsequent methane explosion killed 16 people on a visit to the installation. Yet, in the community near the tunnel entrance, many people knew informally about coal measures near the surface, and about regular natural emissions of "marsh gas".

[12] For this argument, in various versions, see Tribe, L.H. (1973), Technology assessment and the fourth discontinuity, *Southern California Law Review,* **46**, 609–646; Winner, L. (1972), On criticizing technology, *Public Policy,* **20**, 35–59; Winner, L. (1977), *Autonomous Technology* (MIT Press, Cambridge, MA). On the idea of interference effects, see Hacking, I. (1986), in D. MacLean (Ed), *Values at Risk* (Rowman and Allenheld, Totowa, NJ).

[13] Fiksel, J. (1985), Quantitative risk analysis for toxic chemicals in the environment, *Journal of Hazardous Materials,* **10**, 227–240.

[14] Black, D. (Chairman) (1984), *Investigation of the Possible Increased Incidence of Cancer in West Cumbria* (HMSO, London). See also Crouch, D. (1985), *Science, Teleology and Society: Child Cancer Around the Sellafield Nuclear Reprocessing Plant,* unpublished M.Sc. Thesis, University of Sussex, UK; MacGill, S.M., Ravetz, J.R., and Funtowicz, S. (1985), *Scientific Reassurance as Public Policy: The Logic of the Black Report,* draft, University of Leeds, UK.

[15] Ricci, P. and Cirillo, M.C. (1985), Uncertainty in health risk analysis, *Journal of Hazardous Materials,* **10**, 433–448.

[16] On this point, see especially Collins, H.M. (1985), *Changing Order* (London: Sage).

[17] See Wynne, *op. cit.*, [6]; MacGill *et al.*, *op. cit.*, [14].

[18] See Black, *op. cit.*, [14] and Crouch, *op. cit.*, [14].

[19] Thompson, M. (1983), Postscript: A cultural basis for comparison, in H. Kunreuther and J. Linnerooth (Eds), *Risk Analysis and Decision Processes: The Siting of Liquefied Energy Gas Facilities in Four Countries* (Springer, Berlin). On the UK and political ritual, see Lukes, S. (1977), Political ritual and social integration, in *Essays in Social Theory* (Macmillan, London).

[20] See also Douglas, M. and Wildavsky, A. (1982), *Risk and Culture* (University of California Press, Berkeley, CA). The critical evaluation of environmentalism in that book is irrelevant to the validity of the sociological analysis.

[21] Vogel, D. (1983), Cooperative Regulation: Environmental regulation in Great Britain, *The Public Interest,* **72**, 88–106; Wildavsky, A. (1984), *Doing More and Using Less: The Utilization of Research as a Result of Regime*, draft mss. For a balanced comparison, see Brickmann, R., Jasanoff, S., and Ilgen, T. (1985), *Controlling Chemicals: A Comparative Study of Policy and Politics* (Cornell University Press, Ithaca, NY). Some of the US criticism of its own adversarial style seems to be inspired by the belief that it is essentially anti-enterprise, yet there is little evidence that US industry is hampered relative to European industry. On the contrary, one could say that the adversarial hullaballoo of US regulation is an integral part of an enterprise culture.

[22] Dyson, K. (1982), West Germany: The search for a rationalist consensus, in J. Richardson (Ed), *Policy Styles in Western Europe* (Allen and Unwin, London).

[23] Personal communication.

[24] See Brickmann *et al.*, *op. cit.* [21]; Dyson, *op. cit.*, [22].

[25] Edelman, M. (1976), *Political Language: Words That Succeed and Policies That Fail* (Heinemann, London).

[26] For further discussion of this, see Brickmann *et al.*, *op. cit.*, [21]; Wynne, B. and O'Riordan, T. (1987), Comparative regulatory styles, in H. Kunreuther and P. Kleindorfer (Eds). (1987), *Insuring and Managing Hazardous Risks: From Seveso to Bhopal and Beyond,* Springer, Berlin.

[27] Wilson, J.Q. (Ed) (1981), *The Politics of Regulation* (Free Press, New York). Also McGarity, T. (1979), Substantial and procedural discretion in administrative resolution of science policy questions: Regulating carcinogens in EPA and OSHA, *Georgetown Law Journal,* **81**, 724–810.

[28] McCray, L. (1983), *An anatomy of risk assessment,* Working Paper for the US National Research Council's Committee on the Institutional Means for Assessment of Risks to Public Health, *Risk Assessment in the Federal Government: Managing the Process* (National Academy Press, Washington, DC).

[29] For example, in the disputes following World War II about the freedom or planning of science in the UK, left-wing scientists led by J.D. Bernal expressed a rationalist, "accountability" model of science, while conservatives led by M. Polyani, expressed an intuitive craft, or "tacit factors" model of science. The association of these models with specific political positions varies according to historical context.

[30] Storer, N. (Ed) (1973), *The Sociology of Science: Theoretical and Empirical Investigations* (University of Chicago Press, Chicago, IL). Popper, K.R. (1959), *The Logic of Scientific Discovery* (Routledge and Kegan Paul, London).

[31] Kuhn, T.S. (1970), *The Structure of Scientific Revolutions*, 2nd edn. (University of Chicago Press, Chicago, IL). See also note [8]; Mulkay, M.J. (1980), *Science and the Sociology of Knowledge* (Allen and Unwin, London); Ravetz, J.R. (1972), *Scientific Knowledge and its Social Problems* (Clarendon Press, Oxford, UK); Polanyi, M. (1958), *Personal Knowledge* (Routledge and Kegan Paul, London); Hesse, M.B. (1980), *Revolutions and Reconstructions in the Philosophy of Science* (Harvester Press, Hassocks, UK). The journal *Social*

Studies of Science (Sage) is the seminal publication for work in this area.

[32] Kuhn, *op. cit.*, [31]; and Kuhn, T.S. (1977), *The Essential Tension: Selected Studies in Scientific Tradition and Change* (University of Chicago Press, Chicago, IL).

[33] Habermas, J. (1975), *Legitimation Crisis* (Beacon Press, New York, NY). See also McCarthy, T. (1976), A theory of communicative competence, in P. Connerton (Ed), *Critical Sociology*, pp. 470–497 (Penguin, Harmondsworth, UK) See also Bernstein, R.J. (1983), *Beyond Objectivism and Relativism* (Blackwell, Oxford, UK).

[34] In addition to the earlier references, see *Policy Studies Journal,* special issue, 1982, on environmental policy implementation and enforcement. See also Fromm, G. (Ed) (1981), *Studies in Public Regulation* (MIT Press,Cambridge, MA); Edelman, M. (1966) *The Symbolic Uses of Politics* (University of Illinois Press, Chicago, IL).

[35] Diver, *op. cit.*, [3].

[36] If the "answer" provided by the "tightly specified" rules is sufficiently discordant with dominant values, a new answer can always be found. When saccharin fell foul of the Delaney Amendment of the US Food, Drugs and Cosmetics Act, which banned any food additive found to be associated with cancers in laboratory animals, administrators redefined saccharin as a drug, which fortunately allowed it to remain legal, because public opinion was solidly in favor of using saccharin. A current example is the attempt by various regulatory agencies to incorporate factors for public "risk aversion" into optimization methods for acceptable routine plant emissions standards (for example, by the UK National Radiological Protection Board and the Health and Safety Executive).

[37] In personal discussions at various times throughout the research.

[38] Ezrahi, Y. (1980), Utopian and pragmatic rationalism: The political context of scientific advice, *Minerva,* **18**(1), 111–131.

[39] Tribe, *op. cit.*, [12].

[40] Forrester, J. (1982), The policy analysis–critical theory affair, *Journal of Public Policy,* **2**, 145–164.

[41] Latour, B. and Woolgar, S. (1982), *Laboratory Life* (Sage, Beverley Hills, CA) Knorr-Cetina, K.D. (1982), *The Manufacture of Knowledge* (Addison-Wesley, New York, NY).

[42] See, e.g., Weinberg's representation, *op. cit.*, [7].

[43] For a hint of the complexity and richness of this question, see Karin Knorr-Cetina's (1985) review, entitled "Germ Warfare", of Latour, B., Les Microbes: guerre et paix suivi de Irreductions, *Social Studies of Science,* **15**, 577–585.

[44] Ezrahi, *op. cit.*, [38]; Ravetz, J. (1987), Uncertainty, ignorance, and policy, and B. Wynne, (1987), Uncertainty – technical and social, in *Science for Public Policy,* unpublished proceedings of a conference at the International Institute for Applied Systems Analysis, Laxenburg, Austria; Brooks, H. and Cooper, C. (Eds.) *Science for Public Policy,* (Pergamon, Oxford), Rip, A. (1986), Experts in public arenas, in H. Otway and M. Peltu (Eds), *Regulating Industrial Risks*, pp. 94–110 (Butterworth, London).

[45] See, e.g., Edelman, *op. cit.* [25].

[46] Habermas and Connerton, both *op. cit.*, [33].

[47] von Winterfeldt, D. (1987), Value-tree analysis, in Kunreuther and Kleindorfer, *op. cit.*, [26].

[48] Checkland, P. (1985), From optimizing to learning: A development of systems thinking for the 1990s, *Journal of the Operations Research Society*, **36**, 757–767; Checkland, P. (1981), *Systems Thinking, Systems Practice* (Wiley, Chichester, UK).

[49] Vickers, G. (1983), *Human Systems are Different* (Harper and Row, London).

[50] Forrester, *op. cit.*, [40].

[51] Majone, *op. cit.*, [4].

[52] Of course, this is an idealized view of positivist legal processes. They can be extremely socially exclusive, and use rationalistic ideologies of discovery to conceal and privatize negotiation of values.

[53] MacNeill, W. (1981), The care and repair of public myths, *Foreign Affairs*, **61**, 1–13.

[54] Bonhoeffer, D. (1964), *Ethics*, pp. 363–372 (Collins, London).

Index

Back and regulation, 26, 56, 111, 197, 211, 228, 238, 264, 349, 372, 398, 429
Best practicable environmental option (BPEO), 203, 303
Black-box,
  judgment, 276, 371
  technology as, 20, 277
Booy Clean, 93-100
Centre-periphery relationships, 45, 400, 424
Centralization of facilities, 156, 180, 190, 260
Concentration thresholds, 40, 58, 66, 85, 86, 119, 123, 236, 332, 405, 415
Concern, 25, 29, 34, 40, 49, 74, 88, 92, 101, 116, 126, 139, 150, 159, 167, 233, 245, 250-253, 266, 313, 341, 356-391, 397, 405, 414, 426-438
Costs of disposal, 164, 165, 182-184, 187
Cradle-to-grave waste control, 62, 67, 72, 135, 139, 150, 155, 167, 197, 203, 211, 229
Credibility, 1-16, 69, 80, 107-109, 237, 260, 287, 299, 312, 318, 322, 338, 348-351, 365, 390, 397, 412-414, 419-426, 429
  public, 1, 6, 31, 227, 301, 305, 339, 368, 390, 398, 407, 412-414, 424
Cross-border transportation of wastes, *see* Transfrontier movement of wastes
Decision-analysis, 356-391, 413, 426
Decision rules, 10, 12, 87, 238, 270, 332, 338-351, 400, 405
Decision tree, 275-280, 286, 314, 315, 364
Degree-of-hazard system, 114, 122, 130-133, 135, 141, 161, 217, 318, 321, 326-331
Deskilling, *see* Political deskilling
Detoxification, 220
Downstream regulation, 259, 349
Elutriate procedure test (EP), 123, 137, 331-341, 346, 423
Enforcement, 8, 24, 31, 70, 86, 98, 111, 154, 213, 221-233, 240, 255, 305, 329, 402, 424, 437
  gap, 3, 4, 185-187
Environmental regulation, 26, 50, 67, 121, 139, 145, 252
Export of wastes, 23, 56, 87, 93, 159, 165, 176, 195, 263
  ban, 23, 33, 64, 177, 404, 418
Groundwater contamination, 25, 59, 66, 125, 152, 159, 218, 253, 262, 333
Hazard ranking, 85, 255, 318
  see also Degree-of-hazard system
Hazardous waste,
  characteristics, 60, 124
  classification, 32-34, 54, 65, 71-73, 110, 114-148, 197, 216, 240, 259, 271, 296, 332, 402, 407
  definition, 8, 20, 34, 46-48, 55, 65, 114-148, 197-199, 203, 212-214, 233, 255, 290, 297, 402-404, 409, 415, 424
  exemptions from regulation, 48, 209, 257, 327
  life cycles, *see* Life cycles of wastes
  listing, 114-148, 157, 171, 178, 197, 203, 204, 209-215, 233, 255, 257, 296, 327, 328, 332, 415
  registration, *see* Trip-ticket system
  regulatory framework, 28, 46-48, 50, 57, 60, 67, 71, 106, 110, 125, 156, 162, 163, 195-197, 201-211, 225, 228, 232, 247, 251-257, 296, 311-351, 402-404, 432, 437

Ignorance in risk knowledge, 6, 10-14, 27, 31, 52, 58, 78-80, 102, 104, 271, 283, 298-303, 312-314, 318, 322-326, 331, 374, 384, 390, 400, 407-417, 422
Illegal dumping, 23, 24, 94, 132, 151, 166, 174, 184, 208, 225, 231, 238, 293, 417
Implementation gaps, 4, 6, 8, 27, 31, 402, 424
Incineration, 46, 61, 70, 87, 107, 131, 139, 151, 157-166, 169, 173, 178, 180-182, 188-190, 199-201, 227, 236, 259-263, 329
Inference rules, 342-347, 431
  see also Decision rules *and* Observation rules
Inspection of sites, 90, 207, 224, 226, 233, 255, 257-260, 295, 303, 406, 438
Institutions
  conflict, 336, 407
  credibility and risk perceptions, *see* Credibility
  trust, 227, 234, 350, 406
Interpretation rules, 341-342
Landfill codisposal, 25, 35, 135, 153, 196-198, 200, 217-220, 233-235, 329, 406
Leaching/leachates, 46, 62, 66, 124, 137, 152, 159, 207, 219, 240, 250, 303, 332-336
Legitimation, 10, 13, 30, 40, 85, 110, 145, 235, 338, 368, 374, 397-399, 419-425, 432
Life-cycles of wastes, 1, 8, 12, 27, 32, 39, 49, 54-56, 60-73, 79, 86, 114, 118, 141, 150, 179, 195-198, 202, 211, 223, 228-233, 236, 240, 246, 297, 302, 311, 319, 322, 326, 329, 351, 400-402, 408, 437
Liquid energy gas (LEG), 3, 271, 281-283, 298
Love Canal, 27, 46, 103, 118, 300, 316, 319, 369, 373
No observable effects levels (NOELs), 312
Nuclear,
  power, "dread dimension", 360-363
  reactors/power plants, 3, 39, 75, 271, 276, 280, 288, 302, 326
  technology, 285
  waste, 47, 58, 73, 76, 127, 155, 164, 399
Observation rules, 332, 339-341
Pesticides, 49, 110, 247, 286-289, 296, 306, 330, 336, 364, 409, 415
Political cultures, 35-41, 84-86, 101, 110, 115, 143, 150, 154, 169, 190, 266, 304, 337, 386, 397
  and legitimation, 413-425
Political deskilling, 386, 389, 413, 433
Public,
  anxiety, reactions, 3, 14, 27, 37, 63, 73, 79, 89, 100, 126, 132, 163, 196, 240, 252-254, 266, 288, 299, 347-349, 356-391, 400, 414, 433, 437
  concern, *see* Concern
  credibility, *see* Credibility, public
  ownership of facilities, 151, 156, 165, 166, 168, 169, 171-179, 188-191, 404, 414, 437, 438
  perception of risks, 9-16, 64, 74, 79, 261, 356-391, 398, 426, 435
Radiation, effects of, 312-318
Radioactive wastes, *see* Nuclear waste
Rationality, 84-112, 288, 299, 382, 388, 397, 407, 411, 414, 421-436, 439
Recycling of wastes, 48, 54, 77, 152, 161, 186-189, 212, 220, 230, 247, 265, 298, 438
Regulatory styles, 5, 29, 35-41, 143, 150, 269, 412, 419-422
  see also Political cultures
Risk,
  analysis, 1-13, 26, 45, 56, 65, 69-72, 78-80, 214, 240, 266, 270-391, 397, 407-413, 426
  assessment, formal models, 65-73, 321-325, 330, 351
  attributes, 14, 274, 357-364, 374-377, 384, 430
  definition, 272-274, 375
  intrinsic-situational, 13, 56, 65-70, 272, 280, 287, 297, 326-333, 400, 412
  management of, 2, 54, 67, 75, 85, 108, 151, 235, 240, 251, 300, 314, 326, 331, 341, 349, 359, 363, 374, 397, 422
  perceptions, 3, 10, 270, 356-391, 426
Scientific,
  ignorance, *see* Ignorance in risk knowledge
  knowledge, 2, 6-9, 12, 30, 36, 56, 70, 123, 153, 317, 325, 341-349, 372, 403, 413, 423, 430-432
  uncertainty, *see* Uncertainty
Sellafield, *see* Windscale
Seveso, 27, 33, 63, 411
Site licensing, 24, 58, 71, 90, 115, 140, 144, 195-202, 205-207, 210-212, 215-217, 221-226, 232, 238-240, 259, 329, 404-406
Social learning, 14, 371, 386, 413, 429-438
Soil clean-up/sanitation, 89, 101-109, 111, 418
Technological,
  problems, 337
  risks, 269-306
    see also Risk, attributes and Risk, definition
Toxic chemicals/waste, 2, 24-28, 46-50, 58-60, 87, 92, 102, 107, 115, 119, 161, 186, 197, 233, 253-255, 275, 311-351, 406, 410, 423

Toxicological,
test protocols, 340, 399
testing, 67, 121, 141, 212-215, 317, 327, 337, 342, 346, 431
tier-testing, 120-124, 137, 318-325
Transcience, 409, 431
Transfrontier movement of wastes, 25, 28, 32-34, 63, 115, 131, 137, 144, 176, 186, 401
Treatment and disposal (T&D), 25-29, 33, 45, 53-58, 62, 67-72, 76-78, 86, 91, 96, 99, 111, 130, 158, 166, 168, 173, 195-201, 205-207, 210-212, 216, 221-223, 228-233, 235, 238, 246, 254, 259-268, 290, 295, 303, 327, 400, 404, 407, 414, 417, 428, 437
Trip-ticket system, 72, 119, 128, 130, 134, 155, 228, 233, 238, 257, 265
Uncertainty,
institutional, 9, 14, 29, 56, 59, 74, 80, 85, 101, 109, 116, 143, 216, 227, 234-236, 269-306, 339, 349, 405, 408, 418
public, 251, 261
scientific knowledge, and, 6-9, 14, 26, 39, 103, 106, 115, 124, 141-145, 300, 304, 333-338, 341, 347-351, 423
social, 251, 346, 367, 369, 372, 374, 378, 416, 422, 430
structural, mistaken for technical uncertainty, 7, 79, 269-306, 349, 407, 413, 429
technical imprecision, and , 5, 7-9, 56, 74, 90, 101, 109, 153, 199, 215, 234, 269-306, 356, 408, 422
Upstream,
production decisions, 50, 259, 414, 430
production intervention, 38
regulation, 69, 237, 240, 408, 429
Waste-environment-technology (WET), 70, 303
Waste reduction, 55, 187, 261, 267, 404
Windscale (Sellafield), 285, 306, 362, 373, 376, 410, 413
Worst-case accident scenario, 137, 273, 326, 329, 330, 335